Singularities in PDE and the Calculus of Variations

Volume 44

CRM PROCEEDINGS & LECTURE NOTES

Centre de Recherches Mathématiques
Montréal

Singularities in PDE and the Calculus of Variations

Stanley Alama
Lia Bronsard
Peter J. Sternberg
Editors

The Centre de Recherches Mathématiques (CRM) of the Université de Montréal was created in 1968 to promote research in pure and applied mathematics and related disciplines. Among its activities are special theme years, summer schools, workshops, postdoctoral programs, and publishing. The CRM is supported by the Université de Montréal, the Province of Québec (FCAR), and the Natural Sciences and Engineering Research Council of Canada. It is affiliated with the Institut des Sciences Mathématiques (ISM) of Montréal, whose constituent members are Concordia University, McGill University, the Université de Montréal, the Université du Québec à Montréal, and the École Polytechnique. The CRM may be reached on the Web at www.crm.umontreal.ca.

American Mathematical Society
Providence, Rhode Island USA

The production of this volume was supported in part by the Fonds pour la Formation de Chercheurs et l'Aide à la Recherche (Fonds FCAR) and the Natural Sciences and Engineering Research Council of Canada (NSERC).

2000 *Mathematics Subject Classification.* Primary 35J50, 47J30.

Library of Congress Cataloging-in-Publication Data

Singularities in PDE and the calculus of variations / Stanley Alama, Lia Bronsard, Peter J. Sternberg, editors.
p. cm. — (CRM proceedings & lecture notes, ISSN 1065-8580 ; v. 44)
Includes bibliographical references.
ISBN 978-0-8218-4350-5 (alk. paper)
1. Singularities (Mathematics) 2. Differential equations, Partial. 3. Calculus of variations. I. Alama, Stanley, 1962– II. Bronsard, Lia, 1963– III. Sternberg, Peter J., 1959– IV. Title. V. Series.

QC20.7.S54 S565 2008
515′.353—dc22 2008060012

∞ The paper used in this book is acid-free and falls within the guidelines
established to ensure permanence and durability.
This volume was submitted to the American Mathematical Society
in camera ready form by the Centre de Recherches Mathématiques.
Visit the AMS home page at `http://www.ams.org/`

10 9 8 7 6 5 4 3 2 1 13 12 11 10 09 08

Contents

Preface

This is the companion volume to the *Workshop on Singularities in PDE and the Calculus of Variations* held at the CRM from July 17 to July 21, 2006. The main theme of the meeting was the formation of geometrical singularities in PDE problems with a variational formulation. These equations typically arise in some applications — to physics, engineering, or biology, for example — and their resolution often requires a combination of methods coming from such areas as functional and harmonic analysis, differential geometry and geometric measure theory.

A recurrent theme in the meeting involved the Cahn – Hilliard model for phase transitions, which is ubiquitous in the description of domain walls (codimension-one singularities). Several talks addressed phase transitions in various contexts, including dynamics of crystal facets, stochastically driven switching between minima, and the effects of nonlocal terms (in micromagnetic and copolymer models) in producing minimizers exhibiting unexpected geometric structures.

A second major theme of the workshop was the family of Ginzburg – Landau models for superconductors, superfluids, and Bose – Einstein condensates. In these models, the singularities are typically "vortices" which are quantized codimension-two defects. Results were presented on the interaction dynamics between vortices, the existence of local minimizers with vortex-line structure, and the phase transitions which occur in increasing magnetic or electric fields. We also heard about interesting results on related gauge field models (Chern – Simons and d-wave models) with tools derived for the Ginzburg – Landau functional.

Many fascinating talks fell outside these broad categories, illustrating the breadth of the field. We heard results on singular perturbation problems with concentrations along curves or surfaces, on variational versus geometrical methods in image processing, and on variational problems in biology. Some talks concerned the analytic foundations behind the variational methods, including Monge – Ampère functions, the lifting of Sobolev functions with values in S^1, and weak forms of the Gauss – Green theorem.

Participants at the meeting included researchers and students from several countries in North and South America, Europe, and Asia, many of whom met for the first time. We were truly impressed by the high level of the talks, the range and depth of the results presented, and the remarkably high level of congeniality and camaraderie in the meeting.

The high level of interaction and congeniality shared by the participants was certainly due in part to the outstanding organization and facilities of the CRM (including the wine and cheese reception and morning croissants) and to its remarkable location in the wonderful city of Montreal. We are grateful to the CRM for its generous funding of the meeting and also to the National Science Foundation for providing a generous grant to support young mathematicians coming from the US.

Stan Alama, Lia Bronsard, Peter Sternberg

Centre de Recherches Mathématiques
CRM Proceedings and Lecture Notes
Volume **44**, 2008

Variational Problems Arising in Biology

Elena R. Alvarez-Buylla, Mariana Benítez, Álvaro Chaos, Yuriria Cortés, Gerardo Escalera-Santos, Carlos Espinosa, and Pablo Padilla

ABSTRACT. The problem of understanding the emergence of biological structures and their changes is very complex and involves several temporal and spatial scales. At the biochemical level, it is grounded, to a great extent, on the interaction of the genes responsible for the differentiation of individual cells. These genetic regulatory networks are in turn coupled, at a larger scale, with external factors which include intercellular signaling, the diffusion of hormones, mechanical effects, etc. In this paper we propose a systematic way of deriving equations starting from experimental information on the architecture of the underlying regulatory network. We first construct an associated discrete dynamical system whose equilibrium points can be identified with cell types. We also analyze the behavior of the system when it is subject to random perturbations and relate it to the observed temporal sequence of cell differentiation. Based on this system, we can obtain a set of partial differential equations of Turing type. In this context, the study of pattern formation becomes a mathematical question: the formation of coherent structures and singularities and whether some variational or optimization principle can explain the way in which they interact and their overall organization. As a specific example we consider the genetic network involved underlying cell fate determination during early stages of flower development in *Arabidopsis thaliana*.

1. Introduction

In the preface to his book *On Growth and Form* [7] D'Arcy Thompson writes:

> The terms Growth and Form, which make up the title of this book, are to be understood, as I need hardly say, in their relation to the study of organisms. We want to see how, in some cases at least, the forms of living things, and of the parts of living things, can be explained by physical considerations, and to realize that in general no organic form exists save such as are in conformity with physical and mathematical laws.

The development of molecular genetics in the past ten years or so has made possible to start fulfilling D'Arcy's project. We are now able, in some cases, to

---

2000 *Mathematics Subject Classification.* 35Q35, 35Q53.
This is the final form of the paper.

determine causal relationships among the genes responsible for specific developmental processes in the cell and to establish links between these processes and the formation of patterns and structures. In biological terms, we are beginning to understand in detail the relationship between genotype and phenotype. This has important implications, not only because developmental questions can be posed in a more precise way, but also because, from the mathematical point of view, it is possible to implement models in which the parameters have a clear interpretation and can be, at least in principle, measured. The patterns obtained from the mathematical models could be in some sense optimal. In particular, for some reaction – diffusion equations exhibiting the formation of singularities or coherent structures (peaks, transition layers, spiral waves, etc.) it is a natural question to see if there is an underlying variational problem that determines the global behavior of the system (for a specific example in which a sphere packing problem arises in the context of pattern formation see [4]). In many cases, it is possible to identify an underlying geometric problem, specially if a small parameter is present. In the case of reaction – diffusion systems of the activator – inhibitor type, the ratio of the diffusion coefficient of the activator to that of the inhibitor can be taken in many applications as a small quantity. We mention [5] and references therein for more details on models arising in diverse biological contexts leading to closed geodesics and minimal surface problems.

In this paper we present a systematic methodology that allows us to construct mathematical models grounded on experimental information about genetic regulatory networks. It has already been applied to specific systems. In [3] we proposed a network architecture for the cell fate determination of the organs during early flower development in *Arabidopsis thaliana*. We intend to give a detailed account that is accessible to mathematicians, keeping the biological technicalities to a minimum. This methodology can be described in the following way. Based on experimental data we first construct a network reflecting the interactions of the genes which are relevant. Then we build a discrete and deterministic dynamical system whose dynamics is determined by the architecture of the network. The equilibrium points of this system can be interpreted as the different cell types. In fact, our system recovers all of the observed cell types and has no spurious equilibrium points.

We may regard this discrete system as the "skeleton" of a continuous dynamical system. In other words, we interpolate the discrete system to obtain a continuous vector field. We argue that once this system is randomly perturbed, the solution of the associated Fokker – Planck equation contains relevant biological information. In particular, the temporal sequence of appearance of singular structures or peaks in the solution of the FP equation corresponds to the order in which floral organ differentiation actually takes place in nature.

Later on, we take the vector field, or rather a two-dimensional projection and use it as the reactive term of a Turing system. In this way we are able to explore the space – time dynamics of developmental processes controlled by the network. Our first numerical simulations seem to indicate that we can recover a well established biological model, called the ABC model, using this approach. Moreover, according to this view, it is possible to explain why the morphogens originally proposed by

Turing are in many cases so difficult to identify. Indeed, what comes out of the analysis is that it is not the activity of individual genes (or the concentrations of the proteins they synthesize) but actually nontrivial combinations of them that should be taken as the effective morphogenes.

2. From genetic regulatory networks to Turing systems

A regulatory genetic network can be represented in a simplified way as a directed graph. The nodes are the genes and the arrows the interactions. A green arrow represents the fact that the gene from which it starts tends to activate the target gene. On the other hand, a blue arrow is used to indicate that the interaction is inhibitory. As an example, consider the network shown in Figure 1.

It represents one of the most basic, yet most important networks, that is, an activator – inhibitor system. The node B activates itself, as in an autocatalytic reaction and this is characterized by the arrow pointing to itself. At the same time, this gene favors the activation of A, which in turns inhibits B. As simple as it may seem, this network appears in real biological systems and has been used to explain some features of the hair formation in roots and leaves in *Arabidopsis* [2].

In order to construct a dynamical system, a more precise description of the interaction is needed. First, and for the sake of simplicity, we assume that the genes can be in two states only, 0 (off) and 1 (on). A complete description of the behavior of a gene can be summarized in a table, listing all possible inputs of the genes that affect a particular gene and the corresponding outputs for the latter. In the activator – inhibitor example, these rules can be summarized in a table (see Table 1).

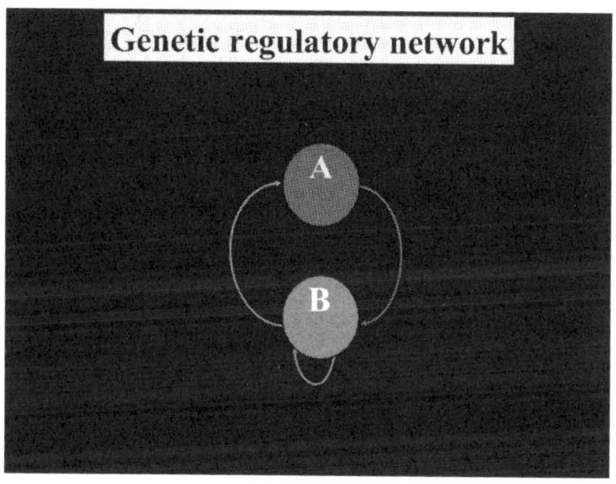

FIGURE 1. Activator – inhibitor network

TABLE 1. Table with the logic rules for the activator – inhibitor system

Input B	Input A	Output B	Output A
0	0	0	0
0	1	0	0
1	0	1	0
1	1	1	0

If there is no activity of the B gene, then both genes remain off. If A and B are on, the self activation of B is assumed to be stronger and the inhibitor to eventually decay. The entry A equals 0, B equals 1 corresponds then to an equilibrium point. In short, it is only when the self-activated gene B is on that the state of the system is nontrivial. In a natural way, these rules determine a discrete dynamical system. More precisely, if we denote by $x(n)$ the state of the system at the nth iteration, where for our example $x(n) = \big(x_1(n), x_2(n)\big)$, $x_1(n)$ representing the activity of gene B and correspondingly, $x_2(n)$ that of gene A, the dynamics of the system is given by

$$x(n+1) = f\big(x(n)\big).$$

Here f is a shorthand for the transformation defined in Table 1. A graphical representation of this discrete dynamical system gives a phase portrait in $[0,1] \times [0,1]$.

Notice that there are two equilibrium points $(0,0)$ and $(1,0)$, which correspond to two different cell types. The discrete phase space contains all possible configurations of genetic expression. If we linearly interpolate this vector field, implicitly assuming that the genes can now take all the intermediate activation values between 0 and 1, we obtain a vector field on the unit square (the space of genotypes). We denote the interpolated field by F, so the equation we get is

$$\dot{x}(t) = F(x),$$

where $x \in [0,1] \times [0,1]$.

We do this for technical reasons. Indeed, if we interpolate the system and then introduce a random perturbation, let us say, Brownian motion, we can write down the stochastic differential equation and the corresponding Kolmogorov or Fokker – Planck equation that the transition probability satisfies. Figure 2 shows the interpolated vector field in the two gene activator – inhibitor network.

As mentioned in the introduction, it is natural to assume that the biological processes taking place within the cell, in particular those associated with the determination of genetic expression patterns are subject to random fluctuations and perturbations. The transition probability provides us with a quantitative tool and encodes useful biological information. We illustrate this in the next section in a biologically meaningful context. We emphasize the fact that the interpolation procedure is done only for technical reasons, since it allows us to introduce the Fokker – Planck equation. In principle, the discrete dynamical system can also be perturbed with noise and the transition probability analyzed and the same results should be obtained. The comparison between both approaches has been carried out for the particular example of *Arabidopsis*, a plant which is the standard [1].

Once the differential equation is perturbed with noise, we obtain a stochastic differential equation

$$\mathrm{d}x(t) = F(x)\,\mathrm{d}t + \sqrt{2\varepsilon}\,\mathrm{d}W,$$

where for simplicity we have taken a Brownian random perturbation W and the intensity ε to be constant. In fact the role that noise plays in biological processes and specifically in genetic regulatory networks is a subject that is being actively investigated (see for instance [6]). If this setting is chosen, then the transition probability function $u(x,t)$ describing the probability that the system is in the

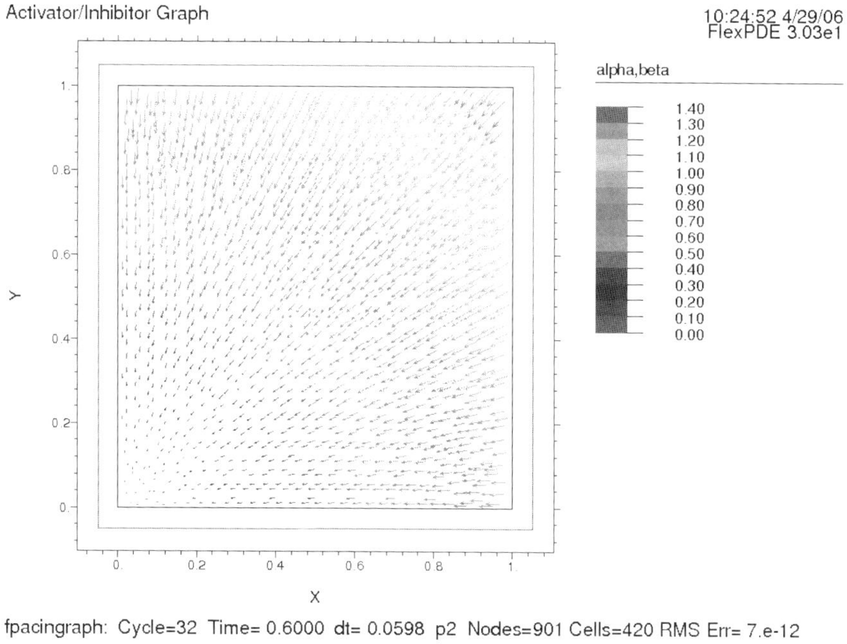

FIGURE 2. Interpolated field

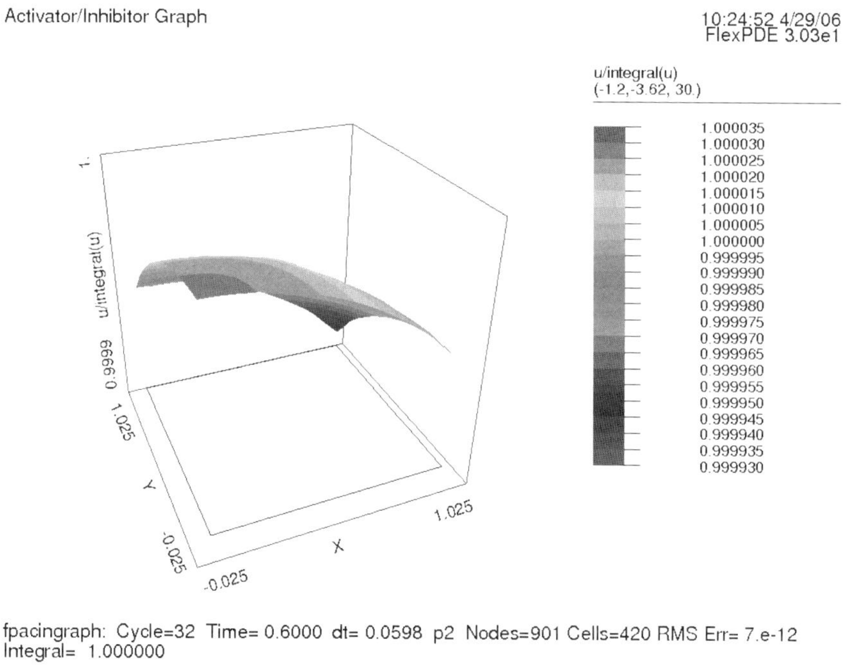

FIGURE 3. Solution to the Fokker – Planck equation

state x at time t satisfies the Fokker – Planck or Kolmogorov equation

$$\frac{\partial u}{\partial t} = -\operatorname{div}(F(x)u) + \varepsilon \Delta u.$$

In Figure 3 we present a numerical solution of the Fokker – Planck equation for the activator – inhibitor network.

3. Cell fate determination for floral organs in *Arabidopsis*

In this section we present an example for a real genetic regulatory network. It is responsible for the determination of the cells fate in the meristem of *Arabidopsis*. More informally, when an *Arabidopsis* flower is to be formed (see Figure 4), all cells in the bud are in the same undifferentiated state. The network is responsible for the "decision" that cells attain a particular state at a given time. The main four cell fates at the early flower developmental stage modelled here correspond to primordial cells of: sepals, petals, stamens and carpels.

Based on decades of experimental experimental work by many groups, it has been possible to identify which genes are involved in this process and how they interact. A network architecture was proposed in [3] and it is shown in Figure 5.

The equilibrium points that are obtained for the associated dynamical system are shown in Figure 6.

We notice that all possible initial conditions were systematically tested (approximately 2^{15}; fewer because some genes were fixed to state "1" because they are expressed constitutively) and eventually, in no more than ten iterations, each of them ends up in one of these equilibria. Moreover, the equilibrium points are fixed-point attractors and correspond to configurations of genetic expression, either for vegetative cells or for floral organs and, in fact, they recover all the observed configurations during early stages of flower development in *Arabidopsis*. In other words, our the proposed network architecture and the associated dynamical system describes all the observed cell types and only them. Besides, the network is robust under perturbations and also capable of correctly predicting mutants. In order to carry out a similar process to the one described for the activator inhibitor system, we first made a simplification. That is, instead of solving the Fokker – Plank equation associated with the fifteen-dimensional interpolated vector field, we projected

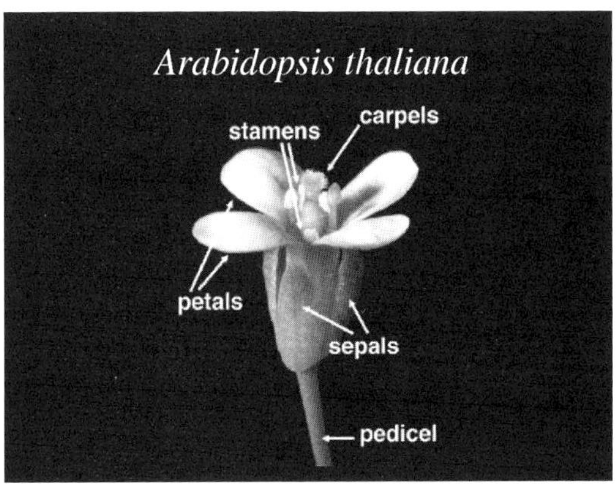

FIGURE 4. *Arabidopsis thaliana*

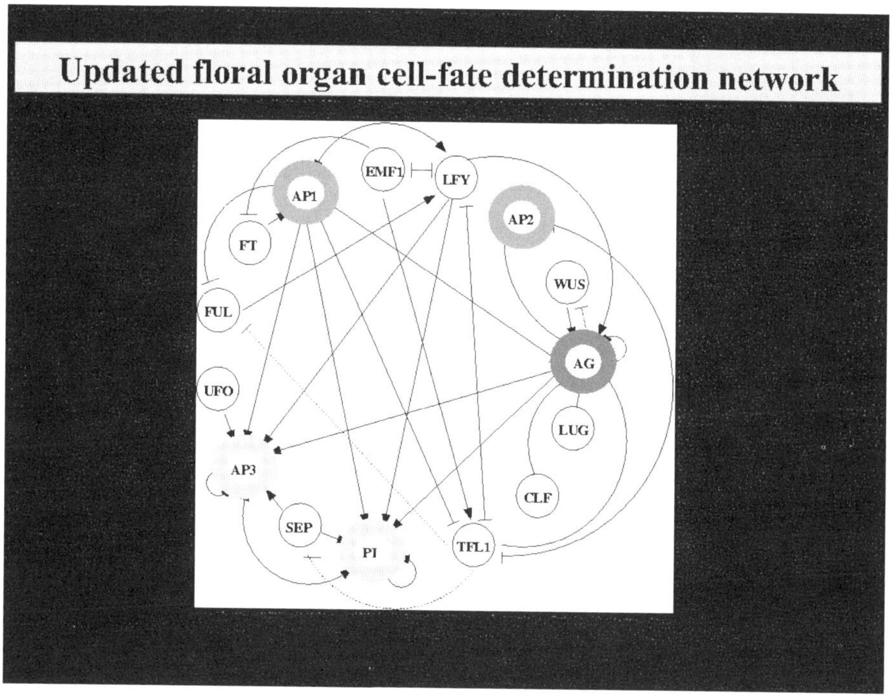

FIGURE 5. *Arabidopsis thaliana* floral organ determination network

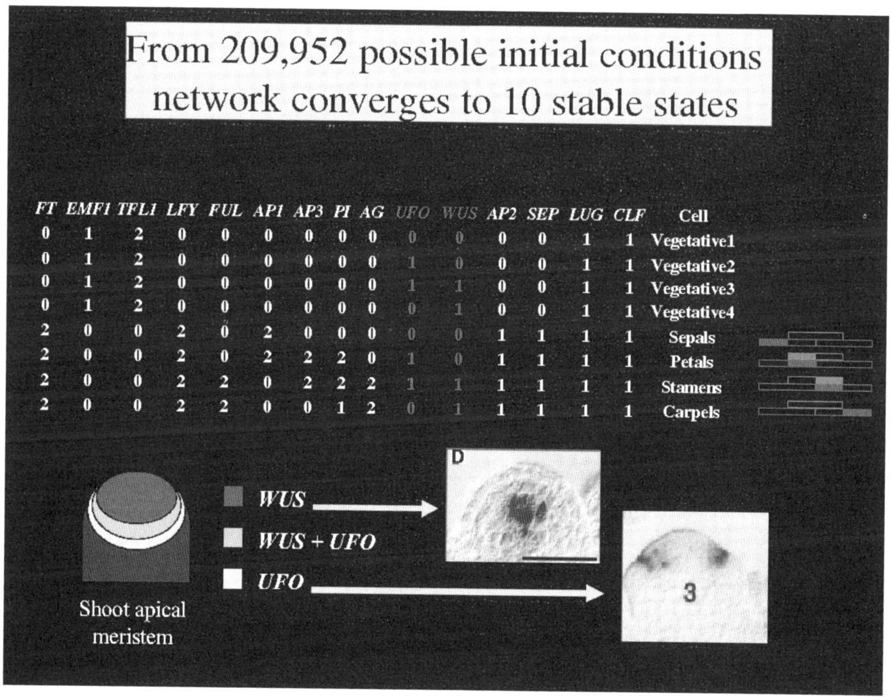

FIGURE 6. Equilibrium points found for the network

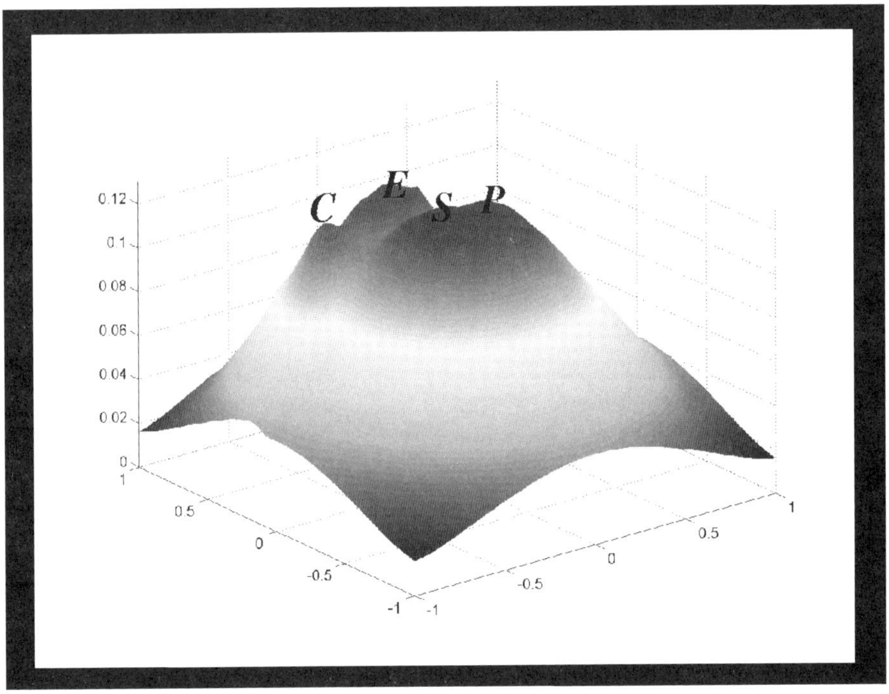

FIGURE 7. Equilibrium points found for the network (S — sepal primordial cells, P — petal primordial cells, E — stamen primordial cells and C — carpel primordial cells)

the dynamics onto a two-dimensional plane. More precisely, we considered, by using least squares, the plane that best fitted the location of the equilibrium points. Then we assumed that the dynamics could be well represented by a gradient system and a potential function was constructed as a "well" around each equilibrium point, whose depth was taken to reflect the number of initial conditions that are attracted to that particular point. The final probability distribution that results from solving the corresponding FP equation is shown in Figure 7. Notice that we are only showing the points corresponding to floral organs.

The initial condition corresponds to the state of undifferentiated cells and we observe that the peaks that are formed above each of the attractors appear in the same temporal order that is normally observed during flower development, namely, first the cells corresponding to sepals are differentiated, then those of petals, and then stamens and carpels.

It is questionable whether all of the simplifications described above maintain the same features of the system. In particular the temporal sequence described above. So we also studied the discrete fifteen dimensional system when it is perturbed randomly. The results coincide and this fact supports the use of the two dimensional vector field as a good approximation.

Given that, it is also natural trying to understand the spatio-temporal dynamics, so we used this field as the reactive term in a Turing system on a circle, in order to investigate whether the gene expression patterns could be recovered. The results are shown in Figure 8.

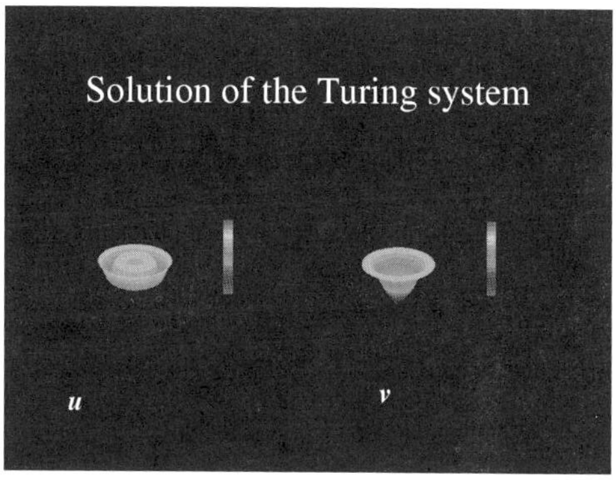

FIGURE 8. Equilibrium points found for the network

The results are still preliminary and further research needs to be done. However, at this stage, it is interesting to notice that, at least qualitatively, some features from the observed patterns are recovered. In particular the fact that the organization occurs in concentric rings, which is the basic spatial way in which the spatial areas of differentiation are observed in real flowers.

4. Conclusions

We have presented a methodology that allows us to develop systematically reaction–diffusion equations for which the connection with experimental data is clear. At this stage, there are several mathematical questions that can be addressed. Besides fundamental questions related to existence and well posedness in general, there are various others that are also interesting. In particular, it is likely that the system does not have a variational structure. However, arguing along the same lines as in the introduction, it is to be expected that an asymptotic variational problem can be identified. Moreover, other important physical and biological effects can be readily incorporated once a partial differential equation has been proposed. These include active transport, growth, mechanical considerations as well as the effects of a nontrivial geometry.

References

1. E. R. Alvarez-Buylla, M. Benítez, Á. Chaos, Y. Cortés-Poza, C. Espinosa-Soto, R. B. Lotto, D. Malkin, Escalera Santos. G. J., and P. Padilla-Longoria, *Floral morphogenesis: stochastic explorations of a gene network epigenetic landscape*, PLoS ONE, submitted.
2. M. Benítez, C. Espinosa-Soto, P. Padilla-Longoria, J. Díaz, and E. R. Alvarez-Buylla, *Equivalent genetic regulatory networks in different contexts recover contrasting spatial cell patterns that resemble those in* Arabidopsis *root and leaf epidermis: a dynamic model*, The International Journal of Developmental Biology **51** (2007), no. 2, 139–155.
3. C. Espinosa-Soto, P. Padilla-Longoria, and E. R. Alvarez-Buylla, *a gene regulatory network model for cell-fate determination during* Arabidopsis thaliana *flower development that is robust and recovers experimental gene expression profilles*, The Plant Cell **16** (2004), no. 11, 2923–2939.
4. C. E. Garza-Hume and P. Padilla, *A pattern formation problem on the sphere*, Proceedings of the Sixth Mississippi State–UBA Conference on Differential Equations and Computational Simulations (Starkville, MS, 2005) (J. Graef, H. Lim, R. Shivaji, B. Soni, and J. Zhu, eds.), Elecon. J. Differ. Equ. Conf., vol. 15, Southwest Texas State Univ., San Marcos, TX, 2007, pp. 97–106.

5. P. Padilla, *Variational, dynamical and geometric aspects of some nonlinear problems*, Variational Methods: Open Problems, Recent Progress, and Numerical Algorithms (Flagstaff, AZ, 2002) (J. M. Neuberger, ed.), Contemp. Math., vol. 357, Amer. Math. Soc., Providence, RI, 2004, pp. 151–169.
6. J. M. Pedraza and A. van Oudenaarden, *Noise propagation in gene networks*, Science **307** (2005), no. 5717, 1965–1969.
7. D. W. Thompson, *On growth and form*, Dover, New York, NY, 1992.

Instituto de Ecologia, Universidad Nacional Autónoma de México, Circuito Escolar 04510, México D.F., México
E-mail address, E. Alvarez-Buylla: eabuylla@gmail.com
E-mail address, M. Benítez: marianabk@gmail.com
E-mail address, A. Chaos: achaos@miranda.ecologia.unam.mx
E-mail address, G. Escalera-Santos: jsantos@buzon.uaem.mx
E-mail address, C. Espinosa: c.espinosas@gmail.com

Facultad de Ciencias, Universidad Nacional Autónoma de México, Circuito Escolar 04510, México D.F., México
E-mail address: yuriria@matem.unam.mx

Instituto de Investigaciones en Matemáticas Aplicadas y en Sistemas, Universidad Nacional Autónoma de México, Circuito Escolar 04510, México D.F., México
E-mail address: pablo@mym.iimas.unam.mx

Centre de Recherches Mathématiques
CRM Proceedings and Lecture Notes
Volume **44**, 2008

On the Cauchy Problem for Phase and Vortices in the Parabolic Ginzburg–Landau Equation

Fabrice Béthuel, Giandomenico Orlandi, and Didier Smets

Abstract. The purpose of this note is twofold. Firstly, we survey some of our recent works concerning the dynamics of vortices in the two dimensional parabolic Ginzburg–Landau equation, emphasizing in particular the vortex-phase interaction. Secondly, we wish to supplement the analysis carried out in these work with a discussion on the initial value problem both for the vortices and the phase.

1. The equation

In this paper we wish to survey some of our recent work [6, 8, 9] on the two-dimensional complex-valued parabolic Ginzburg–Landau equation

$$(\mathrm{PGL})_\varepsilon \qquad \frac{\partial u_\varepsilon}{\partial t} - \Delta u_\varepsilon = \frac{1}{\varepsilon^2} u_\varepsilon (1 - |u_\varepsilon|^2) \quad \text{on } \mathbb{R}^2 \times \mathbb{R}^+,$$

where $0 < \varepsilon < 1$ is a small parameter. The emphasis is put there on the asymptotic limit as the parameter ε goes to 0, which is a rather usual point of view in the topic. We will also provide some new results, and additional remarks and comments, in particular concerning the initial value problem for phase and vortices. Finally we will address some open problems for the higher dimensional case suggested by our analysis.

Recall that $(\mathrm{PGL})_\varepsilon$ is the heat-flow for the Ginzburg–Landau energy E_ε, given, for $v \colon \mathbb{R}^2 \to \mathbb{C}$, by

$$E_\varepsilon(v) = \int_{\mathbb{R}^2} e_\varepsilon(v) = \int_{\mathbb{R}^2} \frac{|\nabla v|^2}{2} + \frac{(1 - |v|^2)^2}{4\varepsilon^2}. \tag{1}$$

In particular, for any $0 \le T_1 \le T_2$, we have the energy identity

$$E_\varepsilon\big(u_\varepsilon(\cdot, T_2)\big) + \int_{T_1}^{T_2} \int_{\mathbb{R}^2} |\partial_t u_\varepsilon|^2 \, \mathrm{d}x \, \mathrm{d}t = E_\varepsilon\big(u_\varepsilon(\cdot, T_1)\big). \tag{2}$$

Our only assumption throughout is on the energy of the initial datum $u_\varepsilon^0(\cdot) \equiv u_\varepsilon(\cdot, 0)$. We assume that

$$(\mathrm{H}_0) \qquad E_\varepsilon(u_\varepsilon^0) \le M_0 |\log \varepsilon|,$$

2000 *Mathematics Subject Classification.* 35B40, 35K55, 35Q40.
This is the final form of the paper.

where $M_0 > 0$ is some given constant, independent of the parameter ε. The energy regime given by (H_0) allows for the formation of topological defects called vortices, which we describe next.

2. Vortices and phase

Let Ω be a simply connected domain in $\mathbb{R}^2$, and let $v\colon \Omega \to \mathbb{C}$ be continuous. If v does not vanish on Ω, then one may decompose v as

$$v = \rho \exp(\mathrm{i}\varphi) \quad \text{on } \Omega,$$

where $\rho = |v|$ is the modulus and $\varphi\colon \Omega \to \mathbb{R}$, the phase, is continuous and uniquely determined up to a multiple of 2π. It is in general not always possible to perform the above decomposition for maps having zeroes. A typical example is given, on $\mathbb{R}^2$ by maps having the form

$$v(r,\theta) = f(r)\exp(\mathrm{i}d\theta) \quad \text{on } \mathbb{R}^2, \tag{3}$$

where $d \in \mathbb{Z}^*$, and where f is assumed to be real-valued, continuous, vanishing at zero and positive elsewhere. For such maps, the phase, given by $d\theta$, is not continuous. The somewhat vague terminology "vortex," refers to these kind of discontinuity in the phase. In the example (3) the origin is a zero of v, hence a vortex: since the winding number around 0 is d, such a vortex is termed a vortex of degree d.

Going back to the Ginzburg – Landau energy (1), it is worthwhile noticing that the potential term $(1-|v|^2)/\varepsilon^2$ forces the map v to be close to the circle S^1 when ε is small, and therefore, if a bound on the energy is imposed, vortices should be scarce. To fix ideas, we consider the following minimization problem: let $\Omega = D^2 = \{z \in \mathbb{C} \simeq \mathbb{R}^2 : |z| = 1\}$, and let $g\colon \partial\Omega = S^1 \to S^1$ be defined by $g(z) = z^d$, i.e., $g(\theta) = \exp(\mathrm{i}d\theta)$. Consider the minimization problem

$$\mu_\varepsilon = \inf\{E_\varepsilon(v) : v \in H^1_g(D^2;\mathbb{C})\}. \tag{4}$$

Since the winding number d at the boundary is assumed to be nonzero, any test map $v \in H^1_g(D^2,\mathbb{C})$ has to vanish somewhere, i.e., to have vortices. It is natural, due to the symmetries in the problem, to seek first solutions of the form

$$w_\varepsilon(z) = f_\varepsilon(r)\exp(\mathrm{i}d\theta) = f_\varepsilon(r)\frac{z^d}{|z^d|}, \tag{5}$$

where $z = r\exp(\mathrm{i}\theta)$ in polar coordinates, and $f_\varepsilon : \mathbb{R}^+ \to \mathbb{R}$ is smooth, with $f_\varepsilon(0) = 0$. Choosing f_ε such that $f_\varepsilon(r) = 1$ for $r \geq \varepsilon$, and $|f'| \leq 2$, a simple computation shows that

$$\mu_\varepsilon \leq E_\varepsilon(w_\varepsilon) = \pi d^2|\log\varepsilon| + O(1). \tag{6}$$

Indeed, since $\partial w_\varepsilon/\partial\theta = \mathrm{i}df(r)\exp(\mathrm{i}d\theta)$,

$$\begin{aligned}\int_{D_2}|\nabla w_\varepsilon|^2 &= \int_{D_2}\left|\frac{\partial w_\varepsilon}{\partial r}\right|^2 + \frac{1}{r^2}\left|\frac{\partial w_\varepsilon}{\partial\theta}\right|^2 = 2\pi\int_0^1\left[|f'(r)|^2 + d^2\frac{|f(r)|^2}{r^2}\right]r\,\mathrm{d}r \\ &= 2\pi\int_\varepsilon^1 \frac{d^2}{r^2}r\,\mathrm{d}r + 2\pi\int_0^\varepsilon\left[|f'(r)|^2 + d^2\frac{|f(r)|^2}{r^2}\right]r\,\mathrm{d}r.\end{aligned} \tag{7}$$

Our assumptions on f imply that the second term in the r.h.s. of (7) is $O(1)$, so that

$$\int_{D_2} |\nabla w_\varepsilon|^2 = 2\pi \int_\varepsilon^1 \frac{d^2}{r}\,\mathrm{d}r + O(1) = 2\pi d^2 |\log \varepsilon| + O(1).$$

On the other hand,

$$\int_{D_2} \frac{(1-|w_\varepsilon|^2)^2}{\varepsilon^2} = O(1),$$

and the claim (6) follows. It turns out that the bound given in (6) is only optimal in the case $d = \pm 1$. For general $d \in \mathbb{Z}$, it can be proved (see [4]) that

$$\mu_\varepsilon = \pi |d|\,|\log \varepsilon| + O(1),$$

so that if $|d| \neq 1$ and ε is small, the minimizers are not radially symmetric. On the other hand, for $|d| = 1$, it has been proved that the minimizers u_ε, for small ε, do have radial symmetry [16, 17]. The case $|d| = 1$ and the corresponding minimizing radial solution plays the role of a ground state in the theory. Indeed, as in the scalar case, we may define an optimal profile (although it is not given by an explicit formula). More precisely, there exists a unique function $f\colon \mathbb{R}^+ \to \mathbb{R}$ satisfying

$$\begin{cases} -f'' - \dfrac{1}{r} f' + \dfrac{1}{r^2} f = f(1-f^2) & \text{on } [0, +\infty) \\ f(0) = 0. \end{cases}$$

Then we have

$$u_\varepsilon(z) \simeq f\Big(\frac{|z|}{\varepsilon}\Big) \exp(\mathrm{i}\theta)$$

and

$$u_\varepsilon(z) \to u_*(z) = \frac{z}{|z|} \quad \text{as } \varepsilon \to 0, \text{ in } W^{1,p},\ p < 2\ , \text{ and in } C^k_{\mathrm{loc}}(D^2 \setminus \{0\}).$$

The map $u_*(z) = z/|z|$ realizes thus the prototypical singularity that can appear in the asymptotics for minimization problems. The convergence above shows that the set where the minimizer is close to zero concentrates near the origin.

More generally, it can be proved that solutions u_ε to the minimization problem (4) converge, up to a subsequence to maps u_* of the form (if $d > 0$)

$$u_\varepsilon \rightharpoonup u_* = \exp(\mathrm{i}\varphi) \prod_{i=1}^{d} \frac{z - a_i}{|z - a_i|} \quad \text{in } W^{1,p},\ p < 2, \tag{8}$$

where φ is a harmonic function on Ω, $a_1, \dots, a_d$ are d points in Ω. Moreover the convergence is in C^k_{loc} off the vortices. If the points $a_1, \dots, a_d$ are known, the phase φ is then completely determined, in view of the boundary condition. As before, the set where u_ε is close to zero concentrates near the points a_i. These points are the singularities (or defects) of the map u_*, and are termed vortices as well, although u_* itself does not vanish.

The discussion on the minimization problem (4) shows that the energy cost of a singularity of degree d is (at least) $\pi |d|\,|\log \varepsilon|$. A natural question is to investigate if this energy regime alone allows to define vortices, at least in this asymptotic sense. Therefore, we are led to consider the energy level set W^M_ε defined by

$$W^M_\varepsilon = \{v \in H^1(\Omega) : E_\varepsilon(v) \leq M |\log \varepsilon|\}.$$

In contrast with the prescribed boundary minimization problem, no general compactness result for reasonable norms can be derived in W_ε^M, due to possible divergences in the phase. Indeed, take for instance the sequence

$$w_\varepsilon(z) = \exp\big(\mathrm{i}\varphi(z)\sqrt{|\log\varepsilon|}\big), \tag{9}$$

with $\varphi\colon \Omega \to \mathbb{R}$ a nonconstant smooth function chosen so that $\|\varphi\|_{L^2}^2 \leq 2M$. We have $|w_\varepsilon| = 1$, hence

$$E_\varepsilon(w_\varepsilon) = \frac{1}{2}\int_\Omega |\nabla w_\varepsilon|^2 = \frac{|\log\varepsilon|}{2}\int_\Omega |\nabla\varphi|^2 \leq M|\log\varepsilon|.$$

On the other hand, $|\nabla w_\varepsilon| = O(|\log\varepsilon|^{1/2})$, so that any norm of the gradient will diverge as $\varepsilon \to 0$.

For arbitrary sequences in W_ε^M it is however possible to identify the possible diverging part of the phase, and to split it from the contribution of the vortices, as the next result shows (see [3,5]).

Theorem 1. *Let $M > 0$ and $(v_\varepsilon)_{\varepsilon>0}$, $v_\varepsilon : \Omega \to \mathbb{C}$ such that*

$$E_\varepsilon(v_\varepsilon) \leq M|\log\varepsilon|.$$

Let $G \subset\subset \Omega$ be a smooth open simply connected set. Then, there exists a subsequence $\varepsilon_n \to 0$, l points $a_1, \dots, a_l \in G$, integers $d_1, \dots, d_l \neq 0$, with $\sum_1^l |d_i| \leq M'$, for some constant M' depending only on M, and functions $\varphi_{\varepsilon_n} : G \to \mathbb{R}$ such that

$$\int_G |\nabla\varphi_{\varepsilon_n}|^2 \leq M|\log\varepsilon|$$

and

$$v_{\varepsilon_n} \cdot \exp(-\mathrm{i}\varphi_{\varepsilon_n}) \to \prod_{i=1}^{l}\left(\frac{z-a_i}{|z-a_i|}\right)^{d_i} \quad \textit{in } H^s(G),\ s<1. \tag{10}$$

In example (9) we have $l = 0$, (i.e., there are no vortices) and taking $\varphi_\varepsilon = \varphi \cdot \sqrt{|\log\varepsilon|}$, one may write, as above,

$$w_{\varepsilon_n} \cdot \exp(-\mathrm{i}\varphi_{\varepsilon_n}) \equiv 1.$$

In contrast with the convergence (8), the degrees d_i may be different from ± 1, for instance, one may check that the sequence w_ε defined in (5) converges to $w_* = (z/|z|)^d$. Moreover, the phase φ_ε in (10) is not constrained, whereas it has to be harmonic in (8).

Theorem 1 shows that the possible lack of compactness is merely due to the phase φ_ε (which is a real-valued function). On the other hand, the "topological" contribution due to the vortices is essentially compact.

A related but conceptually different approach for locating the vorticity for maps v_ε satisfying the bound $\mathcal{E}_\varepsilon(v_\varepsilon) \leq M|\log\varepsilon|$ has been proposed in [1,14]. The main idea there is to look at the Jacobians of v_ε, which allows to characterize its topological part. More precisely, for $v = (v^1, v^2)\colon \Omega \to \mathbb{R}^2$ a smooth map, its Jacobian Jv is the 2-form defined by

$$Jv = \mathrm{d}v^1 \wedge \mathrm{d}v^2 = \tfrac{1}{2}\mathrm{d}(v^1\,\mathrm{d}v^2 - v^2\,\mathrm{d}v^1).$$

In two dimensions, it may be identified with a scalar function, namely

$$Jv = \det(\nabla v) = v_x \times v_y,$$

where, for $a, b \in \mathbb{R}^2$, $a \times b = a^1b^2 - a^2b^1$. Note that $v_x \times v_y = 0$ whenever v_x and v_y are collinear. Hence, when $|v| = 1$, we have $Jv \equiv 0$. On the other hand, topological singularities yield a nontrivial contribution to the Jacobian. For instance, we have

$$J\left(\prod_{i=1}^{l}\left(\frac{z-a_i}{|z-a_i|}\right)^{d}\right) = \pi\sum_{i=1}^{l} d_i\delta_{a_i},$$

so that the Jacobian detects the presence of vortices as well as their degree. In contrast, oscillations in the phase of v are not seen by its Jacobian Jv. It is then proved [14] that

Theorem 2. *Let $v_\varepsilon : \Omega \to \mathbb{R}^2$ such that $E_\varepsilon(v_\varepsilon) \leq M|\log\varepsilon|$. Then there exists a subsequence $\varepsilon_n \to 0$, l points $a_1, \ldots, a_l \in \Omega$, and integers $d_1, \ldots, d_l \neq 0$, with $\sum_1^l |d_i| \leq M'$, for some constant M' depending only on M, such that*

$$Jv_{\varepsilon_n} \rightharpoonup J_* = \pi\sum_{i=1}^{l} d_i\delta_{a_i} \quad \text{in } [\mathcal{C}_c^{0,\alpha}(\Omega)]^*, \text{ for any } \alpha > 0. \tag{11}$$

The above result above may be specified at level ε. If v_ε belongs to W_ε^M then, there exists points a_i^ε, integers d_i^ε, such that

$$\left\| Jv_\varepsilon - \pi\sum_{i=1}^{l} d_i^\varepsilon\delta_{a_i^\varepsilon}\right\|_{[C_c^{0,\alpha}]^*} \leq C(M)\varepsilon^\mu, \tag{12}$$

for some $0 < \mu < 1$. In particular, since the coefficients d_i^ε are integers, vortices and their degrees are also well-defined (up to a precision of order ε^μ) also for an arbitrary map v_ε in W_ε^M.

3. Evolution law for phase and vortices

By assumption (H_0) and the energy identity (2) we have, for all $t > 0$, $E_\varepsilon\big(u_\varepsilon(\cdot,t)\big) \leq M_0|\log\varepsilon|$, so that the result in the previous section apply to $u_\varepsilon(\cdot,t)$ for all $t > 0$, and we are in position to give a meaning to phase and vortices, i.e.,

$$u_\varepsilon(\cdot,t) \simeq \exp\big(i\varphi_\varepsilon(\cdot,t)\big)\prod_{i=1}^{l(t)}\left(\frac{z-a_i(t)}{|z-a_i(t)|}\right)^{d_i(t)}.$$

The central question is to determine the evolution in time of the phase φ_ε as well as of the points $a_i(t)$. A simple heuristic argument show that the evolution of the phase is related to the heat equation. Indeed consider for instance solutions to $(\mathrm{PGL})_\varepsilon$ of the form

$$u_\varepsilon = \rho_\varepsilon \exp(i\varphi_\varepsilon),$$

so that vortices are not present. The system for ρ_ε and φ_ε is then

$$\begin{cases} \rho_\varepsilon^2\partial_t\varphi_\varepsilon - \operatorname{div}(\rho_\varepsilon^2\nabla\varphi_\varepsilon) & = 0 \\ \partial_t\rho_\varepsilon^2 - \Delta\rho_\varepsilon^2 + 2|\nabla u_\varepsilon|^2 & = \dfrac{2\rho_\varepsilon^2(1-\rho_\varepsilon^2)}{\varepsilon^2}. \end{cases}$$

The second equation suggests $\rho_\varepsilon^2 \simeq 1$, so that the first equation yields

$$\partial_t\varphi_\varepsilon - \Delta\varphi_\varepsilon \simeq 0. \tag{13}$$

In particular, the original time scale t is appropriate to study the evolution of the phase.

The analysis of the evolution of the vortices raises a number of difficulties: one of the most important ones is the fact that their number and their degrees may vary in time, so that their labeling is not straightforward.

In contrast with the phase, it can be shown that, at first order of approximation, vortices do not move in the original time scale [12,18]. It has been recognized that in the two-dimensional case that we study here, accelerating time, e.g., considering the time scale $s = t/|\log\varepsilon|$, is appropriate to investigate the dynamics of the vortices. Therefore we consider the map $\mathfrak{u}_\varepsilon$, defined on $\mathbb{R}^2 \times \mathbb{R}^+$ by

$$\mathfrak{u}_\varepsilon(z,s) = u_\varepsilon(z, s|\log\varepsilon|). \tag{14}$$

The next theorem, proved in [6, 8] after earlier results in [13, 15], gives a precise meaning to the notion of vortices in this setting, as well as a first description of their dynamics.

Theorem 3. *Assume $(u_\varepsilon)_{\varepsilon>0}$ verifies* (PGL)$_\varepsilon$ *and* (H$_0$). *Then, for a subsequence $\varepsilon_n \to 0$ we have, for any $s > 0$,*

$$\mathfrak{u}_{\varepsilon_n}(z,s) \to \mathfrak{u}_*(z,s) = \prod_{i=1}^{l(s)} \left(\frac{z - a_i(s)}{|z - a_i(s)|} \right)^{d_i(s)} \exp[\mathrm{i}(\langle \vec{c}(s), z\rangle + b(s)], \tag{15}$$

where, for $i = 1,\ldots,l(s)$, $a_i(s) \in \mathbb{R}^2$, $d_i(s) \in \mathbb{Z}$, $b(s) \in [0,2\pi)$ and $\vec{c}\colon \mathbb{R}^+ \to \mathbb{R}^2$ is a Lipschitz function. The convergence in (15) *is uniform on every compact subset of $\mathbb{R}^2 \times \mathbb{R}^+_* \setminus \Sigma_\mathfrak{v}$, where $\Sigma_\mathfrak{v} = \bigcup_{s>0}\bigcup_{i=1}^{l(s)}\{(a_i(s),s)\}$ is a closed, 1-dimensional rectifiable subset of $\mathbb{R}^2 \times \mathbb{R}^+_*$.*

We proved moreover that the numbers $l(s)$ and $d_i(s)$ are uniformly bounded by a constant depending only on M_0. The set $\Sigma_\mathfrak{v}$ corresponds, at least in a weak sense, since vortices are not yet labeled, to the trajectory set. The information on its Hausdorff dimension shows that (15) is not void. It can be proved that convergence off $\Sigma_\mathfrak{v}$ is in C^k for space variables, and therefore also in $W^{1,p}$ norm locally on $\mathbb{R}^2$, for any time $s > 0$. Comparing convergence (15) with the general result stated in convergence (10), we would like to stress the following improvements. First, the convergence holds uniformly in the accelerated time s. Second, the phase shift φ_ε in (10) may be taken as

$$\varphi_\varepsilon(z,s) = \langle \vec{c}(s), z\rangle + b(s). \tag{16}$$

Since the functions $\vec{c}$ and b depend only on the accelerated time variable s but not on the space variable z, the phase φ_ε corresponds to a plane wave, with wavenumber $\vec{c}(s)$. Since $\vec{c}$ is Lipschitz and locally bounded in time, we have gained compactness for the phase φ_ε. In contrast, in the original time scale t there is no compactness for the phase, due to possible oscillations of order $O(\sqrt{|\log\varepsilon|})$ in the phase of the initial datum. After a time of order $O(|\log\varepsilon|)$, i.e., in the accelerated time, these oscillations have been damped to order $O(1)$. As we will see later, only low frequencies persist, contributing to the phase term (16). The possible presence of low frequencies is of course related to the fact that the domain $\mathbb{R}^2$ is unbounded. This term is not present for bounded domains and appropriate boundary conditions.

Concerning the dynamics of vortices, we have

Theorem 4. *There exists a finite number of times $0 = \tau_0 < \tau_1 < \cdots < \tau_q < \tau_{q+1} = +\infty$ such that*

(i) *The number of vortices* $l(s) \equiv l_k$ *is constant on each interval* (τ_k, τ_{k+1}), *for* $k = 0, \ldots, q$.

(ii) *The restriction of* $\Sigma_{\mathfrak{v}}$ *to* $\mathbb{R}^2 \times (\tau_k, \tau_{k+1})$ *is a disjoint union of* l_k *smooth one-dimensional graphs. More precisely, relabeling possibly the points* $a_1(s), \ldots, a_{l_k}(s)$, *their degrees* $d_i(s) = d_i$ *are constant in* (τ_k, τ_{k+1}), *are distinct from zero, and their trajectories are given by the system of ordinary differential equations*

$$d_i^2 \frac{\mathrm{d}a_i}{\mathrm{d}s}(s) = -\nabla_{a_i} W(a_1, \ldots, a_{l_k}) + d_i c(s)^{\perp}, \quad i = 1, \ldots, l_k, \tag{17}$$

where W *is the Kirchhoff–Onsager function defined as*

$$W(a_1, \ldots, a_{l_k}) = -2 \sum_{i \neq j = 1}^{l_k} d_i d_j \log |a_i - a_j|. \tag{18}$$

Notice that formula (17) shows that in the accelerated time scale considered here, vortices and phase do interact via the drift term $d_i c(s)^{\perp}$. In particular, the induced motions have opposite directions for vortices of opposite degrees.

The motion law (17) stated in Theorem 4(ii) was already proved in [13, 15, 19, 21] for bounded domains with suitable boundary conditions, in the case of "well-prepared" initial data and before the first singular time τ_1. In these works the initial datum is assumed to have l vortices of degree ± 1 and an energy $E_\varepsilon(u_\varepsilon^0) = \pi l |\log \varepsilon| + O(1)$. As mentioned before, this energy is actually the minimal energy required for such a vortex configuration. In [20], this well-prepared assumption was somewhat relaxed to $E_\varepsilon(u_\varepsilon^0) \leq \pi l |\log \varepsilon| + |\log \varepsilon| / (\log |\log \varepsilon|)^\beta$ for some $\beta > 1$. Since the domain is bounded, the term $\vec{c}$ vanishes and the drift term is hence not present, so that (17) is then the gradient flow of the Kirchhoff–Onsager function (18) suitably modified according to the imposed boundary conditions.

Theorem 4 handles the general situation in which multiple degree vortices and singularities in the dynamics for vortices are present. One major point in its statement is that the number of singularities in the dynamics is finite. The times τ_k, for $k \geq 1$, are precisely the singular times and correspond to branchings of $\Sigma_{\mathfrak{v}}$ (the issues related to $\tau_0 = 0$ will be addressed in Section 5). A branching point is either a splitting or a recombination point, or possibly both. More precisely, we say that a point $(a_i(s_0), s_0) \in \Sigma_{\mathfrak{v}}$ is a splitting (resp. recombination) point if and only if there there are at least two branches emanating from (resp. converging to) that point, or in other words if there exists a sequence $s_n \to s_0^+$ (resp. $s_n \to s_0^-$) such that for any $r > 0$ one has $\sharp\{a_j(s_n) | a_j(s_n) \in B(a_i(s_0), r)\} \geq 2$ for n large enough. Simple topological considerations show that if $(a_i(\tau_k), \tau_k)$ is a branching point, then the quantity $\sum_j d_j(s)$ is constant through the branching, and therefore equal to $d_i(\tau_k)$, whereas the analysis of the energy shows that the quantity $\pi \sum_j d_j(s)^2$ decreases at least by 2π. As a consequence of this last fact, splitting may only occur for multiple degree vortices (i.e., $|d_i| \geq 2$).

A typical example of recombination is a collision. Collisions are in fact an unavoidable aspect of the ordinary differential equation (17), in particular when vortices of opposite degrees are present.

For instance, if u_ε^0 has two vortices of degree $+1$ and -1 located at the points $a_{-1}(0) = -1$ and $a_1(0) = 1$, if $\vec{c} \equiv 0$, then in view of (17) and the result in the Section 5, where the Cauchy problem is addressed, the limiting map $\mathfrak{u}_*$ has two

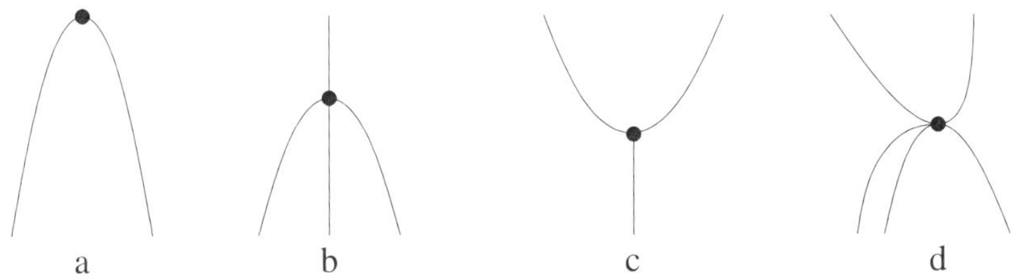

FIGURE 1. Four classes of branching points.

vortices given by

$$a_i(s) = (-1)^i\sqrt{1-2s}, \quad i=-1,1,\ \forall\, s<\tfrac{1}{2}.$$

These two vortices will collide at time $s=\frac{1}{2}$ and annihilate each other, so that after collision time the map $\mathfrak{u}_*$ is constant (see Figure 1a). Similarly, we may construct a collision without total annihilation as follows. Consider an initial datum with three vortices given by $a_k(0)=k-2$, $d_k=(-1)^{k+1}$, for $k=1,2,3.$[1] If $\vec{c}\equiv 0$, then the solution of (17) with initial datum $a_k(0)=k-2$ and $d_k=(-1)^{k+1}$, for $k=1,2,3$, is given by

$$a_k(s) = (2-k)\sqrt{1-2s}, \quad \text{for } 0\le s<\tfrac{1}{2} \text{ for } k=1,2,3.$$

From Theorem 4 we deduce that for $s>\frac{1}{2}$, a_2 and a_3 have disappeared and that

$$a_1(s)=0 \quad \text{for } s\ge \tfrac{1}{2}.$$

In particular, the branching point $(0,\frac{1}{2})$ is as in Figure 1b.

Whereas collisions are determined by singularities in the ordinary differential equation (17), splittings are not, and Theorem 4 does not settle the question of the occurrence of such splittings. In this direction, our next result [9, Theorem 5.1] provides an example of splitting of a degree $+2$ vortex into two vortices of degree $+1$ occurring at any arbitrary positive time.

Proposition 1. *Let $s_0>0$ be given. There exists M_0 and a sequence of solutions $(u_\varepsilon)_{\varepsilon>0}$ of $(\mathrm{PGL})_\varepsilon$ such that $\Sigma_{\mathfrak{v}}^s$ is reduced to a single vortex located at the origin and of degree $d=+2$, for $s<s_0$, and which splits at time $s=s_0$ into two distinct vortices of degree $+1$.*

It is an open problem to construct similar examples of splitting of vortices with higher degrees.

4. Energy quantization

As seen in the proof of Proposition 2, arguments based on the energy are central. In view of assumption (H_0), the normalized energy density

$$\mathfrak{v}_\varepsilon^s(x) \equiv \frac{e_\varepsilon\big(\mathfrak{u}_{\varepsilon_n}(x,s)\big)}{|\log\varepsilon|}\,\mathrm{d}x$$

verifies $|\mathfrak{v}_\varepsilon^s|(\mathbb{R}^2)\le M_0$ and hence is uniformly bounded. It turns out that this energy concentrates exactly at the vortices and that moreover the density is an integer multiple of π. This is a consequence of the properties of the equation and

[1]Since the total degree near the origin is different from zero, one has to add a "vortex at infinity" in order to have total degree zero at infinity so that (H_0) is verified, see [9, Section 5.2].

is not true for arbitrary maps, as considered in Section 2. The times $\tau_1, \dots, \tau_q$ are identified in [8] as the only times of dissipation in an appropriate asymptotic sense. More precisely, the following was proved there [8, Theorem 4 and Corollary 3.1].

Theorem 5. *For* $s \notin \{\tau_0, \dots, \tau_q\}$,

$$\mathfrak{v}^s_{\varepsilon_n}(x)\frac{e_{\varepsilon_n}(\mathfrak{u}_{\varepsilon_n}(x,s))}{|\log \varepsilon_n|}\,\mathrm{d}x \rightharpoonup \mathfrak{v}^s_* = \pi \sum_{i=1}^{l(s)} d_i^2(s)\delta_{a_i(s)} \tag{19}$$

in the sense of measures on $\mathbb{R}^2$, *and*

$$|\partial_t u_{\varepsilon_n}|^2\,\mathrm{d}x\,\mathrm{d}s \rightharpoonup \omega_* = \pi \sum_{k=0}^{q}\sum_{i=1}^{l(\tau_k)} \beta_i^k \delta_{(a_i(\tau_k),\tau_k)}, \tag{20}$$

in the sense of measures on $\mathbb{R}^2 \times \mathbb{R}^+$, *where* $\beta_i^k \in \mathbb{N}$.

Whereas the proof of (19) is rather involved, and is inspired by the corresponding result for the elliptic equation [11], the proof of (20) is an elementary consequence of (19) and the nonincreasing property of the energy. Indeed, since the total energy $|\mathfrak{v}^s_*|(\mathbb{R}^2) = \pi \sum d_i^2(s)$ is quantized and nonincreasing, it is also piecewise constant. The times $\tau_1, \dots, \tau_q$ correspond therefore to the times of energy loss, where dissipation concentrates. The points $(a_i(\tau_k), \tau_k)$ for which $\beta_i^k \neq 0$ are called the *dissipation points.*

Theorem 6. *A point* $(a_i(\tau_k), \tau_k)$ *is a branching point if and only if it is a dissipation point. In this case, we have*

$$\sum_{j=1}^{l_i^-} d_j^- = d_i(\tau_k) = \sum_{j=1}^{l_i^+} d_j^+ \qquad \textit{(Conservation of the degree)} \tag{21}$$

$$\sum_{j=1}^{l_i^-} (d_j^-)^2 \geq d_i^2(\tau_k) \geq \sum_{j=1}^{l_i^+} (d_j^+)^2 \qquad \textit{(Energy decrease)} \tag{22}$$

where the first (resp. second) inequality in (22) *is strict whenever* $l_i^- \geq 2$ *(resp.* $l_i^+ \geq 2$*). In particular,*

$$\sum_{j=1}^{l_i^-} (d_j^-)^2 > \sum_{j=1}^{l_i^+} (d_j^+)^2.$$

In the above formulas, for a given point $(a_i(\tau_k), \tau_k)$, the integers $d_i^\pm$ are defined as follows. Let $\mathcal{C}_1^-, \dots, \mathcal{C}_{l_i^-}^-$ and $\mathcal{C}_1^+, \dots, \mathcal{C}_{l_i^+}^+$ denote the vortices trajectories respectively in $\mathbb{R}^2 \times (\tau_{k-1}, \tau_k)$ and $\mathbb{R}^2 \times (\tau_k, \tau_{k+1})$ for which $(a_i(\tau_k), \tau_k)$ is an endpoint. Then, $d_1^-, \dots, d_{l_i^-}^-$ and $d_1^+, \dots, d_{l_i^+}^+$ are the degrees of the corresponding vortices. Notice that it follows from (19), (20) and the formula for the evolution of the energy in localized form (see, e.g., [8, Proposition 3.3]), that

$$\beta_i^k = \sum_{j=1}^{l_i^-} (d_j^-)^2 - \sum_{j=1}^{l_i^+} (d_j^+)^2. \tag{23}$$

Notice that (21) and (22) impose some restriction on the possible nature of the branching points.

The description of the asymptotic dynamics of phase and vortices presented so far holds only on $\mathbb{R}^2 \times (0, +\infty)$. As a matter of fact, convergence (15) cannot be extended to time $s = 0$, in view of the large freedom for the phase, as discussed in Section 2. In the next sections, we wish to complete the results in Theorems 3 and 4, and analyze the relationship of the trajectories for $s > 0$ with the asymptotic behavior of the initial datum, i.e., at time $s = \tau_0 = 0$. This issue was not addressed directly in our previous works [6, 8, 9].

5. The Cauchy problem for vortices

The results presented in this section and in the next one are new.

In view of Theorem 4, the set $\Sigma_{\mathfrak{v}} \cap \mathbb{R}^2 \times (0, \tau_1)$, i.e., the restriction of the trajectories to $\mathbb{R}^2 \times (0, \tau_1)$ is a union of l_0 smooth curves, and we may label the vortices as $a_1(s), \dots, a_{l_0}(s)$, $\forall\, 0 < s < \tau_1$, and their degrees by $d_1, \dots, d_{l_0}$. We have

Proposition 2. *For $i = 1, \dots, l_0$, the limit*

$$a_i(0) \equiv \lim_{s \to 0^+} a_i(s)$$

exists and is related to the Jacobian of the initial datum by

$$Ju^0_{\varepsilon_n} \rightharpoonup \pi \sum_{i=1}^{l_0} d_i \delta_{a_i(0)}, \quad \text{as } n \to \infty,$$

where $(\varepsilon_n)_{n \in \mathbb{N}}$ is the sequence in Theorem 3.

In the above statement, the points $a_i(0)$ need not be necessarily distinct. The collection of distinct points in the collection $\{a_k(0)\}$ corresponds to the vortices $\{a_i^0\}_{1 \le i \le l}$ of the initial data, in the sense of Theorem 1, the corresponding degrees d_i^0, $i = 1, \dots, l$ being the sum of the degrees of the points $a_i(s)$ converging to a_i^0.

Outline of the proof of Proposition 2. We briefly show how the various constructions in [6] can be pasted together to yield the desired result. In view of the compactness result for Jacobians (see Theorem 2), we know, that up to a subsequence,

$$Ju^0_{\varepsilon_n} \rightharpoonup \pi \sum_{i=1}^{\tilde{l}_0} d_i^0 \delta_{\mathfrak{a}_i^0}, \quad \text{as } n \to \infty, \tag{24}$$

where the points $\{\mathfrak{a}_i^0\}_{i=1,\dots,\tilde{l}_0}$ are in $\mathbb{R}^2$, and the numbers d_i^0 are integers in $\mathbb{Z}^*$. To prove Proposition 2, if suffices to show that for any given radius $\lambda > 0$, there exists an accelerated time s_λ such that, for any $0 < s < s_\lambda$ we have

$$\bigcup_{i=1}^{\tilde{l}_0} \{\mathfrak{a}_i^0\} \subset \bigcup_{i=1}^{l_0} B(a_i(s), \lambda). \tag{25}$$

and

$$\bigcup_{i=1}^{l} \{a_i(s)\} \subset \bigcup_{i=1}^{\tilde{l}_0} B(\mathfrak{a}_i^0, \lambda). \tag{26}$$

In order to prove (25) and (26), it is appropriate to distinguish short time dynamics and long time dynamics. The short time dynamics is provided by the following.

Step 1. *There exists a time T_λ such that, passing possibly to a subsequence, we have*

$$Ju_{\varepsilon_n}(T_\lambda) \rightharpoonup \pi \sum_{i=1}^{l_\lambda} d_i^\lambda \delta_{\mathfrak{b}_i}, \quad \text{as } n \to \infty, \tag{27}$$

and

$$\bigcup_{i=1}^{l_\lambda} \{\mathfrak{b}_i\} \subset \bigcup_{i=1}^{\tilde{l}_0} B\left(\mathfrak{a}_i^0, \frac{\lambda}{2}\right) \quad \text{and} \quad \bigcup_{i=1}^{\tilde{l}_0} \{\mathfrak{a}_i^0\} \subset \bigcup_{i=1}^{l_\lambda} B\left(\mathfrak{b}_i, \frac{\lambda}{2}\right). \tag{28}$$

PROOF. We consider on $\mathbb{R}^2 \times \mathbb{R}^+$, the space-time Jacobian of u_ε,

$$\mathcal{J} u_\varepsilon = \sum_{0 \le i < j \le 2} (\partial_{x_i} u_\varepsilon \times \partial_{x_j} u_\varepsilon)\, \mathrm{d}x_i \wedge \mathrm{d}x_j,$$

with the convention that $x_0 \equiv t$. In view of the space-time bounds on the Ginzburg–Landau energy, we may invoke the work of Jerrard and Soner [14] (see also [2]), to assert that

$$\mathcal{J} u_\varepsilon \rightharpoonup \mathcal{J}_* \quad \text{in } \left(C_c^{0,\alpha}(\mathbb{R}^N \times \mathbb{R}^+)\right)^*,$$

where $\mathcal{J}_*$ is an 1-rectifiable vector-valued measure. Moreover, it is shown in [2, 14], that $(1/\pi)\mathcal{J}_*$ can be identified with an integer multiplicity 1-current, whose boundary is exactly $\mathcal{J}_*^0$ (the slice at time zero). Here, for $t \ge 0$, $\mathcal{J}_*^t$ denotes the slice of the current $\mathcal{J}_*$ on $\mathbb{R}^N \times \{t\}$ (which is well-defined), and coincides (see [7]) with the limit in the sense of currents of the space Jacobians $Ju_\varepsilon(\cdot, t)$. In particular,

$$\mathcal{J}_*^0(z,t) = \pi \sum_{i=1}^{\tilde{l}_0} d_i^0 \delta_{\mathfrak{a}_i^0}(z) \otimes \delta_0(t),$$

and if for some time $T > 0$,

$$Ju_{\varepsilon_n}(\cdot, T) \rightharpoonup \pi \sum_{i=1}^{l_T} d_i^T \delta_{\mathfrak{b}_i^T}, \tag{29}$$

then $\mathcal{J}_*^T(z,t) = \pi \sum_{i=1}^{l_0} d_i^0 \delta_{\mathfrak{b}_i^T}(z) \otimes \delta_T(t)$, and

$$\partial[\mathcal{J}_{*|\mathbb{R}^2 \times (0,T)}] = \mathcal{J}_*^T(z) \otimes \delta_T(t) - \mathcal{J}_*^0(z) \otimes \delta_0(t) \tag{30}$$

It was proved in [7, Part II, inequality (43)] that

$$|\mathcal{J}_*^T|\left(\mathbb{R}^2 \times (0,T)\right) \le C(M_0) T^{1/2}. \tag{31}$$

It follows from (30) that the minimal connection (in the sense of [10]) between the configurations $\{\mathfrak{a}_i^0, d_i^0\}$ and $\{\mathfrak{b}_i^T, d_i^T\}$ is less than $|\mathcal{J}_*^T|(\mathbb{R}^2 \times (0,T))$ and hence by (31) less than $CT^{1/2}$. In particular if the time $T \equiv T_\lambda$ is chosen sufficiently small, this implies (28). □

We emphasize that the time T_λ is taken in the original time scale, not in the accelerated one. In particular the time interval $(0, T_\lambda)$ in the original time scale corresponds to a vanishing (as ε goes to zero) boundary layer in the accelerated time scale. The first step described above allows to consider the time T_λ instead of the initial time $t = 0$. At time T_λ, we may take advantage of some regularization properties of $(\mathrm{PGL})_\varepsilon$. In particular at the the ε-level the map $u_\varepsilon(., T)$ is smooth for every $T > 0$ and it can be shown using the clearing-out lemma (see [6, 7]) that, for some constant $\mu_0 > 0$ depending only on M_0, there exist l_ε points $x_i^\varepsilon(t)$ such that, for any time $T \geq \sqrt{\varepsilon}$

$$|u_\varepsilon(x,T)| \geq \tfrac{1}{2} \quad \text{on } \mathbb{R}^2 \setminus \bigcup_{i=1}^{l_\varepsilon} B(x_i^\varepsilon(T), \mu_0\varepsilon), \tag{32}$$

with

$$B(x_i^\varepsilon(T), 2\mu_0\varepsilon) \cap B(x_i^\varepsilon(T), 2\mu_0\varepsilon) = \varnothing \quad \text{if } i \neq j, \tag{33}$$

and $|l_\varepsilon| \leq C(M_0)$. In particular, one may assign a degree $d_i^\varepsilon(T)$ around each of the points $x_i^\varepsilon(T)$. Passing possibly to a subsequence, we may assume that $x_i^\varepsilon(T) \to \tilde{\mathfrak{b}}_i^T$. The point $\mathfrak{b}_i^T$ in (29) correspond to the points $\tilde{\mathfrak{b}}_i^T$ with nonzero degree.

Thanks to the clearing-out lemma, the vortices $x_i^\varepsilon(t)$ are identified, in a suitable sense with the concentration sets for the energy $\Omega_\delta^\varepsilon(t)$, defined, for $\delta > 0$ and $t > 0$, by

$$\Omega_\delta^\varepsilon(t) = \left\{ x \in \mathbb{R}^2, \int_{B(x,\delta)} e_\varepsilon\big(u(\cdot,t)\big)\,\mathrm{d}x \geq \frac{\eta_0}{2}|\log\varepsilon| \right\}, \tag{34}$$

where η_0 is some positive constant appearing in the clearing-out lemma (see [6, Proposition 2.1]). A straightforward covering argument shows that there exist l points $x_i^\varepsilon(t)$ such that

$$\Omega_\delta^\varepsilon(t) \subset \bigcup_{i=1}^{l} B(x_i^\varepsilon(t), 2\delta),$$

where the number l of points is uniformly bounded by $l \leq C\frac{M_0}{\eta_0}$. The central point is that as $\varepsilon \to 0$ an $\delta \to 0$ the points $\{x_i^\varepsilon(s|\log\varepsilon|)\}$ converge to set of vortices $\{a_i(s)\}_{i=1,\dots,l(s)}$, as described in Theorem 3.

We next turn to the long time dynamics, that is the dynamics on time intervals of size of order $|\log\varepsilon|$. A major tools is provided by the cylinders lemma, which bounds the speed of concentration sets.

Proposition 3 (Cylinders lemma). *Let $t_0 \geq 1$ and $r > 0$ be given. There exists positive constants σ_0, γ_0 and r_0 depending only on M_0 such that, if*

$$\Omega^\varepsilon(t_0) \equiv \Omega_{r/4}^\varepsilon(t_0) \subset \bigcup_{i=1}^{l} B(x_i^\varepsilon, r), \tag{35}$$

for some $|\log\varepsilon|^{-1/6} \leq r \leq r_0$ and some points $(x_i^\varepsilon)_{1\leq i\leq l}$ verifying

$$|x_i^\varepsilon - x_j^\varepsilon| \geq \sigma_0 r, \quad \forall\, i \neq j, \tag{36}$$

then

$$\Omega^\varepsilon(t) \subset \bigcup_{i=1}^{l} B\Big(x_i^\varepsilon, \frac{\sigma_0 r}{8}\Big), \tag{37}$$

for every $t_0 \leq t \leq t_0 + \gamma_0 r^2|\log\varepsilon|$.

We are now in position to go to

Step 2. *There exists a time $s_\lambda > 0$ such that for every $0 < s \leq s_\lambda$, we have*

$$\bigcup_{i=1}^{l}\{a_i(s)\} \subset \bigcup_{i=1}^{\tilde{l}_\lambda} B\Big(\mathfrak{b}_i, \frac{\lambda}{2}\Big). \tag{38}$$

IDEA OF THE PROOF OF STEP 2. It was proved in [6, Section 3] that concentration of energy occurs also in the original time scale, i.e,. we may assume that

$$\mu_{\varepsilon_n}^{T_\lambda} \equiv \frac{e_\varepsilon(u_\varepsilon(\cdot, T_\lambda))}{|\log \varepsilon|} \rightharpoonup \mu_*^{T_\lambda}$$

in the sense of measures, where $\mu_*^{T_\lambda}$ has the form

$$\mu_*^{T_\lambda} = \sum_{i=1}^{\tilde{l}_\lambda} \sigma_i(T_\lambda)\delta_{b_i} + \frac{|\nabla \Phi_*|^2}{2}(\cdot, T_\lambda)\,\mathrm{d}x,$$

where $\sigma_i(T_\lambda) \geq \eta_0$ and the function Φ_* satisfies the heat equation on $\mathbb{R}^2 \times \mathbb{R}_+$, and hence is smooth at time T_λ. On the other hand we know that

$$\|J_*^{T_\lambda}\| \leq \mu_*^{T_\lambda},$$

so that

$$\bigcup_{i=1}^{l_\lambda}\{\mathfrak{b}_i\} \subset \bigcup_{i=1}^{\tilde{l}_\lambda}\{b_i\}. \tag{39}$$

In order to relate the points b_i to the points $a_i(s)$ we invoke the following direct consequence of the cylinders lemma.

Lemma 1. *Let $r > 0$ be given. There exist constants σ_0, γ_0 depending only on M_0 such that if*

$$\bigcup_{i=1}^{\tilde{l}_\lambda}\{b_i\} \subset \bigcup_{i=1}^{n} B(x_i, r),$$

where the points $x_1, \ldots, x_n \in \mathbb{R}^2$ verify

$$|x_i - x_j| \geq \sigma_0 r \quad \forall\, i \neq j$$

then

$$\bigcup\{a_i(s)\} \subset \bigcup_{i=1}^{n} B\left(x_i, \frac{\sigma_0 r}{8}\right)$$

for every $0 < \sigma < \gamma_0 r^2$.

The proof of this lemma is similar to the proof of [8, Lemma 2.4] and therefore we omit it. We choose in Lemma 1 $x_i = b_i$, take

$$r = \min\left\{\frac{|b_i - b_j|}{\sigma_0}\right\} \cup \left\{\frac{4\lambda}{\sigma_0}\right\}$$

and then set $s_\lambda = \gamma_0 r^2$. It follows that for $0 < s < s_\lambda$

$$\bigcup_{i=1}^{l}\{a_i(s)\} \subset \bigcup_{i=1}^{\tilde{l}_\lambda} B\left(b_i, \frac{\lambda}{2}\right). \tag{40}$$

For $0 < s < s_\lambda$ we consider the set of indices $I_i(s)$ defined by $j \in I_i(s)$ if $a_i(s) \in B(b_i, \lambda/2)$. it follows from Lemma 1 that $I_i(s)$ is independent of s. We claim that, if $I_i \neq \varnothing$, then

$$D_i(s) \equiv \sum_{i \in I_i} a_i(s) \neq 0. \tag{41}$$

To prove the claim, we argue by contradiction. First notice that by conservation of the degree, $D_i(s)$ is independent of s. Next assume $I_i \neq \varnothing$ and $D_i(s) = 0$. Set $R = 2r$ and $0 < \kappa < 1$ be a small number. We invoke Lemma 1 to infer that

$$\bigcup_{i \in I_i} \{a_i(s)\} \subset B(b_i, \lambda/2) = \varnothing$$

for $0 < s < cr^2$. A contradiction, so that the claim (41) is established.

Finally, to establish (38) it suffices to combine (39), (40) and (41). □

6. The Cauchy problem for $\vec{c}$

As already mentioned, the presence of the plane wave $\vec{c}$ in Theorems 3 and 4 is related to persistency of low frequencies for the heat equation. In order to describe this phenomenon, we first consider the linear homogeneous heat equation.

6.1. Persistency of low frequencies for the linear heat equation. We consider a family $(g_n)_{n\in\mathbb{N}}$ of solutions to the linear homogeneous heat equation

$$\partial_t g_n - \Delta g_n = 0 \quad \text{on } \mathbb{R}^N \times \mathbb{R}_+,\ N \geq 1,$$

where $g_n(\cdot, 0)$ belongs to $L^2(\mathbb{R}^N)$. We set

$$\Gamma_n \equiv \|g_n(\cdot, 0)\|_{L^2(\mathbb{R}^N)}$$

and we assume that

$$\Gamma_n \to +\infty \quad \text{as } n \to +\infty.$$

in view of the properties of the heat equation, for fixed n, $g_n(\cdot, t)$ converges locally to zero, at $t \to +\infty$. Our aim is to show next

Lemma 2. *Let $\gamma_n \geq \gamma_n^0 \equiv \Gamma_n^{4/N}$. For every $s > 0$, the sequence $\big(g_n(\cdot, \gamma_n s)\big)_{n\in\mathbb{N}}$ is bounded in $L^\infty(\mathbb{R}^N)$.*

Proof. We introduce the Fourier transform $\hat{g}_n(\xi, t) \equiv \int_{\mathbb{R}^N} \exp(-\mathrm{i}x\xi) g_n(x, t)\, \mathrm{d}x$ with respect to spatial variables, so that, in view of Plancherel formula,

$$\int_{\mathbb{R}^N} |\hat{g}_n(\xi, 0)|^2 = (2\pi)^N \Gamma_n^2.$$

Since g_n is a solution to the heat equation,

$$\hat{g}_n(\xi, t) = \exp(-t|\xi|^2)\hat{g}_n(\xi, 0),$$

so that we are led to

$$|g_n(\cdot, t)|_{L^\infty(\mathbb{R}^N)} \leq (2\pi)^{-N} \int_{\mathbb{R}^N} |\hat{g}_n(\xi, t)|\, \mathrm{d}\xi = (2\pi)^{-N} \int_{\mathbb{R}^N} \exp(-t|\xi|^2)|\hat{g}_n(\xi, 0)|\, \mathrm{d}\xi.$$

If $t = \gamma_n s$, we may introduce the change of variable $u = \gamma_n^{1/2}\xi$, so that the previous computation leads to, using Cauchy–Schwarz inequality,

$$\begin{aligned} \|g_n(\cdot, \gamma_n s)\|_{L^\infty(\mathbb{R}^N)} &\leq (2\pi)^{-N}\gamma_n^{-N/2} \int_{\mathbb{R}^N} \exp(-s|u|^2) \left|\hat{g}_n\left(\frac{u}{\sqrt{\gamma_n}}, 0\right)\right| \mathrm{d}u \\ &\leq (2\pi)^{-N}\gamma_n^{-N/2} \left(\int_{\mathbb{R}^N} \exp(-2s|u|^2)\, \mathrm{d}u\right)^{1/2} \left(\int_{\mathbb{R}^N} \left|\hat{g}_n\left(\frac{u}{\sqrt{\gamma_n}}, 0\right)\right|^2 \mathrm{d}u\right)^{1/2}. \end{aligned} \tag{42}$$

Since

$$\int_{\mathbb{R}^N}\left|\hat{g}_n\left(\frac{u}{\sqrt{\gamma_n}},0\right)\right|^2 \mathrm{d}u = \gamma_n^{N/2}\int_{\mathbb{R}^N}|\hat{g}_n(\xi,0)|^2\,\mathrm{d}\xi = (2\pi)^N\gamma_n^{-N/2}\Gamma_n^2 \tag{43}$$

we obtain

$$\|g_n(\cdot,\gamma_n s)\|_{L^\infty(\mathbb{R}^N)} \le C\gamma_n^{-N/4}\Gamma_n,$$

and the conclusion follows from the condition $\gamma_n \ge \gamma_n^0 \equiv \Gamma_n^{4/N}$. □

The previous result shows that compactness may be achieved in the accelerated time scale $t = \gamma_n s$, and we may therefore perform an asymptotic analysis in that time scale. The previous computations suggest to consider the scaled maps $\tilde{g}_n^0$ defined by

$$\tilde{g}_n^0(x)\} = g_n(\gamma_n^{1/2}x,0) \quad \forall x\in\mathbb{R}^N.$$

We first notice that, if $\gamma_n \ge \Gamma_n^{4/N}$, then

$$\|\tilde{g}_n^0\|_{L^2(\mathbb{R}^N)} \le 1 \qquad \forall n\in\mathbb{N}.$$

Therefore, passing possibly to a subsequence, we may assume that

$$\tilde{g}_n^0(x) \rightharpoonup \vec{\sigma}_*(x) \quad \text{as } n\to+\infty, \tag{44}$$

weakly in $L^2(\mathbb{R}^N)$. This leads to

Lemma 3. *For every $s>0$, if $(\gamma_n)_{n\in\mathbb{N}}$ and $\vec{\sigma}_*$ are as above, then*

$$g_n(\cdot,\gamma_n s) \to \frac{1}{(4\pi s)^{N/2}}\int_{\mathbb{R}^N}\exp\left(-\frac{|x|^2}{4s}\right)\vec{\sigma}_*(x)\,\mathrm{d}x = \frac{1}{(2\pi)^N}\int_{\mathbb{R}^N}\exp(-s|\xi|^2)\hat{\vec{\sigma}}_*(\xi)\,\mathrm{d}\xi$$

uniformly on compact sets.

Proof. In view of the scale invariance of the heat equation we have

$$g_n(x,\gamma_n s) = h_n\left(\frac{x}{\sqrt{\gamma_n}},s\right),$$

where h_n denotes the solution of the homogeneous heat equation with initial datum $\tilde{g}_n^0$. In view of (44), h_n converges uniformly on compact sets of $\mathbb{R}^N\times\mathbb{R}_+^*$ to h_*, the solution of the homogeneous heat equation with initial datum $\vec{\sigma}_*$. It follows that

$$g_n(x,\gamma_n s) \to h_*(0,s),$$

which yields the result. □

Example. Let φ_0 be a smooth map with compact support in $\mathbb{R}^N$, and let $\varphi(\cdot,\cdot)$ be the solution to the heat equation with initial datum φ_0. Let $\alpha > -N/2$ be given. We consider the map g_λ^0 defined on $\mathbb{R}^N$ by

$$g_\lambda^0(x) = \lambda^\alpha\varphi_0\left(\frac{x}{\lambda}\right)$$

and g_λ the solution to the heat equation with initial datum g_λ^0. By scaling, we have

$$\Gamma_\lambda \equiv \|g_\lambda^0\|_{L^2(\mathbb{R}^N)} = \lambda^{\alpha+(N/2)}\|\varphi_0\|_{L^2(\mathbb{R}^N)}$$

and therefore

$$\gamma_\lambda^0 = c_0\lambda^{2+(4\alpha/N)}.$$

It follows from the proof of Lemma 3 that, if $-N/2<\alpha<0$, then

$$g_\lambda(x,\gamma_\lambda^0 s) \to 0 \quad \text{as } \lambda\to+\infty.$$

If $\alpha = 0$, then

$$g_\lambda(x, \gamma_\lambda^0 s) \to \varphi(0, c_0 s) \quad \text{as } \lambda \to +\infty.$$

finally, if $\alpha > 0$, one checks, using for instance an integral representation of φ that

$$g_\lambda(x, \gamma_\lambda^0 s) \to 0 \quad \text{as } \lambda \to +\infty.$$

6.2. Computing $\vec{c}$. As we have seen in Section 3, if vortices are not present, i.e., if $u_\varepsilon = \rho_\varepsilon \exp(i\varphi_\varepsilon)$, then the phase φ_ε is essentially governed by the heat equation (see (13)), and so does its gradient. In that situation, $\int |\nabla\varphi_\varepsilon|^2 \leq 2M_0|\log\varepsilon|$, in view of assumption (H_0). We may therefore apply the results of the previous subsection to $\nabla\varphi_\varepsilon$ with $\Gamma_\varepsilon = \sqrt{|\log\varepsilon|}$ and $N = 2$. It follows that for any $s > 0$, $\nabla\varphi_\varepsilon$ becomes bounded at the time $\gamma_\varepsilon s$, where

$$\gamma_\varepsilon = \Gamma_\varepsilon^2 = |\log\varepsilon|.$$

The accelerated time scale is precisely the one given by (14). In the case considered above where vortices are not present, the phase φ_ε is globally well-defined and uniquely determined up to an integer multiple of 2π. Its gradient $\nabla\varphi_\varepsilon$ can be deduced from u_ε by the formula

$$u_\varepsilon \times \nabla u_\varepsilon = \rho_\varepsilon^2 \nabla\varphi_\varepsilon = |u_\varepsilon|^2 \nabla\varphi_\varepsilon.$$

When vortices are present, it is no longer possible to define a phase φ_ε globally: however, the quantity $u_\varepsilon \times \nabla u_\varepsilon$ is still defined, and a good alternate to the phase φ_ε is provided by the scalar function Φ_ε given by a Hodge – de Rham decomposition of $u_\varepsilon \times \delta u_\varepsilon$ on $\mathbb{R}^2 \times \mathbb{R}_+$,

$$u_\varepsilon \times \delta u_\varepsilon = \delta\Phi_\varepsilon + \delta^*\Psi_\varepsilon, \tag{45}$$

where δ and δ^* denote respectively the exterior differentiation operator for differential forms on $\mathbb{R}^2 \times \mathbb{R}$ and its formal adjoint (while we use the standard notation d and d^* when restricting to time slices $\mathbb{R}^2 \times \{t\}$). Taking the exterior product of $(\mathrm{PGL})_\varepsilon$ with u_ε, we obtain

$$u_\varepsilon \times \partial_t u_\varepsilon - \operatorname{div}(u_\varepsilon \times \nabla u_\varepsilon) = 0 \quad \text{on } \mathbb{R}^2 \times \mathbb{R}_+. \tag{46}$$

Inserting equality (45) into (46), we obtain

$$\partial_t \Phi_\varepsilon - \Delta\Phi_\varepsilon = L(\Psi_\varepsilon) \quad \text{on } \mathbb{R}^2 \times \mathbb{R}_+, \tag{47}$$

where L is a second order differential operator defined on $\mathbb{R}^2 \times \mathbb{R}$ (it is defined in [6, equation (4.20)]). Applying the operator δ to (45) and using the fact that $\delta\Psi_\varepsilon = 0$, we have

$$-\Delta_{x,t}\Psi_\varepsilon = J_{x,t} u_\varepsilon, \tag{48}$$

so that Ψ_ε may be deduced from the space-time Jacobian $J_{x,t}u_\varepsilon$, and therefore essentially from the location of vortices. In particular, as ε goes to zero, if vortices are not present, then Ψ_ε vanishes asymptotically. As already mentioned, Φ_ε is a substitute for the phase φ_ε when vortices are present: the asymptotic limit of $\nabla\Phi_\varepsilon$ gives rise to the term $\vec{c}$. It is coupled to the vortices through the heat equation (47) with source term $L(\Psi_\varepsilon)$, but it turns out that this coupling is weak and actually vanishing in the limit considered here.

To complete these introductory remarks, it is worthwhile to emphasize that Hodge – de Rham decompositions as (45) are not unique and that different boundary conditions may be imposed. As a matter of fact, different decompositions are used in [6, 7].

Remark 1. A possible different approach to equation (47) is provided by the evolution equation for $u_\varepsilon \times \nabla u_\varepsilon$

$$\partial_t(u_\varepsilon \times \nabla u_\varepsilon) - \Delta(u_\varepsilon \times \nabla u_\varepsilon) = \nabla^\perp(Ju_\varepsilon) + 2\partial_t u_\varepsilon \times \nabla u_\varepsilon. \tag{49}$$

This equation is intrinsic and does not rely on any Hodge – de Rham decomposition. It relates $u_\varepsilon \times \nabla u_\varepsilon$ with the spatial and space-time components of the Jacobian. We will use (49) in the proof of Proposition 4.

Going back to the discussion in Section 6.1 and find, in view of (44), a subsequence $\varepsilon_n \to 0$ such that

$$u^0_{\varepsilon_n} \times \nabla u^0_{\varepsilon_n}(\sqrt{|\log\varepsilon|}x) \rightharpoonup \vec{\sigma}_* \quad \text{as } n \to +\infty \tag{50}$$

weakly in L^2 (we assume for simplicity that $|u^0_\varepsilon| \leq 3$, so that $u^0_{\varepsilon_n} \times \nabla u^0_{\varepsilon_n}(\cdot\sqrt{|\log\varepsilon|})$ is bounded in L^2). Combining Lemma 3 with our previous arguments, we may prove

Proposition 4. *Let $\vec{\sigma}_*$ and $(\varepsilon_n)_{n\in\mathbb{N}}$ as in (50). Then, the function $\vec{c}\colon \mathbb{R}^+ \to \mathbb{R}^2$ provided in Theorem 3 can be computed as*

$$\vec{c}(s) = \frac{1}{4\pi s}\int_{\mathbb{R}^2} \exp\Big(-\frac{|x|^2}{4s}\Big)\vec{\sigma}_*(x)\,\mathrm{d}x. \tag{51}$$

Identity (51) shows that the phase term $\vec{c}(s)$ can be completely and uniquely inferred, at any time $s > 0$, from the initial datum u^0_ε through the limiting function $\vec{\sigma}_*$. This function does not depend on the vortices nor on their degrees, as the next example shows. Let l points $a_1, \dots, a_l$, l integers $d_1, \dots, d_l$ be given, and consider initial data of the form

$$u^0_\varepsilon(z) = \exp\big(\mathrm{i}\varphi^0_\varepsilon(z)\big)\prod_{i=1}^{l} f\Big(\frac{|z-a_i|}{\varepsilon}\Big)\Big(\frac{z-a_i}{|z-a_i|}\Big)^{d_i}.$$

Assume that

$$\nabla\varphi^0_\varepsilon(\sqrt{|\log\varepsilon|}x) \rightharpoonup \vec{\sigma}_* \quad \text{as } \varepsilon \to 0,$$

weakly in L^2. We leave it as an exercise to the reader to prove that, in that situation, we have also

$$u^0_\varepsilon \times \nabla u^0_\varepsilon(\sqrt{|\log\varepsilon|}x) \rightharpoonup \vec{\sigma}_* \quad \text{as } \varepsilon \to 0$$

weakly in L^2. In particular, the limiting functions $\vec{\sigma}_*(x)$ and $\vec{c}(s)$ depend only on φ^0_ε, not on the location of vortices, nor on their degree. If one takes for instance, as in [6], the specific choice

$$\varphi^0_\varepsilon(z) = \sqrt{|\log\varepsilon|}\varphi_0\Big(\frac{z - \sqrt{|\log\varepsilon|}x_0}{\sqrt{|\log\varepsilon|}}\Big),$$

where φ_0 is a smooth function in $\mathcal{S}(\mathbb{R}^2)$ and $x_0 \in \mathbb{R}^2$ is fixed, then the limiting phase function $\vec{c}$ is given, in view of the discussion in Section 6.1, by

$$\vec{c}(s) = \nabla\varphi(x_0, s),$$

where φ is the solution to the homogeneous heat equation with initial datum φ_0.

Remark 2. Formula (51) shows that $\vec{c}$ is smooth on $\mathbb{R}_+$. Moreover, applying Cauchy – Schwarz inequality to (51) we obtain the bound

$$|\vec{c}(s)| \leq \frac{C(M_0)}{\sqrt{s}} \quad \text{for any } s > 0.$$

Remark 3. Propositions 2 and 4 allow to integrate the trajectories in the case when at time zero all the vortices a_i^0 have degree $d_i^0 = \pm 1$. Indeed, since splitting is excluded for degree one vortices, the number l_0 is exactly the number l of the vortices a_i^0, and the first singular time τ_1 corresponds to the first singularity in the o.d.e. (17), i.e., the first collision time of (17). In this situation, the time τ_0 is not singular. Examples of such integrations with collisions have been provided in the previous sections.

More generally, a natural question is the following: does the knowledge of positions and degrees of the vortices (together with the drift term $\vec{c}$) at time $\tau_0 = 0$ completely determines their positions and degrees at future times? Whereas collisions are determined by singularities in the ordinary differential equation (17), splittings are not, and clearly Theorem 4 does not settle the question of the occurrence of such splittings. More precisely, the singular times $\tau_1, \ldots, \tau_q$ as well as the number of vortices and their degrees involved after splittings are not inferred in a constructive way. As a matter of fact, an important part of the relevant information is lost in the limiting procedure. An example of splitting occurring at arbitrary time has been provided in Proposition 1.

Idea of the proof of Proposition 4. The main ingredient in the analysis of $\vec{c}$ is [6, Proposition 4.1], which states that the map u_ε may be decomposed as

$$u_\varepsilon = w_\varepsilon \exp(\mathrm{i}\phi_\varepsilon),$$

where φ_ε satisfies the homogeneous heat equation on $\mathbb{R}^2 \times \mathbb{R}_+^*$. It is then proved in [6] that w_ε converges to $\prod_{i=1}^{l} \big((z - a_i)/|z - a_i|\big)^{d_i}$ in the accelerated time scale, whereas $\nabla\varphi_\varepsilon$ converges to the field $\vec{c}$ (see [6, Proposition 6.1]). In order to derive (51), we will therefore explain next how φ_ε is constructed. First one extends u_ε to $\mathbb{R}^2 \times \mathbb{R}$ by reflection and performs there the Hodge – de Rham decomposition (45). Going back to (47) one writes

$$\Phi_\varepsilon = \phi_\varepsilon + \Phi_{0,\varepsilon} \quad \text{on } \mathbb{R}^2 \times \mathbb{R}_*,$$

where $\Phi_{0,\varepsilon}$ is the solution of the evolution problem

$$\begin{cases} \partial_t \Phi_{0,\varepsilon} - \Delta \Phi_{0,\varepsilon} = L(\Psi_\varepsilon) & \text{on } \mathbb{R}^2 \times \mathbb{R}_+^* \\ \Phi_{0,\varepsilon} = 0 & \text{on } \mathbb{R}^2 \times \{0\} \end{cases} \tag{52}$$

so that ϕ_ε (and hence $\nabla\phi_\varepsilon$) therefore clearly solves the homogeneous heat equation on $\mathbb{R}^2 \times \mathbb{R}_+$. In view of (45) one has on every time slice $\mathbb{R}^2 \times \{t\}$, for $t \geq 0$,

$$\mathrm{d}\phi_\varepsilon = u_\varepsilon \times \mathrm{d}u_\varepsilon - (\delta^* \Psi_\varepsilon)_\top - \mathrm{d}\Phi_{0,\varepsilon}.$$

Using this identity and various estimates it is shown in the proof of [6, Proposition 4.1] that there exists some slice $t_0 \in [\frac{1}{4}, \frac{1}{2}]$ such that

$$\nabla\phi_\varepsilon = u_\varepsilon \times \nabla u_\varepsilon + f_1 + f_2 + f_3 \quad \text{on } \mathbb{R}^2 \times \{t_0\},$$

where

$$\|f_1\|_{L^1(\mathbb{R}^2 \times \{t_0\})} + \|f_2\|_{L^2(\mathbb{R}^2 \times \{t_0\})} + \|f_3\|_{L^5(\mathbb{R}^2 \times \{t_0\})} \leq C(M_0).$$

(see [6, inequality (4.33)]. Actually, these estimates hold for a subset of $[\frac{1}{4}, \frac{1}{2}]$ of measure at least 1/16). Since $\nabla\varphi_\varepsilon$ satisfies the heat equation, we may write

$$\nabla\phi_\varepsilon(x,|\log\varepsilon|s)$$
$$= \frac{1}{4\pi(s|\log\varepsilon|-t_0)}\int_{\mathbb{R}^2}\exp\Big(-\frac{|x-y|^2}{4(s|\log\varepsilon|-t)}\Big)u_\varepsilon\times\nabla u_\varepsilon(y)+[f_1+f_2+f_3](y)\,\mathrm{d}y.$$

Standard $L^p - L^q$ estimates for the heat kernel and a change of variable then lead to

$$\nabla\phi_\varepsilon(x,|\log\varepsilon|s) = \frac{1}{4\pi s}\int_{\mathbb{R}^2}\exp\Big(-\frac{|y|^2}{4s}\Big)u_\varepsilon\times\nabla u_\varepsilon(y\sqrt{|\log\varepsilon|},t_0)\,\mathrm{d}y + o(1)$$

on every compact set of $\mathbb{R}^2$. In order to conclude one needs to compare $u_\varepsilon\times\nabla u_\varepsilon(y\sqrt{|\log\varepsilon|},t_0)$ and $\sigma_*(y)$. We claim that

$$e^{s\Delta}[u_\varepsilon\times\nabla u_\varepsilon(\cdot\sqrt{|\log\varepsilon|},t_0)]\to e^{s\Delta}\sigma_*(\cdot) \tag{53}$$

in L^2 for any $s>0$.

PROOF OF THE CLAIM. For $(y,\tau)\in\mathbb{R}^2\times\mathbb{R}_+$, set

$$f_\varepsilon(y,\tau) = u_\varepsilon\times\nabla u_\varepsilon(\sqrt{|\log\varepsilon|}x,|\log\varepsilon|\tau)$$

so that, in view of (49),

$$\partial_t f_\varepsilon - \Delta f_\varepsilon = g_\varepsilon$$

where

$$g_\varepsilon \equiv |\log\varepsilon|[\nabla^\perp Ju_\varepsilon(\sqrt{|\log\varepsilon|}x,|\log\varepsilon|\tau)+2\partial_t u_\varepsilon\times\nabla u_\varepsilon(\sqrt{|\log\varepsilon|}x,|\log\varepsilon|\tau).$$

By Duhamel's formula

$$\mathrm{e}^{s\Delta}f_\varepsilon(\cdot,t_0/|\log\varepsilon|) = \mathrm{e}^{s+(t_0/|\log\varepsilon|)\Delta}f_\varepsilon(\cdot,0)+\int_0^{t_0/|\log\varepsilon|}|\log\varepsilon|\mathrm{e}^{s\Delta}g_\varepsilon(\cdot,s)\,\mathrm{d}s. \tag{54}$$

Since by assumption $f_\varepsilon(\cdot,0)$ converges to σ_* weakly in L^2,

$$\mathrm{e}^{s+(t_0/|\log\varepsilon|)\Delta}f_\varepsilon(\cdot,0)\to\mathrm{e}^{s\Delta}\sigma_* \quad \text{in } H^k(\mathbb{R}^2)\ \forall k\in\mathbb{N}.$$

On the other hand, by Jacobian estimates and the regularizing properties of $e^{s\Delta}$,

$$\Big\|\int_0^{t_0/|\log\varepsilon|}|\log\varepsilon|\mathrm{e}^{s\Delta}g_\varepsilon(\cdot,s)\,\mathrm{d}s\Big\|_{H^k}\le\frac{C_k}{|\log\varepsilon|},$$

and the conclusion follows. □

7. Some open problems

Problem 1. We believe that the evolution equation (49), that we did not use in our previous works, might be exploited to provide simpler proofs to some results in [6, 7].

Problem 2. Construct examples of splitting of vortices with degree higher than two.

Construct collisions of vortices giving rise to splitting, as in Figure 1 d.

Problem 3 (Quantization of energy in higher dimension). In [7], we showed that for solutions u_ε to $(\mathrm{PGL})_\varepsilon$ in dimension $N\ge 2$, the normalized energy densities $\mu_\varepsilon^t = e_\varepsilon\big(u_\varepsilon(x,t)\big)/|\log\varepsilon|\,\mathrm{d}x$ converge, for any $t>0$, to the measures

$$\mu_*^t = \frac{|\nabla\Phi_*|^2}{2}\,\mathrm{d}x+\Theta_*(x,t)\mathcal{H}^{N-2}\llcorner\Sigma_\mu^t, \tag{55}$$

where Σ_μ^t is a $(N-2)$-dimensional rectifiable set and $\Theta_*(\cdot,t)$ is a bounded function. Prove or disprove that the multiplicity function Θ_* is an integer multiple of π.

A simpler version of this problem is to consider axisymmetric solutions for $N=3$, which reduce to a two-dimensional problem.

Problem 4. In our works we considered initial data verifying (H_0), so they had to depend on ε. Assume instead u_0 is smooth and independent of ε, $u_0 = 1$ outside a compact set, and that $|u_0| \geq \frac{1}{2}$, so that we may write $u_0 = \rho_0 \exp(\mathrm{i}\varphi_0)$. Prove that for any given $t > 0$, $u_\varepsilon(\cdot,t) \to \exp\big(\mathrm{i}\varphi(\cdot,t)\big)$, where φ solves the homogeneous heat equation with initial datum φ_0. Analyze the boundary layer near $t=0$.

Problem 5 (Generation of vortices). In the same setting as before, we assume that u_0 is independent of ε, smooth, equal to 1 outside a compact set and has only a finite number of zeroes, assumed to be nondegenerate. Show that the zeroes give rise to vortices of degree ± 1 and recover their dynamical law.

References

1. G. Alberti, *Un risultato di convergenza variazionale per funzionali di tipo Ginzburg-Landau in dimensione qualunque*, Boll. Unione Mat. Ital. Sez. B Artic. Ric. Mat. (8) **4** (2001), no. 2, 289–310.
2. G. Alberti, S. Baldo, and G. Orlandi, *Variational convergence for functionals of Ginzburg–Landau type*, Indiana Univ. Math. J. **54** (2005), no. 5, 1411–1472.
3. L. Almeida and F. Béthuel, *Topological methods for the Ginzburg–Landau equations*, J. Math. Pures Appl. (9) **77** (1998), no. 1, 1–49.
4. F. Béthuel, H. Brezis, and F. Hélein, *Ginzburg–Landau vortices*, Progr. Nonlinear Differential Equations Appl., vol. 13, Birkhäuser, Boston, 1994.
5. F. Béthuel and G. Orlandi, *Ginzburg–Landau functionals, phase transitions and vorticity*, Noncompact Problems at the Intersection of Geometry, Analysis, and Topology, Contemp. Math., vol. 350, Amer. Math. Soc., 2004, pp. 35–47.
6. F. Béthuel, G. Orlandi, and D. Smets, *Collisions and phase-vortex interaction in dissipative Ginzburg–Landau dynamics*, Duke Math. J. **130** (2005), no. 3, 523–614.
7. ———, *Convergence of the parabolic Ginzburg–Landau equation to motion by mean curvature*, Ann. of Math. (2) **163** (2006), no. 1, 37–163.
8. ———, *Quantization and motion law for Ginzburg–Landau vortices*, Arch. Rational Mech. Anal. **183** (2007), no. 2, 315–370.
9. ———, *Dynamics of multiple degree Ginzburg–Landau vortices*, Comm. Math. Phys. **272** (2007), no. 1, 229–261.
10. H. Brezis, J.-M. Coron, and E. H. Lieb, *Harmonic maps with defects*, Comm. Math. Phys. **107** (1986), no. 4, 649–705.
11. M. Comte and P. Mironescu, *Remarks on nonminimizing solutions of a Ginzburg–Landau type equation*, Asymptotic Anal. **13** (1996), no. 2, 199–215.
12. W. E, *Dynamics of vortices in Ginzburg–Landau theories with applications to superconductivity*, Phys. D **77** (1994), no. 4, 383–404.
13. R. L. Jerrard and H. M. Soner, *Dynamics of Ginzburg–Landau vortices*, Arch. Rational Mech. Anal. **142** (1998), no. 2, 99–125.
14. ———, *The Jacobian and the Ginzburg–Landau energy*, Calc. Var. Partial Differential Equations **14** (2002), no. 2, 151–191.
15. F. H. Lin, *Some dynamical properties of Ginzburg–Landau vortices*, Comm. Pure Appl. Math. **49** (1996), no. 4, 323–359.
16. P. Mironescu, *Les minimiseurs locaux pour l'équation de Ginzburg-Landau sont à symétrie radiale*, C. R. Acad. Sci. Paris Sér. I Math. **323** (1996), no. 6, 593–598.
17. F. Pacard and T. Rivière, *Linear and nonlinear aspects of vortices: The Ginzburg–Landau model*, Progr. Nonlinear Differential Equations Appl., vol. 39, Birkhäuser, Boston, 2000.
18. L. Peres and J. Rubinstein, *Vortex dynamics in* U(1) *Ginzburg–Landau models*, Phys. D **64** (1993), no. 1-3, 299–309.

19. É. Sandier and S. Serfaty, *Gamma-convergence of gradient flows with applications to Ginzburg–Landau*, Comm. Pure Appl. Math. **57** (2004), no. 12, 1627–1672.
20. S. Serfaty, *Vortex collision and energy-dissipation rates in the Ginzburg–Landau heat flow. I: Study of the perturbed Ginzburg–Landau equation*, J. Eur. Math. Soc. (JEMS) **9**, no. 2, 177–217, preprint.
21. D. Spirn, *Vortex dynamics of the full time-dependent Ginzburg–Landau equations*, Comm. Pure Appl. Math. **55** (2002), no. 5, 537–581.

Laboratoire Jacques-Louis Lions, Université Pierre et Marie Curie, 4 place Jussieu, BC 187, 75252 Paris Cedex 05, France
E-mail address: bethuel@ann.jussieu.fr

Dipartimento di Informatica, Università di Verona, Strada le Grazie, 37134 Verona, Italy
E-mail address: orlandi@sci.univr.it

Laboratoire Jacques-Louis Lions, Université Pierre et Marie Curie, 4 place Jussieu, BC 187, 75252 Paris Cedex 05, France
E-mail address: smets@ann.jussieu.fr

Centre de Recherches Mathématiques
CRM Proceedings and Lecture Notes
Volume **44**, 2008

Nonlocal Cahn – Hilliard and Isoperimetric Problems: Periodic Phase Separation Induced by Competing Long- and Short-Range Interactions

Rustum Choksi

1. Introduction

The modeling of pattern morphologies via energy minimization involving long and short-range competitions is well-established and ubiquitous in science (cf. [21, 33, 47] and the references therein). It also provides mathematicians with a wealth of rich variational problems consisting of both local (associated with short-range interactions) and nonlocal (associated with long-range interactions) terms (see for example [29] and the references therein).

Here we are concerned with two nonlocal variational problems which describe the simplest morphology class within this general setting, that of *periodic phase separation*. Many physical systems exhibit a phase separation which, according to experiments, can be roughly described as follows:

$\mathcal{P}$1. The phase separation is (nearly) periodic on some fixed (mesoscopic) scale.

$\mathcal{P}$2. Within a period cell, the structure seems to want to minimize surface area between the two phases.

1.1. The physical paradigm: diblock copolymers. A paradigm for $\mathcal{P}1$ and $\mathcal{P}2$ is provided by microphase separation of diblock copolymers (cf. [5, 6, 50]). A diblock copolymer is a linear-chain molecule consisting of two sub-chains joined covalently to each other. One of the sub-chains is made of monomers of type A and the other of type B. Below a critical temperature, even a weak repulsion between unlike monomers A and B induces a strong repulsion between the sub-chains, causing the sub-chains to segregate. A macroscopic segregation whereby the sub-chains detach from one another cannot occur because the chains are chemically bonded: Rather, a phase separation on a mesoscopic scale with A and B-rich domains emerges. The observed mesoscopic domains, illustrated in Figure 1, are highly regular periodic structures with interfaces strongly resembling surfaces with constant mean curvature; for example lamellar, spheres, cylindrical tubes, and single and double-gyroids

2000 *Mathematics Subject Classification.* Primary 49S05, 74N15; Secondary 74N20.
This is the final form of the paper.

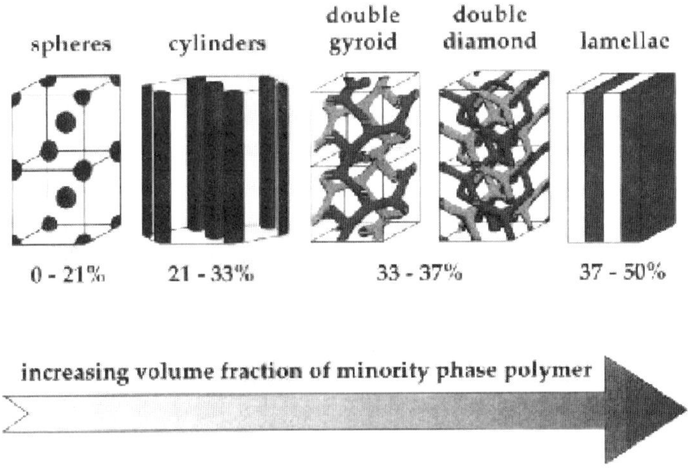

FIGURE 1. Periodic phase separation. Cartoon taken from Edwin Thomas' talk at MSRI 1999 — taken from reference [46]

(see for example, [6]). The connection between observed structures and triply periodic constant mean curvature surfaces (many of which are stable in a suitable sense) has been well-established in the literature (see for example, [3, 27, 50]).

1.2. Mathematical paradigms.

The periodic Cahn–Hilliard and isoperimetric problems. The simplest mathematical model for capturing $\mathcal{P}1$ and $\mathcal{P}2$ would be to a priori enforce the periodicity (a hard constraint) and look at the resulting interfacial problems; thus the long-range interactions no longer play any role. One is thus led to the periodic Cahn–Hilliard and isoperimetric problems. Let

$$E_\epsilon(u) = \int_{\mathbb{T}^n} \frac{(1-u^2)^2}{4} + \frac{\epsilon^2}{2}|\nabla u|^2 \, dx$$

for $u \in H^1(\mathbb{T}^n)$, $\int_{\mathbb{T}^n} u = m$, and

$$E_0(u) = c_0 \int_{\mathbb{T}^n} |\nabla u|$$

for $u \in \mathrm{BV}(\mathbb{T}^n, \pm 1)$, $\int_{\mathbb{T}^n} u = m$. Only can easily extend E_ϵ and E_0 to all function in $L^1(\mathbb{T}^n)$ by defining them to be zero if u is not in $H^1(\mathbb{T}^n)$ and $\mathrm{BV}(\mathbb{T}^n, \pm 1)$, respectively. By the the *periodic Cahn–Hilliard problem* we mean

(PCH) Minimize $E_\epsilon(u)$ over all $u \in L^1(\mathbb{T}^n)$.

By the *periodic isoperimetric problem* we mean

(PIP) Minimize $E_0(u)$ over all $u \in L^1(\mathbb{T}^n)$.

Both problems are clearly well-posed and the rigorous connection between them (for an appropriate choice of constant c_0 and rescaling of E_ϵ) is via Gamma convergence. Our main interest here is *not* to treat the periodicity as a hard constraint, but rather when it results from an interplay between minimizing nonlocal and local terms. However, we do make a few comments concerning (PCH) and (PIP) based upon recent work with P. Sternberg ([14]). Our primary concern in [14] was the wealth of nontrivial stable local minimizers of (PIP) (stable constant mean curvature surfaces), and whether or not these surfaces have diffuse interface analogues (local minimizers of (PCH)).[1] Our primary tool was the notion of a strict (up to translation) L^1-local minimizer in the sense of [30] which carries through to the diffuse problem via Gamma convergence. Hence the relevant question was whether or not stability (i.e., a positive bound on the second variation) was equivalent to being a strict L^1-local minimizer. In 2D the situation is greatly simplified, and we proved the equivalence of the two notions. The 3D case remains open.

The nonlocal Cahn–Hilliard[2] *and isoperimetric problems.* Any model for microphase separation of diblock copolymers can not be based solely on minimization of A-B monomer interfaces, and must take into account the nature of the *joined* A- and B-subchain interactions. Indeed, a density functional theory, first proposed by Ohta and Kawasaki [36], entails the minimization of a *nonlocal* Cahn–Hilliard like energy (cf. [34]) whereby the standard Cahn-Hilliard free energy is augmented by a long-range interaction term — associated with the connectivity of the sub-chains in the diblock copolymer macromolecule.[3] Let Ω be the unit cube $(0,1)^n$ in R^n and fix $m \in (-1,1)$. Minimize

$$\text{(NLCH)} \qquad \text{Minimize } \frac{\epsilon^2}{2}\int_\Omega |\nabla u|^2\,dx + \int_\Omega \frac{(1-u^2)^2}{4}\,dx + \frac{\sigma}{2}\int_\Omega |\nabla v|^2\,dx,$$

over all $u \in \{u \in H^1(\Omega,[-1,1]) \mid \int_\Omega u = m\}$. Here, v is related to u via the BVP

$$(1) \qquad -\Delta v = u - m,$$

in the sense of distributions with *either* Neumann or periodic boundary conditions for v on $\partial\Omega$. The relevant parameter regimes are either $\epsilon \sim \sigma \ll 1$ or $\epsilon \ll \sigma \ll 1$. From the (NLCH), the incentive for pattern formation is clear: The first term wants uniform phases, with the second term penalizing transitions between phases. However the nonlocal term is lowered by oscillations between the two phases. The effect of the three is to set a fine scale (probably periodic) structure.

If one focuses on the strong segregation limit, wherein the interfacial thickness of the phase boundaries tends to zero, one can easily show (cf. [1, 12, 39]) that the relevant sharp interface problem can be written as:

$$\text{(NLIP)} \qquad \text{Minimize } \int_\Omega |\nabla u|\,dx + \gamma\int_\Omega |\nabla v|^2\,dx,$$

over all $u \in \{u \in \mathrm{BV}(\Omega,\{-1,1\}) \mid \int_\Omega u = m\}$. Here γ is a parameter which will determine the length scale of minimizers, and again v is related u via (1). We refer to this problem as *the nonlocal isoperimetric problem.*

We are interested in the following three issues surrounding (NLCH) and (NLIP):

(i) The periodicity resulting from competing local and nonlocal terms

---

[1] We should mention here that Pacard and Ritoré [37] prove a result connecting critical points of the sharp and diffuse interface problems. With a completely different machinery, they address the problem in great generality. However, our question is different as it pertains to structures which are in some sense minimizing (not just critical points).

[2] The term *the nonlocal Cahn–Hilliard equation* has already been extensively used. It pertains to a different problem and should not be confused with either the variational problem (NLCH) or the gradient flow PDE (4).

[3] See [12] for full derivation and [13] for a derivation of a vector-valued analogue describing diblock copolymer-homopolymer blends.

(ii) The geometry of minimizers: the effect of the nonlocal term, perturbation from the phase boundary being area-minimizing/constant mean curvature
(iii) The PDE resulting from the H^{-1} gradient flow for (NLCH): numerical simulation of steady states and the dynamic stages of the phase separation (spinodal decomposition, coarsening etc.)

In Section 2, we describe a rigorous result addressing issue (i) for (NLIP) — joint work with G. Alberti and F. Otto. In Section 3, we provide a first attempt at a general method for addressing issue (ii) via some recent results on the consequences of stability for (NLIP) — joint work with P. Sternberg. Finally in Section 4, we discuss issue (iii) and present some simulations, focusing more on questions than on answers.

2. The nonlocal isoperimetric problem: uniform energy distribution

For dimension $n = 1$, one can show (either for (NLCH) or (NLIP)) that minimizers are periodic (see [32] for the case $m = 0$ and [41] for general m). Heuristically, it is straightforward to see that the scale of the periodicity is set by the nonlocal term; For σ sufficiently large, the periodicity on a scale smaller than the domain size is enforced as a weak constraint via interactions between the perimeter and the nonlocal terms. The natural question is to what extend are minimizers periodic in higher-D. Here we present a rigorous result to support the conjecture that *in higher-D, minimizers are "nearly" periodic* on a length scale set by γ, the dimension n, and volume fraction m. We do so via the energy associated with a minimizer, and its uniform distribution in space.

The length scale for minimizers is set by γ. Since in this section, we are interested in a qualitative "periodicity" result on sufficiently large domains, we may just as well set it to unity. We will however need the following equivalent reformulation of (NLIP). Let Ω be a cube in $\mathbb{R}^n$ and $m \in (-1, 1)$. Then for any $u \in \mathrm{BV}(\Omega_L, \pm 1)$, there exits $\mathbf{b} \in L^2(\Omega, \mathbb{R}^n)$ such that $\nabla \cdot \mathbf{b} = u - m$ in the sense of distributions, with either Neumann (i.e., $\mathbf{b} \cdot \mathbf{n} = 0$) or periodic boundary conditions for $\mathbf{b}$. Minimizing over all such vector fields $\mathbf{b}$, we find

$$\min \int_\Omega |\mathbf{b}|^2 = \int_\Omega |\nabla v|^2,$$

where as usual v solves (1) on Ω, i.e., $-\Delta v = u - m$, in the sense of distributions with the respective boundary conditions (Neumann or periodic). This reformulation has several advantages. For example, it localizes the functional and facilitates the use of *cutting and pasting* arguments. It further allows us to demonstrate that the particular choice of boundary conditions do not influence the structure away from the boundary — Theorem 2.1 and its proof will support this.

We now present the reformulation of (NLIP): Let $\Omega_L = [0, L]^n$, $L \gg 1$. Consider

$$\min_{u \in \mathcal{A}_0} E(u, \mathbf{b}, \Omega_L) = \int_{\Omega_L} |\nabla u| + \int_{\Omega_L} |\mathbf{b}|^2 \, d\mathbf{x}$$

$$\mathcal{A}_0 = \{(u, \mathbf{b}) : u \in \mathrm{BV}(\Omega_L, \pm 1), \mathbf{b} \in L^2(\Omega_L, \mathbb{R}^n),$$
$$\nabla \cdot \mathbf{b} = u - m, \mathbf{b} \cdot \mathbf{n} = 0 \text{ on } \partial \Omega_L\}.$$

Before presenting our result on the uniform distribution of energy for minimizers, we provide some motivation for the result. Suppose the minimizer $(u_p, \mathbf{b}_p)$ on Ω_L ($L \gg 1$) was exactly periodic, say u_p represents a lamellar structure with period

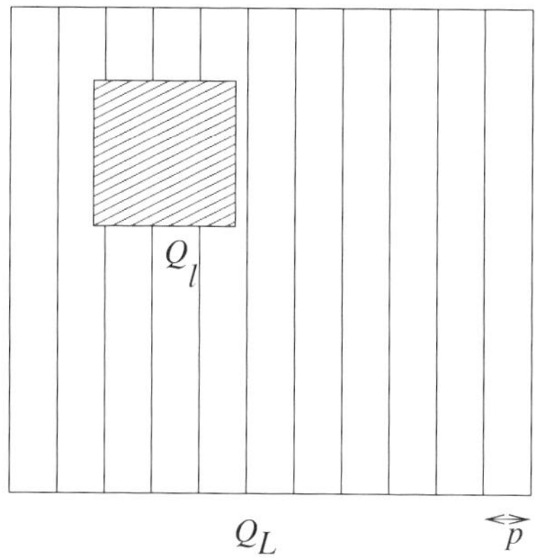

FIGURE 2

p, where $L = kp$ for some integer k — see Figure 2. It is straightforward to see that there exist a constant $\bar{E}_0$ (which depends on p but not on k) such that

$$\frac{E(u_p, \mathbf{b}_p, Q_{kp})}{(kp)^n} = \overline{E}_0.$$

Now consider the structure on some smaller cube, Q_l with size l, $L > l \gg 1$, i.e., consider $(u_p, \mathbf{b}_p)$ restricted to Q_l (see Figure 2). Computing its the energy density, we may not get exactly $\overline{E}_0$, however one finds

$$\left| \frac{E\big(u_p, \mathbf{b}_p, Q_l(x)\big)}{l^n} - \overline{E}_0 \right| \leq \frac{C}{l},$$

for some constant C.

Our result is to prove the same estimate but for *any* minimizer:

Theorem 2.1. *There exist constants $\sigma^* > 0$, $C > 0$ depending only on m, n s.t. if $(u_0, \mathbf{b}_0)$ is a minimizer of E on $\mathcal{A}_0$, then for every $l \leq L$ and $x \in \Omega_L$,*

$$\left| \frac{E\big(u_0, \mathbf{b}_0, Q_l(x)\big)}{l^n} - \sigma^* \right| \leq \frac{C}{l},$$

where $Q_l(x) =$ cube if size l, centered at x. In fact,

$$\sigma^* = \lim_{L \to \infty} \left(\min_{(u_0, \mathbf{b}_0) \in \mathcal{A}_{per}} \frac{E(u_0, \mathbf{b}_0, Q_L)}{L^n} \right).$$

Not surprisingly, the same limit (definition of σ^* above) would be achieved for other boundary conditions (free, etc.). Following arguments similar to those presented in [11], one can also prove a similar estimate for the separate terms in the energy, in particular for just the surface energy.

3. The nonlocal isoperimetric problem: stability and the second variation

The structures which one observes in microphase separation are mostly constant mean curvature (CMC) surfaces. This fact is documented in both the science and mathematics literature (cf. [5, 6, 46, 50]). If one believes that the variational problem

(NLCH), and hence (NLIP), captures the essential physics of the phase separation, then one would expect that minimizers of (NLIP), are close to minimizers of (PIP).

In a series of paper ([40, 42–45]), Ren and Wei consider the stability of lamellar, ring/cylindrical, and spot/spherical solutions (the latter two considered on a radially symmetric domain). The approach here was to linearize the Euler–Lagrange equation (or gradient flow equation) about one of these critical points, and consider the spectrum of the linearized operator. One obtains stability of several structures for small γ and instability for γ sufficiently large. Interestingly, they also prove that certain non-CMC structures (so called *wiggled* lamellar and spot solutions) were stable for a suitable values of γ.

As a first step towards a *global* approach to the consequences of stability, we provide a rigorous statement of the first and second variations on (NLIP). This is joint work with P. Sternberg, see [15] for details. For simplicity let us work with periodic boundary conditions (i.e., $\Omega = \mathbb{T}^n$, the n-dimensional flat torus). See [15] for the analogous results with Neumann boundary conditions. Let

$$\mathcal{E}_\gamma(u) := \int_{\mathbb{T}^n} |\nabla u| + \gamma \int_{\mathbb{T}^n} |\nabla v|^2 \, \mathrm{d}\mathbf{x}$$

We minimize $\mathcal{E}_\gamma$ over all $u \in \mathrm{BV}(\mathbb{T}^n, \pm 1)$ with $\int_{\mathbb{T}^n} u = m$ and $-\Delta v = u - m$ in $\mathbb{T}^n$.

For $u \in \mathrm{BV}(\mathbb{T}^n, \pm 1)$ we can naturally associate a set of finite perimeter $A := \{u = 1\}$ and vice versa. Given any such $u \iff$ set of finite perimeter A, we define an appropriate admissible perturbation A_t (i.e., associated with the volume constraint, see Remark 3.4(i)) and define

$$U(x,t) = \begin{cases} 1 & \text{if } x \in A_t \\ -1 & \text{if } x \in A_t^c. \end{cases}$$

Then we say A is a *critical point* if

$$\left.\frac{\mathrm{d}}{\mathrm{d}t}\right|_{t=0} \mathcal{E}_\gamma\big(U(\cdot,t)\big) = 0,$$

and is further *stable* if:

$$\left.\frac{\mathrm{d}^2}{\mathrm{d}t^2}\right|_{t=0} \mathcal{E}_\gamma\big(U(\cdot,t)\big) \geq 0.$$

Regularity assumption. When $\gamma = 0$, it is well-known (see for example [22]) that in dimensions $n < 8$, the phase boundary associated with any L^1-local minimizer must have constant mean curvature and be an analytic $(n-1)$-dimensional manifold, while in dimensions $n \geq 8$, the same is true off of a (perhaps empty) singular set of Hausdorff dimension at most $n - 8$. While the phase boundary associated with a local minimizer for $\gamma > 0$ will, in general, no longer have constant mean curvature, we strongly suspect that this lower-order, compact perturbation will not destroy regularity, so we expect that the phase boundary associated with a local minimizer of $\mathcal{E}_\gamma$ is still an analytic $(n-1)$-dimensional manifold in dimensions $n < 8$, and in particular is C^2. Where as it quite natural to expect that the nonlocal perturbation does not destroy regularity, a more interesting question is whether or not it can help. That is, can it reduce the Hausdorff dimension of the singular set for $n \geq 8$?

Proposition 3.1 (The first variation). *Let u be a critical point of $\mathcal{E}_\gamma$ such that for $A := \{u = 1\}$, ∂A is C^2 with mean curvature $H(x)$. Then for v solving $-\Delta v = u - m$ in $\mathbb{T}^n$ there exists a constant λ such that*

$$(n-1)\,H(x) + 4\gamma v(x) = \lambda \quad \text{for all } x \in \partial A. \tag{2}$$

Remark 3.2. When $\gamma = 0$, one is simply studying critical points of area subject to a volume constraint so, as is well-known, one gets constant mean curvature as a condition of criticality. For $\gamma > 0$ one sees from (2) that the curvature will not actually be constant unless it happens that ∂A is a level set of v solving (1). This will be the case for a periodic lamellar structure but note, for example, that a sphere (or periodic array of spheres) will never be a critical point — unless one works, as in [42], within the ansatz of radial symmetry.

Next we present the second variation. A formal derivation of essentially the same formula was previously presented in the appendix of [33].

Theorem 3.3 (The second variation). *Let u be a stable critical point of $\mathcal{E}_\gamma$ such that for $A = \{u = 1\}$, ∂A is C^2 with mean curvature $H(x)$. Let ζ be any smooth function on ∂A, with $\int_{\partial A} \zeta(x)\,\mathrm{d}\mathcal{H}^{n-1}(x) = 0$. Then for v solving $-\triangle v = u - m$ in $\mathbb{T}^n$ one has the condition*

$$\begin{aligned} J(\zeta) := &\int_{\partial A} \left(\left|\nabla_{\partial A}\zeta(x)\right|^2(x) - \left\|B_{\partial A}\right\|^2 \zeta^2\right) d\mathcal{H}^{n-1}(x) \\ &+ 8\gamma \int_{\partial A}\int_{\partial A} G(x,y)\,\zeta(x)\,\zeta(y)\,\mathrm{d}\mathcal{H}^{n-1}(x)\,\mathrm{d}\mathcal{H}^{n-1}(y) \\ &+ 4\gamma \int_{\partial A} \nabla v(x)\cdot\nu(x)\,\zeta^2(x)\,\mathrm{d}\mathcal{H}^{n-1}(x) \geq 0. \end{aligned} \tag{3}$$

where, $\nabla_{\partial A}\zeta$ is the gradient of ζ relative to ∂A; $B_{\partial A}$ is the second fundamental form of ∂A, i.e., $\left\|B_{\partial A}\right\|^2 = \sum_{i=1}^{n-1}\kappa_i^2$, $\kappa_1, \dots, \kappa_{n-1}$ principal curvatures; and ν is the unit normal to ∂A pointing out of A.

Remark 3.4. (i) We briefly mention the construction of admissible perturbations. The construction is very similar to [48]). We let $X\colon \mathbb{T}^n \to \mathbb{R}^n$ be C^2 with $\int_{\partial A} X\cdot\nu = 0$, and let $\Psi\colon \mathbb{T}^n \times (-\tau,\tau) \to \mathbb{T}^n$ solve

$$\frac{\partial\Psi}{\partial t} = X(\Psi) \quad \Psi(x,0) = x.$$

Defining $A_t := \Psi(A,t)$, one finds that the perturbation does not (instantaneously) preserve volume up to second order, in fact:

$$\frac{\mathrm{d}^2}{\mathrm{d}t^2}_{|t=0} |A_t| = \int_{\partial A} (\operatorname{div} X(x))(X(x)\cdot\nu)\,\mathrm{d}\mathcal{H}^{n-1}(x).$$

Hence we modify explicitly A_t via distance function to achieve the admissible perturbation.

(ii) The second variation is computed about a critical point of (NLIP) which is in general, a non-CMC surface. Hence perimeter term contributes an additional:

$$(n-1)^2 \int_{\partial A} (H - \overline{H})H\zeta^2\,\mathrm{d}\mathcal{H}^{n-1}(x),$$

where $\overline{H}$ denotes the average of $H(x)$ over ∂A. This term is exactly canceled when computing the second variation of the nonlocal term.

(iii) We note that the term in the second line of (3) is always positive. Hence to apply (3) for structures with CMC, one is simply comparing the gradient term of (3) to the term involving v. Note that the latter generates the instability: it is negative for highly oscillatory ζ. Quantifying this comparison yields certain basic statements about, for example, the stability and instability of periodic lamellar structures (see [15]). What would be more interesting is to apply (3) to structures whose curvature (or mean curvature) is not constant.

4. The modified Cahn – Hilliard equation

It is well-known (see for example [18]) that the Cahn-Hilliard equation can be derived as a gradient flow with respect to the H^{-1} norm of the Cahn – Hilliard energy (the functional in (PCH)). Ignoring for the moment that the Cahn – Hilliard equation can also be derived from more basic physical principles ([9]), one can compute the same H^{-1} gradient flow for (NLCH), obtaining the PDE

$$u_t = \Delta(-\epsilon^2 \Delta u - u + u^3) - \sigma(u - m). \tag{4}$$

We refer to the (4) as the *modified Cahn – Hilliard equation*. Note that with the gradient flow computed with respect to H^{-1}, the nonlocal term in (NLCH) only adds a zeroth order perturbation to the Cahn – Hilliard equation. This new term, however, now drives the solution to have spatial average m. Consider (4) on the *unit* torus (i.e., implement the appropriate periodic boundary conditions). As with the Cahn – Hilliard equation, (4) preserves the total mass $\int_\Omega u$ of the solution provided the initial data u_0 satisfies $\int_\Omega u_0 = m$. Otherwise, one readily sees that the total mass will adjust to m exponentially fast.

4.1. Simulation of steady states. One can readily use (4) to simulate steady states in 2D. Such simulations have appeared through-out the literature, perhaps the earliest ones were in [7]. For a suitable range of m, one can either start with random initial data or with the constant state m, randomly perturbed by noise. One simulates the expected stripes and spots with interesting geometries traversed on the path to steady state. Figures 3 and 4 show a few such experiments using a pseudo spectral code written by J. F. Williams at SFU.

While the situation in 3D is interesting in its own right, it is also important from the point of view of the diblock copolymer application. The phase diagram for microphase separation has received much attention. Here one seeks to predict the general qualitative nature of the structure (spherical, lamellar, gyroid-like) for the given material parameters. There are three such parameters which describe a diblock copolymer melt: the Flory – Huggins interaction parameter χ, the index of polymerization N, and the molecular weight $f = (m+1)/2$. The exact relationship between these dimensionless parameters and ϵ, σ and m is given in [12].

For the phase diagram, the state of art seems to be via the *self-consistent mean field theory* (cf. [6,31]), whereby given an ansatz for the basic structure (e.g., spherical) with one or two degrees of freedom, one is able to compute the free energy and minimize with respect to the degrees of freedom. One then compares the optimal result within each known ansatz to determine the overall winner. From [12], we see that the variational problem (NLCH) is a consequence of the self-consistent mean field theory and a further crucial linearization step. The latter step has created skepticism as to whether or not the essential physics is preserved in the intermediate to strong segregation regime, wherein the interface thickness is

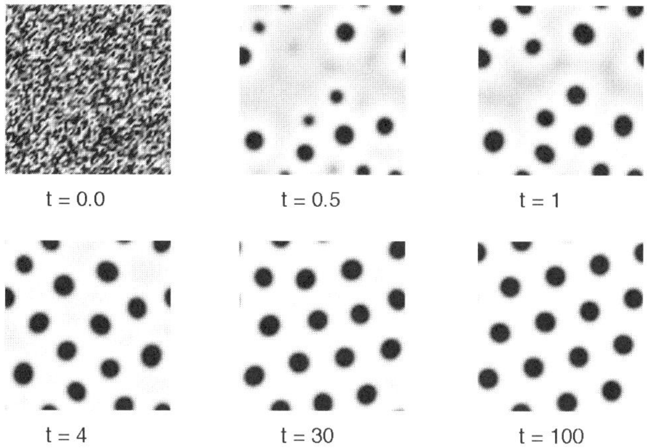

FIGURE 3. 2-D simulation with $\epsilon = 0.8$, $m = 0.5$, $\sigma = 4$ on a torus of size 2π with random initial data

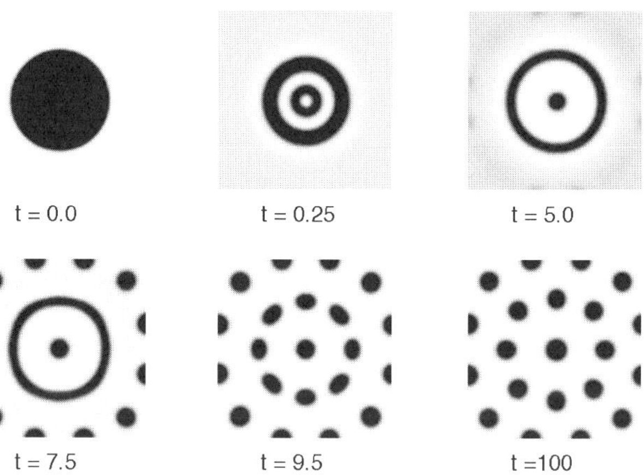

FIGURE 4. 2-D simulation with $\epsilon = 0.8$, $m = 0.5$, $\sigma = 4$ on a torus of size 2π with single disk initial data.

relatively small. My own belief, is that, yes indeed this linearization step throws away a lot of information; however, from the point of view of determination of the phase diagram, all the phases which have been (numerically) predicted purely from the SCMFT (*cf.* [31]), can be (numerically) generated from the (4) theory within a far less ansatz-driven framework. That is, in the end, one only needs to keep a rather crude approximation of the polymer A- and B-chain interactions, in order to determine *the basic qualitative geometry* for a given set of material parameters.

3D simulations for (4) were begun by Teramoto and Nishiura in [49], simulating both gyroids and double-gyroids. However, to my knowledge, no thorough phase diagram calculation has been done via the gradient theory approach of (4), and in

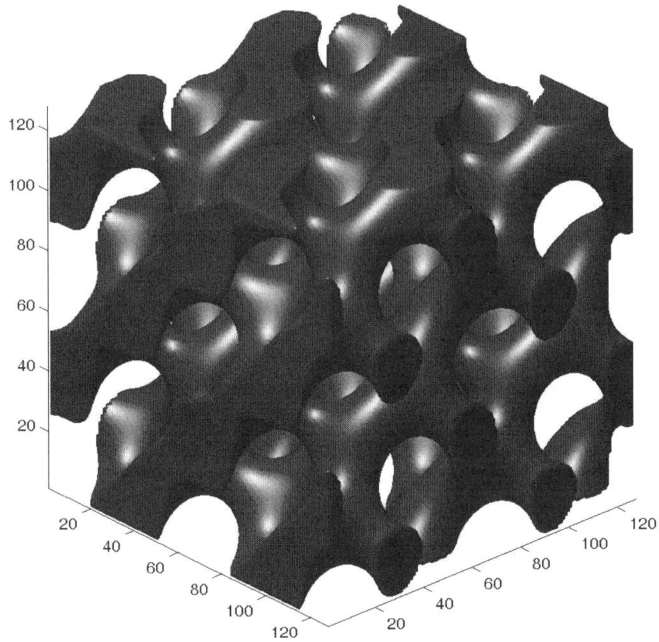

FIGURE 5. The $u = 0$ level set for a numerical simulation of (4) for parameters $\sigma = 10$, $m = 0.3$ cf. Lowengrub group (Irvine).

particular, none within the context of the material parameters χ, N, f so that one can make direct comparison with both the phase diagram of [31], and with experiments (cf. [25]). A preliminary investigation using the code from J. Lowengrub's group (Irvine) suggests that his adaptive, finite difference code may work well here. We give one example in Figure 5, a full investigation is in progress.

Remark 4.1. Following on Remark 3.2, the steady states which one attempts to simulate in Figures 3 and 4 should not be exactly circular: For a rectangular domain, criticality alone precludes interfaces having nonzero constant curvature. This could be either a resolution issue and in fact minimizers look more like the wiggled spots of [42–45], or simply that these perturbations from constant curvature (in general, CMC) are exponentially small. Investigation of these issues is currently in progress with J. F. Williams (SFU). It might also be interesting for 3D simulations (like the gyroid of Figure 5) to place the final structure into the *surface evolver* ([8]) and see qualitatively and quantitatively the perturbation from area-minimizing.

4.2. Dynamics to steady states: questions on spinodal decomposition and coarsening. The dynamics associated with the Cahn – Hilliard equation has received a lot of attention. For an appropriate regime on the mass constraint m, and starting from the constant state, the dynamics is essentially described via two distinct temporal regimes (see for example [38]):

Stage I (short time scale): Nucleation/spinodal decomposition: here a fine grained mixture of the two phases appears and fairly sharp interfaces between the two phases are formed.

Stage II (longer time scale): Interface coarsening via the Mullins – Sekerka law.

In this latter stage, there is a well-established coarsening rate of $t^{1/3}$, i.e., in Stage II, the length scale of the phases coarsens like $t^{1/3}$.

The asymptotic dynamics as ϵ tends to zero are also well understood: solutions converge to the solutions of the Mullins–Sekerka free boundary problem (see for example [4, 38]).

For (4), the third (long-range term) changes the dynamics but analogous issues remain. Starting with a homogeneous state m, there are now three competing mechanisms for the dynamics: (i) Nucleation/spinodal decomposition/formation of phases (still on a fast scale), (ii) setting of the long-range periodicity scale, (iii) coarsening within this scale. A preliminary asymptotic study of first phase was begun in [35]. Naturally the σ term plays a role here when it is sufficiently large — it restricts both the spinodal regime (wherein the homogeneous state is unstable to phase separation) and the extent of the phase separation — no longer will completely pure phases of $u = \pm 1$ form.

The next stage of coarsening is more interesting. A (final) length scale is set by σ preventing the coarsening from going on forever. It would be interesting to see if there is some temporal regime associated with a new a scaling law for the basic length scale as a function of time — this exponent should depend on σ. Certainly for this later stage, one can use the the appropriately scaled asymptotic dynamics (as ϵ tends to zero) which was initially studied by Nishiura and Ohnishi in [34] (see also [17, 19, 26]). Here one obtains the following modified Mullins–Sekerka free boundary problem. That is, setting $A^+ = \{u = 1\}$, $A^- = \{u = -1\}$, and $f = (m+1)/2$, the evolution of ∂A^+ is described via the free boundary problem for ∂A^+ and scalar velocity field v:

$$
\begin{aligned}
\Delta v &= \gamma \begin{cases} 1 - f & \text{in } A^+ \\ -f & \text{in } A^- \end{cases} \\
v &= H \qquad\qquad \text{on } \partial A^+ \\
V_n &= \left[\frac{\partial v}{\partial n}\right]_+^- \qquad \text{on } \partial A^+,
\end{aligned}
\tag{5}
$$

where the normal velocity V_n and mean curvature H are computed with respect to the outward normal to A^+. This free boundary problem can naturally be connected with (NLIP).

References

1. G. Alberti, R. Choksi, and F. Otto, *Uniform energy distribution for minimizers of an isoperimetric problem containing long-range interactions*, submitted.
2. G. Alberti and S. Müller, *a new approach to variational problems with multiple scales*, Comm. Pure Appl. Math. **54** (2001), no. 7, 761–825.
3. A. Aksimentiev, M. Fialkowski, and R. Holyst, *Morphology of surfaces in mesoscopic polymers, surfactants, electrons, or reaction-diffusion systems: Methods, simulations, and measurements* (I. Priogogine and S. A. Rice, eds.), Advances in Chemical Physics, vol. 121, Wiley, New York, NY, 2002, pp. 141–239.
4. N. D. Alikakos, P. W. Bates, and X. Chen, *Convergence of the Cahn–Hilliard equation to the Hele–Shaw model*, Arch. Rational Mech. Anal. **128** (1994), no. 2, 165–205.
5. S. Andersson, S. T. Hyde, K. Larsson, and S. Lidin, *Minimal surfaces and structures: From inorganic and metal crystals to cell membranes and biopolymers*, Chem. Rev. **88** (1988), no. 1, 221–242.
6. F. S. Bates and G. H. Fredrickson, *Block copolymers: Designer soft materials*, Phys. Today **52** (1999), no. 2, 32–38.

7. M. Bahiana and Y. Oono, *Cell dynamical system approach to block copolymers*, Phys. Rev. A (3) **41** (1990), no. 12, 6763–6771.
8. K. A. Brakke, *The surface evolver*, Experiment. Math. **1** (1992), no. 2, 141–165. Software available at: `http://www.susqu.edu/facstaff/b/brakke/evolver/evolver.html`.
9. J. W. Cahn and J. E. Hilliard, *Free energy of a nonuniform system. I: Interfacial free energy*, J. Chem. Phys. **28** (1958), no. 2, 258–267.
10. R. Choksi, *Scaling laws in microphase separation of diblock copolymers*, J. Nonlinear Sci. **11** (2001), no. 3, 223–236.
11. R. Choksi, R. V. Kohn, and F. Otto, *Domain branching in uniaxial ferromagnets: A scaling law for the minimum energy*, Comm. Math. Phys. **201** (1999), no. 1, 61–79.
12. R. Choksi and X. Ren, *On a derivation of a density functional theory for microphase separation of diblock copolymers*, J. Statist. Phys. **113** (2003), no. 1-2, 151–176.
13. ———, *Diblock copolymer/homopolymer blends: Derivation of a density functional theory*, Phys. D **203** (2005), no. 1-2, 100–119.
14. R. Choksi and P. Sternberg, *Periodic phase separation: The periodic Cahn–Hilliard and isoperimetric problems*, Interfaces Free Bound. **8** (2006), no. 3, 371–392.
15. ———, *On the first and second variations of a nonlocal isoperimetric problem*, J. Reine Angew. Math. **611** (2007), 75–108.
16. E. W. Cochran, C. J. Garcia-Cevera, and G. H. Fredrickson, *Stability of the gyroid phase in diblock copolymers at strong segregation*, preprint.
17. J. Escher and Y. Nishiura, *Smooth unique solutions for a modified Mullins–Sekerka model arising in diblock copolymer melts*, Hokkaido Math. J. **31** (2002), no. 1, 137–149.
18. P. C. Fife, *Models for phase separation and their mathematics*, Electron. J. Differential Equations **2000** (2000), no. 48.
19. P. C. Fife and D. Hilhorst, *The Nishiura–Ohnishi free boundary problem in the 1D case*, SIAM J. Math. Anal. **33** (2001), no. 3, 589–606.
20. S. Förster, A. K. Khandpur, J. Zhao, F. S. Bates, I. W. Hamley, A. J. Ryan, and W. Bras, *Complex phase behavior of polyisoprene-polystyrene diblock copolymers near the order-disorder transition*, Macromolecules **27** (1994), no. 23, 6922–6935.
21. R. E. Goldstein, D. J. Muraki, and D. M. Petrich, *Interface proliferation and the growths of labyrinths in a reaction-diffusion system*, Phys. Rev. E (3) **53** (1996), no. 4, part B, 3933–3957.
22. E. Gonzalez, U. Massari, and I. Tamanini, *On the regularity of boundaries of sets minimizing perimeter with a volume constraint*, Indiana Univ. Math. J. **32** (1983), no. 1, 25–37.
23. K. Grosse-Brauckmann, *Stable constant mean curvature surfaces minimize area*, Pacific J. Math. **175** (1997), no. 2, 527–534.
24. D. A. Hajduk, P. E. Harper, S. M. Gruner, C. C. Honeker, G. Kim, E. L. Thomas, and L. J. Fetters, *The gyroid: A new equilibrium morphology in weakly segregated diblock copolymers*, Macromolecules **27** (1994), no. 15, 4063–4075.
25. I. W. Hamley, *The physics of block copolymers*, Oxford Sci. Publ., Oxford Univ. Press, Oxford, 1998.
26. M. Henry, D. Hilhorst, and Y. Nishiura, *Singular limit of a second order nonlocal parabolic equation of conservative type arising in the micro-phase separation of diblock copolymers*, Hokkaido Math. J. **32** (2003), no. 3, 561–622.
27. D. Hoffman, *A new turn for Archimedes*, Nature **384** (1996), 28–29.
28. H. Karcher, *The triply periodic minimal surfaces of Alan Schoen and their constant mean curvature companions*, Manuscripta Math. **64** (1989), no. 3, 291–357.
29. R. V. Kohn, *Energy-driven pattern formation*, International Congress of Mathematicians (Madrid, 2006), to appear.
30. R. V. Kohn and P. Sternberg, *Local minimisers and singular perturbations*, Proc. Roy. Soc. Edinburgh Sect. A **111** (1989), no. 1-2, 69–84.
31. M. W. Matsen and M. Schick, *Stable and unstable phases of a diblock copolymer melt*, Phys. Rev. Lett. **72** (1994), no. 16, 2660–2663.

32. S. Müller, *Singular perturbations as a selection criterion for periodic minimizing sequences*, Calc. Var. Partial Differential Equations **1** (1993), 169 - 204.
33. C. B. Muratov, *Theory of domain patterns in systems with long-range interactions of Coulomb type*, Phys. Rev. E (3) **66** (2002), no. 6, 066108.
34. Y. Nishiura and I. Ohnishi, *Some mathematical aspects of the micro-phase separation in diblock copolymers*, Phys. D **84** (1995), no. 1-2, 31 - 39.
35. I. Ohnishi, Y. Nishiura, M. Imai, and Y. Matsushita, *Analytical solutions describing the phase separation driven by a free energy functional containing a long-range interaction term*, Chaos **9** (1999), no. 2, 329 - 341.
36. T. Ohta and K. Kawasaki, *Equilibrium morphology of block copolymer melts*, Macromolecules **19** (1986), no. 10, 2621 - 2632.
37. F. Pacard and M. Ritoré, *From constant mean curvature hypersurfaces to the gradient theory of phase transitions*, J. Differential Geom. **64** (2003), no. 3, 359 - 423.
38. R. L. Pego, *Front migration with the nonlinear Cahn - Hilliard equation*, Proc. Roy. Soc. London Ser. A **422** (1989), no. 1863, 261 - 278.
39. X. Ren and J. Wei, *On the multiplicity of two nonlocal variational problems*, SIAM J. Math. Anal. **31** (2000), no. 4, 909 - 924.
40. ———, *Concentrically layered energy equilibria of the diblock copolymer problem*, European J. Appl. Math. **13** (2002), no. 5, 479 - 496.
41. ———, *On energy minimizers of the diblock copolymer problem*, Interfaces Free Bound. **5** (2003), no. 2, 193 - 238.
42. ———, *Stability of spot and ring solutions of the diblock copolymer equation*, J. Math. Phys. **45** (2004), no. 11, 4106 - 4133.
43. ———, *Wiggled lamellar solutions and their stability in the diblock copolymer problem*, SIAM J. Math. Anal. **37** (2005), no. 2, 455 - 489.
44. ———, *Existence and stability of spherically layered solutions of the diblock copolymer equation*, SIAM J. Appl. Math. **66** (2006), no. 3, 1080 - 1099.
45. ———, *Droplet solutions in the diblock copolymer problem with skewed monomer composition*, Calc. Var. Partial Differential Equations **25** (2006), no. 3, 333 - 359.
46. A. Ros, *The isoperimetric problem*, Global Theory of Minimal Surfaces (Berkeley , CA, 2001) (D. Hoffman, ed.), Clay Math. Proc., vol. 2, Amer. Math. Soc., Providence, RI, 2005, pp. 175 - 209.
47. M. Seul and D. Andelman, *Domain shapes and patterns: The phenomenology of modulated phases*, Science **267** (1995), no. 5197, 476 - 483.
48. P. Sternberg and K. Zumbrun, *A Poincaré inequality with applications to volume-constrained area-minimizing surfaces*, J. Reine Angew. Math. **503** (1998), 63 - 85.
49. T. Teramoto and Y. Nishiura, *Double gyroid morphology in a gradient system with nonlocal effects*, J. Phys. Soc. Japan **71** (2002), no. 7, 1611 - 1614.
50. E. L. Thomas, D. M. Anderson, C. S. Henkee, and D. Hoffman, *Periodic area-minimizing surfaces in block copolymers*, Nature **334** (1988), 598 - 601.

Department of Mathematics, Simon Fraser University, Burnaby, BC V5A 1S6, Canada

E-mail address: choksi@math.sfu.ca

Centre de Recherches Mathématiques
CRM Proceedings and Lecture Notes
Volume **44**, 2008

On a Generalized Ginzburg–Landau Energy for Superconducting/Normal Composite Materials

Tiziana Giorgi and Hala Jadallah

ABSTRACT. We review and extend some results for a generalized Ginzburg–Landau energy functional, introduced in the physics literature to model composite superconducting/normal body. In this formulation, the order parameter is defined everywhere in the composite system, with the intention to capture the effect of superconducting electrons pairs diffusing in the normal parts. For suitable choices of the materials conductivities, one can show that a superconductor in contact with a normal material may not exhibit surface superconductivity in decreasing applied fields.

1. Introduction

A composite system of superconducting/normal materials is traditionally described within the Ginzburg–Landau theory via the so-called de Gennes boundary condition or, if the normal regions are inclusions or impurities, by a generalized vector potential, where a given smooth function is taken to be nonpositive in the normal parts. These approaches are appropriate in many occasions, but there are situations when they might fail to represent all the possible behaviors; remarkably, for particular choices of the normal state conductivities, as pointed out in physics papers [5, 6, 8, 18].

First introduced in the physics literature in the sixties [8, 18], the model in discussion was re-proposed in greater mathematical details in [7]. Over the years it has been used in numerical (see for example [7, 11, 12]) and, for specific configurations, in analytical works (see as an example [17]), with much of the attention reserved to the numerical analysis of the important phenomena of pinning of vortices and tunneling of the supercurrent. General classical analytical questions were not typically addressed. Inspired by the paper of Hurault [18], we concentrate initially on the case of a bounded superconductor surrounded by a normal material. In our work [13], we address the classical questions of existence of minimizers and estimates on critical points in two and three dimensions. We extend as well to this model the standard result, presented in [15], of loss of superconductivity in high

2000 *Mathematics Subject Classification.* Primary 82D55; Secondary 35Q55.
Funding for the first author was provided by the National Science Foundation Grant #DMS-0604843.
This is the final form of the paper.

magnetic field. The goal was to develop the necessary rigorous set-up needed to explore the claim made in [18] that in decreasing fields and for specific values of the conductivity of the two materials, there would be no surface superconductivity in the transition from the normal state. In other words, to look at the dependence of the critical field H_{c3} upon the conductivities of the involved materials.

In a subsequent work [14], we conduct a rigorous analysis of Hurault work and studied surface nucleation at a normal/superconducting interface. For values of the Ginzburg – Landau parameter κ large but fixed, that is for type-II superconductors, we derive that the third critical field H_{c3} does in fact depend on the ratio of the conductivities, and that there exists a threshold value of this ratio below which $H_{c2} = H_{c3}$. Our results are in agreement with the experimental and theoretical ones presented or predicted in the physics literature, [5, 6, 10, 18]. The approach proposed by Hurault, which is standard when first studying the onset of superconductivity (see for example [25]), consists in considering a semi-infinite superconductor, filling an half-space, and a normal material, which in turn fills the other half-space, in the presence of a constant applied magnetic field parallel to the surface of contact. This simplified model is widely used in the study of surface superconductivity, for via the study of a one-dimensional eigenvalue problem, it supposedly provides in its essential characteristics the behavior of the nucleation field, [10, 18]. More precisely, the nucleation field and the location of nucleation for the standard Ginzburg – Landau energy have been the subject of many papers, and typically the upper and the lower bound estimates for the critical field are obtained using test functions that are constructed using cut-off functions and eigenfunctions of the reduced one-dimensional eigenvalue problem. This means in particular, that the eigenvalue of the one-dimensional problem appears as a leading term in the asymptotic expansion of the critical value of the original problem.

A recent preprint [20] derives the above picture in the so-called high-kappa limit for the two-dimensional situation, assuming the permeability to be the same for both materials. The analysis presented uses the techniques of Helffer and Morame. The author also shows that if the ratio of the conductivities is sufficiently large, the nucleation takes place at the point on the boundary of maximum curvature, as already known for a superconductor in a vacuum, and derives an asymptotic expansion of the critical field in the high kappa limit, using the methods developed for the classical Ginzburg – Landau by Helffer and Pan. To compare with our work, we remark that the author expresses the eigenvalue problem in a different way than we do, so his statements are slightly different, and in particular the one-dimensional problem that he studied is slightly different. However, the main difference in our treatments comes from the fact that [20] assumes the permeability to be the same on both sides of the interface, since the physically accurate situation, where the magnetic permeabilities differ, appears to imply difficulties that the author can not overcome.

The energy density is presented in Section 2. In Section 3 we review existence of minimizers and classical estimate on solutions, as well as loss of superconductivity in high fields. In Section 3, we look at the onset of superconductivity. We review our results for the one-dimensional problem, and we extend them to the case of a two-dimensional superconductor filling the half-space $x < 0$, in contact with a normal material filling the half-space $x > 0$.

2. Energy density

We will distinguish between normal and superconducting quantities with the help of the subscripts n and s, respectively. Hence, for example, $\mathcal{D}_s$ will denote the region occupied by the superconducting material, while $\mathcal{D}_n$ the one occupied by the normal one. We denote in bold vector valued variables.The paper of Chapman, Du and Gunzburger [7] (see also [13, 18]) suggests the following dimensional form for the energy density:

$$(1)\quad \tilde{e}(\tilde{\psi}, \tilde{\mathbf{A}}, \widetilde{\mathbf{H}}_{\mathbf{a}})$$
$$= \left\{ \tilde{f}_s + a_s|\tilde{\psi}|^2 + \frac{b_s}{2}|\tilde{\psi}|^4 + \frac{1}{2m_s}|(i\hbar\nabla + \frac{e_s}{c}\tilde{\mathbf{A}})\tilde{\psi}|^2 + \frac{\mu_s}{8\pi}|\widetilde{\mathbf{H}}|^2 - \frac{\mu_s}{4\pi}\widetilde{\mathbf{H}}\cdot\widetilde{\mathbf{H}}_{\mathbf{a}} \right\}\chi_{\mathcal{D}_s}$$
$$+ \left\{ \tilde{f}_n + a_n|\tilde{\psi}|^2 + \frac{1}{2m_n}|(i\hbar\nabla + \frac{e_n}{c}\tilde{\mathbf{A}})\tilde{\psi}|^2 + \frac{\mu_n}{8\pi}|\widetilde{\mathbf{H}}|^2 - \frac{\mu_n}{4\pi}\widetilde{\mathbf{H}}\cdot\widetilde{\mathbf{H}}_{\mathbf{a}} \right\}\chi_{\mathcal{D}_n},$$

We restrict the problem to the case of one superconductor and one normal material, together filling $\mathbb{R}^n$, $n \leq 3$. However, it is clear how more general cases could be modeled. As in the classical Ginzburg–Landau theory, superconductivity is described by the complex order parameter $\tilde{\psi}$ and the magnetic potential $\tilde{\mathbf{A}}$, which is related to the induced magnetic field $\widetilde{\mathbf{H}}$ by the formula:

$$\widetilde{\mathbf{H}} = \begin{cases} \dfrac{1}{\tilde{\mu}_s}\operatorname{curl}\tilde{\mathbf{A}} & \text{in } \mathcal{D}_s, \\ \dfrac{1}{\tilde{\mu}_n}\operatorname{curl}\tilde{\mathbf{A}} & \text{in } \mathcal{D}_n. \end{cases}$$

The supercurrent density is given by the relation:

$$\tilde{\mathbf{j}} = \begin{cases} \dfrac{e_s}{m_s}\left(-\dfrac{i\hbar}{2}(\tilde{\psi}^*\nabla\tilde{\psi} - \tilde{\psi}\nabla\tilde{\psi}^*) - \dfrac{e_s}{c}\tilde{\mathbf{A}}|\tilde{\psi}|^2\right) + \dfrac{c}{4\pi}\operatorname{curl}\widetilde{\mathbf{H}}_{\mathbf{a}} & \text{in } \mathcal{D}_s, \\ \dfrac{e_n}{m_n}\left(-\dfrac{i\hbar}{2}(\tilde{\psi}^*\nabla\tilde{\psi} - \tilde{\psi}\nabla\tilde{\psi}^*) - \dfrac{e_n}{c}\tilde{\mathbf{A}}|\tilde{\psi}|^2\right) + \dfrac{c}{4\pi}\operatorname{curl}\widetilde{\mathbf{H}}_{\mathbf{a}} & \text{in } \mathcal{D}_n. \end{cases}$$

The applied magnetic field is $\widetilde{\mathbf{H}}_{\mathbf{a}}$. For fixed temperature T, we have $a_s \equiv a_s(T) = a_s(0)(1 - T/T_{cs})$ and $a_n \equiv a_n(T) = a_n(0)(1 - T/T_{cn})$, where $a_s(0) < 0, a_n(0) < 0$ are constant depending on the materials, and T_{cs}, T_{cn} are the so-called zero-field critical temperatures of the materials. We consider situations for which $T_{cn} < T_{cs}$ and look at range of temperatures $T_{cn} < T < T_{cs}$, so that $a_s < 0$ and $a_n > 0$. The positive constants m_s, m_n depend as well on physical parameters, and their ratio, m_n/m_s, is proportional to the ratio, σ_s/σ_n, of the normal conductivities of the materials. The constant $b_s > 0$ is characteristic of the superconducting body, and $\mu_s, \mu_n > 0$ denote the permeability densities of the superconducting and the normal part, respectively. The remaining constants are as follows: $e_s = e_n = 2e$, where e is the electron charge, are the superconducting charge carries, $\tilde{f}_s, \tilde{f}_n$ are the free energy densities of the materials in the nonsuperconducting state in the absence of an applied field, and $2\pi\hbar$ is Planck's constant.

In the mathematical literature, one typically works with nondimensional variables, where the quantities chosen depend on the nature of the problem under consideration. An interesting case has been made for using temperature-independent variables (see [24]). Nevertheless, for various reasons we decided to work with temperature-dependent ones. In [13], to compare our results with the ones for the

Ginzburg–Landau energy presented in [15] we keep their nondimensionalization; while in [14] since we are interested in the onset of superconductivity, we select the second critical field in a vacuum as the reference field, in this work we will employ this second choice. In particular, for lengths we use the coherence length at temperature T of the superconducting material, $\xi_{\mathrm{s}}(T)$, for the order parameter the ratio $|a_{\mathrm{s}}(T)|/b_{\mathrm{s}}$, and for magnetic fields the second critical field of the superconducting material, $H_{\mathrm{c}2} = H_{\mathrm{c}2}(T)$. The corresponding nondimensional energy density

$$
\begin{aligned}
(2)\quad & e(\psi, \mathbf{A}, \mathbf{H_a}) \\
&:= \frac{|a_{\mathrm{s}}|^2}{b_{\mathrm{s}}}\{f_{\mathrm{s}} - |\psi|^2 + \tfrac{1}{2}|\psi|^4 + |(\mathrm{i}\nabla + \mathbf{A})\psi|^2 + \kappa_{\mathrm{s}}^2|\mathbf{H}|^2 - 2\kappa_{\mathrm{s}}^2\,\mathbf{H}\cdot\mathbf{H_a}\}\chi_{\mathcal{D}_{\mathrm{s}}} \\
&+ \frac{|a_{\mathrm{s}}|^2}{b_{\mathrm{s}}}\left\{f_{\mathrm{n}} + \frac{a_{\mathrm{n}}}{|a_{\mathrm{s}}|}|\psi|^2 + \frac{m_{\mathrm{s}}}{m_{\mathrm{n}}}|(\mathrm{i}\nabla + \mathbf{A})\psi|^2 + \frac{\mu_{\mathrm{n}}}{\mu_{\mathrm{s}}}\kappa_{\mathrm{s}}^2|\mathbf{H}|^2 - 2\frac{\mu_{\mathrm{n}}}{\mu_{\mathrm{s}}}\kappa_{\mathrm{s}}^2\mathbf{H}\cdot\mathbf{H_a}\right\}\chi_{\mathcal{D}_{\mathrm{n}}},
\end{aligned}
$$

leads to the generalized Ginzburg–Landau functional:

$$
\begin{aligned}
(3)\quad G(\psi, \mathbf{A}, \mathbf{H_a}) &= \int_{\mathcal{D}_{\mathrm{s}}} \frac{1}{2}(1 - |\psi|^2)^2\,\mathrm{d}\mathbf{x} + \int_{\mathcal{D}_{\mathrm{n}}} a|\psi|^2\,\mathrm{d}\mathbf{x} \\
&+ \int_{\mathbb{R}^n} \frac{1}{m}|\nabla_{\mathbf{A}}\psi|^2\,\mathrm{d}\mathbf{x} + \int_{\mathbb{R}^n} \mu\kappa_{\mathrm{s}}^2\left|\frac{1}{\mu}\operatorname{curl}\mathbf{A} - \mathbf{H_a}\right|^2\mathrm{d}\mathbf{x},
\end{aligned}
$$

where $\nabla_{\mathbf{A}} = (\mathrm{i}\nabla + \mathbf{A})$, $a = a_{\mathrm{n}}/|a_{\mathrm{s}}| \geq 0$, and κ_{s} is the Ginzburg–Landau parameter of the superconductor. The functions $m > 0$ and $\mu > 0$ are piecewise constants, their value in $\mathcal{D}_{\mathrm{s}}$ is 1, while $m = m_{\mathrm{n}}/m_{\mathrm{s}}$ and $\mu = \mu_{\mathrm{n}}/\mu_{\mathrm{s}}$ in $\mathcal{D}_{\mathrm{n}}$.

Critical states are derived as solutions to the Euler-Lagrange equations of (3). We look at applied fields $\mathbf{H_a} \in L^2_{\mathrm{loc}}(\mathbb{R}^n;\mathbb{R}^3)$, and consider

$$\psi \in H^1_{\mathrm{loc}}(\mathbb{R}^n;\mathbb{C}) \equiv \mathcal{H}^1_{\mathrm{loc}}(\mathbb{R}^n), \quad \mathbf{A} \in H^1_{\mathrm{loc}}(\mathbb{R}^n;\mathbb{R}^n)$$

weak solutions to

$$
\begin{aligned}
&\nabla_{\mathbf{A}}\left(\frac{1}{m}\nabla_{\mathbf{A}}\psi\right) - (1 - |\psi|^2)\psi\chi_{\mathcal{D}_{\mathrm{s}}} + a\psi\chi_{\mathcal{D}_{\mathrm{n}}} = 0 \quad \text{in } \mathbb{R}^n, \\
(4)\quad &\operatorname{curl}\left(\frac{1}{\mu}\operatorname{curl}\mathbf{A} - \mathbf{H_a}\right) + \frac{1}{m\kappa_{\mathrm{s}}^2}\left(\frac{\mathrm{i}}{2}(\psi^*\nabla\psi - \psi\nabla\psi^*) + \mathbf{A}|\psi|^2\right) = 0 \quad \text{in } \mathbb{R}^n, \\
&\left(\frac{1}{\mu}\operatorname{curl}\mathbf{A} - \mathbf{H_a}\right) \in L^2(\mathbb{R}^n;\mathbb{R}^3).
\end{aligned}
$$

As in the classical Ginzburg–Landau theory, the solutions to (4) are invariant under the gauge transformation $(\psi, \mathbf{A}) \to (\psi', \mathbf{A}')$, where

$$\psi' = \psi e^{\mathrm{i}\eta}, \quad \mathbf{A}' = \mathbf{A} + \nabla\eta$$

for an arbitrary real-valued function $\eta \in H^2_{\mathrm{loc}}(\mathbb{R}^n)$.

A *normal state* is represented by a weak solution $(\psi, \mathbf{A}) = (0, \mathbf{A}_N)$ of (4), where $\mathbf{A}_N$ satisfies weakly

$$
\begin{aligned}
&\operatorname{curl}\left(\frac{1}{\mu}\operatorname{curl}\mathbf{A}_N - \mathbf{H_a}\right) = 0 \quad \text{in } \mathbb{R}^n, \\
(5)\quad &\left(\frac{1}{\mu}\operatorname{curl}\mathbf{A}_N - \mathbf{H_a}\right) \in L^2(\mathbb{R}^n;\mathbb{R}^3).
\end{aligned}
$$

To investigate the onset of superconductivity, in [14], we took $\mathbf{H_a} = h\mathbf{e}_3$ with $h > 0$, $\mathcal{D}_{\mathrm{s}} = \{(x, y, z), x < 0\}$ and $\mathcal{D}_{\mathrm{n}} = \{(x, y, z), x > 0\}$. Since, in this set-up, if

h is close to or greater than $h_{c_2} \equiv 1$, the induced magnetic field should be parallel to the applied field, and the modulus of the order parameter should depend only on x, [25], the standard simplified assumption is to consider only states of the form:

$$\mathbf{A}(x,y,z) = (0, A(x), 0) \quad \psi(x,y,z) = f(x)\mathrm{e}^{\mathrm{i}k_2 y}\mathrm{e}^{\mathrm{i}k_3 z}, \tag{6}$$

with $f(x)$ real-valued, and k_2, k_3 constants to be determined. The supercurrent is then given by $-(1/m)(A-k_2)f^2$, so that $A-k_2$ is the physically significant quantity, while individually A and k_2 are determined up to a constant. We follow [3], and use this degree of freedom to have a real order parameter, that is we consider $k_2 = 0$. Therefore, if we substitute (6) in (2), and notice that, for f and A fixed, the choice $k_3 = 0$ lowers the energy, we are reduced to consider $(f, A) \in H^1(\mathbb{R}) \times H^1_{\mathrm{loc}}(\mathbb{R};\mathbb{R})$ minimizers of the energy:

$$\mathcal{G}(f,A,h) = \int_{-\infty}^{\infty} \Big[\frac{1}{m}(f')^2 - f^2\chi_- + \frac{1}{2}f^4\chi_- + af^2\chi_+ + \frac{1}{m}A^2 f^2 + \mu\kappa^2\Big(\frac{1}{\mu}A' - h\Big)^2\Big]\,\mathrm{d}x, \tag{7}$$

where, χ_- denotes the characteristic function of the interval $(-\infty, 0)$ and χ_+ the one of $(0,\infty)$. Note that the normal states are pairs of form $\big(0, h(\mu x - \omega)\big)$ for $\omega \in \mathbb{R}$ fixed. In the following sections when referring to the case $n = 1$, we will be talking about the energy functional (7).

3. Global minimizers

The proof of existence of global minimizers resembles the one for the classical Ginzburg–Landau energy of a superconductor filling the whole space. In general, the magnetic potential component of finite energy states does not belong to any Sobolev space $W^{p,r}(\mathbb{R}^n;\mathbb{R}^n)$, but it only has bounded curl in the L^2–norm. In the situation where the order parameter is defined everywhere, the natural space to consider for the physically relevant magnetic potentials is related to the one needed to solve the Laplace equation in $\mathbb{R}^n$, [1]. In particular, existence of global minimizers in dimension $n = 1,2,3$ can be established in the weighted Sobolev spaces, $W^{1,2}_{0,0}(\mathbb{R}^n;\mathbb{R}^n)$, studied in [1]; see also [16].

The case of a superconducting material filling the whole space $\mathbb{R}^n, n = 2,3$ in the presence of an applied field $\mathbf{H_a} \in L^2(\mathbb{R}^n;\mathbb{R}^3)$ is studied by Y. Yang in [26,28]. In three dimensions, [26], existence is proven for magnetic potentials in the closure of $C_0^\infty(\mathbb{R}^3;\mathbb{R}^3)$ with respect to the semi-norm $\|\nabla\ \|_{L^2(\mathbb{R}^3;\mathbb{R}^3)}$, which is the same as the weighted Sobolev space $W^{1,2}_{0,0}(\mathbb{R}^n;\mathbb{R}^n)$ defined in [1]. For the two-dimensional case, [28], to overcome some difficulties peculiar to dimension two, a direct variational method is used to prove existence in a subspace of an ad hoc defined Hilbert space. In fact in [16], we prove that this Hilbert space is equivalent to $W^{1,2}_{0,0}(\mathbb{R}^2;\mathbb{R}^2)$, thus showing that essentially the physically relevant states live in the same space as the solutions of the Laplace operator in $\mathbb{R}^n$.

The main difference in the existence proof between dimensions is due to the fact that for $n = 1,2$ the L^2-norm of the curl in general controls the norm in $W^{1,2}_{0,0}(\mathbb{R}^n;\mathbb{R}^n)$ of divergence free vectors only up to constants, and therefore some extra care is needed.

We consider, in [13, 14], existence of minimizers for $\mathbf{H_a} = h\mathbf{e}$ if $n \le 3$, where h is a positive constant and $\mathbf{e} \in \mathbb{R}^3$ is a fixed unit vector, and for $\mathbf{H_a} \in L^2(\mathbb{R}^n;\mathbb{R}^3)$, if $n = 2, 3$. In particular, we prove the following results:

If $\mathbf{H_a} \in L^2(\mathbb{R}^n;\mathbb{R}^3)$, $n = 2, 3$, *then* $G(\psi, \mathbf{A}, \mathbf{H_a})$ *has a global minimizer in* $\mathcal{H}^1_{\text{loc}}(\mathbb{R}^n) \times W^{1,2}_{0,0}(\mathbb{R}^n;\mathbb{R}^n)$. (See [13, Proposition 2.6])

If $\mathbf{H_a} = h\mathbf{e}$, $n = 2, 3$, *there exists a global minimizer* $(\psi, \mathbf{A})$ *of* $G(\psi, \mathbf{A}, \mathbf{H_a})$ *with* $\psi \in \mathcal{H}^1_{\text{loc}}(\mathbb{R}^n)$, *and* $\mathbf{A} - \mathbf{A}_N \in W^{1,2}_{0,0}(\mathbb{R}^n;\mathbb{R}^n)$. (See [13, Lemma 2.4 and Theorem 2.7])

For the case $n = 1$ we have a stronger statement:

Let $\kappa \ge \sqrt{2}/2$ *be fixed. If* $m_{\text{n}}/m_{\text{s}} \le 1$ *and* $h > 1$ *or if* $m_{\text{n}}/m_{\text{s}} > 1$ *and* $h > m_{\text{n}}/m_{\text{s}}$ *then normal states are the only global minimizers in the class of function* $(f, A) \in H^1(\mathbb{R}) \times H^1_{\text{loc}}(\mathbb{R};\mathbb{R})$, *with* $(1/\mu)\big(A(x) - A(0)\big) - hx \in W^{1,2}_{0,0}(\mathbb{R};\mathbb{R})$.

Let $\kappa \ge 4$ *be fixed. If* $m_{\text{n}}/m_{\text{s}} > 1$ *and* $1 \le h \le m_{\text{n}}/m_{\text{s}}$, *there exists a global minimizer* $(f, A) \in H^1(\mathbb{R}) \times H^1_{\text{loc}}(\mathbb{R};\mathbb{R})$, *with* $(1/\mu)\big(A(x) - A(0)\big) - hx \in W^{1,2}_{0,0}(\mathbb{R};\mathbb{R})$. (See [14, Theorem 6.3])

The last theorem tells that for large enough, constant, applied magnetic fields for $n = 1$ the normal states are the only global minimizers. This is true also for $n = 2$, since we were able to prove, in the spirit of the results obtained for the classical Ginzburg–Landau energy [15], the existence of an upper critical field, for which the only finite energy weak solutions of (4) are the normal states. In other words, if we denote by $\bar{h}$ the so-called *upper critical field*, defined as

$$\bar{h} := \inf\{h' : \text{normal states are the only finite energy weak solutions to (4)} \quad \text{for all } h > h'\};$$

we have the following theorem:

Let $n = 2$, *then* $\bar{h} = \bar{h}(m_{\text{n}}/m_{\text{s}}, \kappa_{\text{s}}, a, \mu_{\text{n}}/\mu_{\text{s}}, \boldsymbol{\mathcal{D}}_{\text{s}}) < \infty$. ([13, Theorem 4.6])

In the three-dimensional case, we can prove $\bar{h} < \infty$, only if the permeability is constant in $\mathbb{R}^3$, that is $\mu \equiv 1$ in $\mathbb{R}^3$ (see [13, Section 4.2]), while in the general case we recover a result analogous to the one for superconducting films presented in [27]:

Let $n = 3$, *for* $m, \kappa_{\text{s}}, a, \mu$ *fixed as* $h \to \infty$, *any finite energy weak solution tends to the normal state.* (See [13, Theorem 4.7])

Notice how in three-dimensions the seemingly simple hypothesis $\mu \ne$ constant can not be treated once again using the same approach valid for the constant case.

We recall that $W^{1,2}_{0,0}(\mathbb{R}^n;\mathbb{R}^n)$ denotes the space of vector functions with components in the space

$$W^{1,2}_{0,0}(\mathbb{R}^n) = \{u \in \boldsymbol{\mathcal{D}}'(\mathbb{R}^n); w(\mathbf{x})\, u \in L^2(\mathbb{R}^n); \lambda \in \mathcal{N}^n : |\lambda| = 1, D^\lambda u \in L^2(\mathbb{R}^n)\},$$

where

$$w(\mathbf{x}) = \begin{cases} \dfrac{1}{(1 + |\mathbf{x}|^2)^{1/2} \ln(2 + |\mathbf{x}|^2)} & \text{if } n = 2, \\ \dfrac{1}{(1 + |\mathbf{x}|^2)^{1/2}} & \text{if } n = 1, 3, \end{cases}$$

here $\boldsymbol{\mathcal{D}}'(\mathbb{R}^n)$ is the dual space of $C_0^\infty(\mathbb{R}^n)$. (See [1, 14] for details)

Note that $W^{1,2}_{0,0}(\mathbb{R}^3;\mathbb{R}^3) \equiv \mathring{W}^{1,2}(\mathbb{R}^3) \equiv \check{H}^1(\mathbb{R}^3)$, and $W^{1,2}_{0,0}(\mathbb{R}^2;\mathbb{R}^2) \equiv \mathcal{H}$; with $\mathring{W}^{1,2}(\mathbb{R}^3)$ defined as in [26], $\check{H}^1(\mathbb{R}^3)$ as in [15], and $\mathcal{H}$ as in [28] (see [16]).

The existence proofs use the fact that the space $W^{1,2}_{0,0}(\mathbb{R}^n;\mathbb{R}^n)$ is a reflexive Banach space with respect to the norm

$$\|\mathbf{B}\|_{W^{1,2}_{0,0}(\mathbb{R}^n;\mathbb{R}^n)} \equiv \left\{\|w(\mathbf{x})\mathbf{B}\|^2_{L^2(\mathbb{R}^n;\mathbb{R}^n)} + \|\nabla\mathbf{B}\|^2_{L^2(\mathbb{R}^n;\mathbb{R}^n)}\right\}^{1/2},$$

and an Hilbert space with respect to the induced scalar product. As well as, more fundamentally, that

$$|\mathbf{B}|_{W^{1,2}_{0,0}(\mathbb{R}^n;\mathbb{R}^n)} \equiv \|\nabla\mathbf{B}\|^2_{L^2(\mathbb{R}^n;\mathbb{R}^n)}$$

defines a seminorm, for which a Poincaré-type inequality holds, that is

$$\|\mathbf{B}\|_{W^{1,2}_{0,0}(\mathbb{R}^3;\mathbb{R}^3)} \leq C|\mathbf{B}|_{W^{1,2}_{0,0}(\mathbb{R}^3;\mathbb{R}^3)},$$

$$\inf_{\mathbf{b}\in\mathbb{R}^2} \|\mathbf{B}+\mathbf{b}\|_{W^{1,2}_{0,0}(\mathbb{R}^2;\mathbb{R}^2)} \leq C|\mathbf{B}|_{W^{1,2}_{0,0}(\mathbb{R}^2;\mathbb{R}^2)},$$

$$\|B - B(0)\|_{W^{1,2}_{0,0}(\mathbb{R};\mathbb{R})} \leq C|B|_{W^{1,2}_{0,0}(\mathbb{R};\mathbb{R})}.$$

4. Onset of superconductivity

In our nondimensional setup, when studying the nucleation of superconductivity in decreasing fields, we are looking at fields $h \geq h_{c2} \equiv 1$. A comparison of (1) and in [18, equation (4)] gives $m_n/m_s = \sigma_s/\sigma_n$, hence the normal state should become unstable at a field $h^n_{c3}(m_n/m_s) > 1$ when $m_n/m_s > 1$, and remain stable down to $h_{c2} \equiv 1$ when $m_n/m_s \leq 1$. Additionally, from previous works (see for example [2, 22]), we know that the nucleation field in a vacuum verifies $H_{c3} = H_{c2}/\beta_0^*$, with β_0^* defined as in [21, Lemma 5.2]. Therefore, we also expect $h^n_{c3}(m_n/m_s) < h_{c3} \equiv 1/\beta_0^*$ for every m_n/m_s, with $h_{c3}(m_n/m_s) \to 1/\beta_0^*$ for $m_n/m_s \to \infty$ and $h^n_{c3}(m_n/m_s)$ increasing in $m_n/m_s > 1$.

We obtain for the energy (7) the above picture in our paper [14], where we define, following physical intuition, the *nucleation field* $h^n_{c3}(m_n/m_s)$ as the field below which there are normal states that are not local minimizers, and above which every normal state is a local minimizer.

If we now look at the two-dimensional case, and we repeat the analysis in [14], to study the critical field where the normal state loses stability and other critical points emerge, we need to analyze the second variation of (3). We let $\phi \in H^1_{\mathrm{loc}}(\mathbb{R}^2)$ and $B \in W^{1,2}_{0,0}(\mathbb{R}^2;\mathbb{R}^2)$ be arbitrary, and compute the second variation around $(\psi \equiv 0, \mathbf{A} = (0, h\mu x))$:

$$\frac{\mathrm{d}^2 G}{\mathrm{d}\epsilon^2}(\psi+\epsilon\phi, A+\epsilon B)\bigg|_{\epsilon=0} = 2\int_{\mathbb{R}^2}\left(\frac{1}{m}|\nabla_{\mathbf{A}}\phi|^2 - |\phi|^2\right)\mathrm{d}x + 2\int_{\mathcal{D}_n}(a+1)|\phi|^2\,\mathrm{d}x + 2\int_{\mathbb{R}^2}\mu\kappa_s^2\left|\frac{1}{\mu}\operatorname{curl} B\right|^2\mathrm{d}x.$$

We find that the normal state is stable for fields h such that $\sigma(h, m_n/m_s) > 1$, where

$$\sigma\left(h, \frac{m_n}{m_s}\right) = \inf_{H^1(R^2)} \frac{\int_{\mathbb{R}^2}(1/m)|\nabla_{\mathbf{A}}\phi|^2 + (a+1)\chi_{\mathcal{D}_n}|\phi|^2\,\mathrm{d}x\,\mathrm{d}y}{\int_{\mathbb{R}^2}|\phi|^2\,\mathrm{d}x\,\mathrm{d}y}. \tag{8}$$

Heuristically speaking, in a thin strip near the superconducting/normal interface one may consider a domain that is simply half superconducting and half normal. If one straighten the boundary so that the curvature is zero everywhere, we are lead to study the variational problem where the full space is split into two parts with the interface a hyperplane. In this situation, the leading term of the variational problem (8) should be given by the formula:

$$\sigma\left(h, \frac{m_{\mathrm{n}}}{m_{\mathrm{s}}}\right) = \inf_{H^1(\mathbb{R}^2)} \frac{\int_{R^2} |(1/m)(\mathrm{i}\nabla + \mathbf{A})\psi|^2 + (a+1)\chi_+ |\phi|^2 \, \mathrm{d}x \, \mathrm{d}y}{\int_{\mathbb{R}^2} |\psi|^2 \, \mathrm{d}x \, \mathrm{d}y}, \tag{9}$$

where $m = 1$ for $\{x < 0\}$, $m = m_{\mathrm{n}}/m_{\mathrm{s}}$ for $\{x > 0\}$, χ_+ is the characteristic function of $\{(x, y) : x > 0\}$, and the magnetic potential is $\mathbf{A} = (0, hx)$ for $\{x < 0\}$, and $\mathbf{A} = (0, h(\mu_{\mathrm{n}}/\mu_{\mathrm{s}})x)$ for $\{x > 0\}$.

Notice the difference between (8) and (9) in the definition of the superconducting and normal domains. In particular, in (8) the superconducting region is bounded.

Critical points of the variational eigenvalue problem (9) satisfy

$$(\mathrm{i}\nabla + \mathbf{A})\frac{1}{m}(\mathrm{i}\nabla + \mathbf{A})\phi + (a+1)\chi_+\phi = \sigma\phi \quad \text{in } \mathbb{R}^2, \tag{10}$$

with the choice of $\mathbf{A} = (0, h\mu x)$ the equation more explicitly appears as

$$-\partial_x \frac{1}{m}\partial_x\psi + \frac{1}{m}(-\partial_{yy}\psi + 2\mathrm{i}h\mu x\partial_y\psi + h^2\mu^2x^2)\psi + (a+1)\chi_+\psi = \sigma\psi \tag{11}$$

The critical value below which the normal state loses stability is then

$$H^*\left(\frac{m_{\mathrm{n}}}{m_{\mathrm{s}}}\right) \equiv \inf\left\{h > 0 : \sigma\left(h, \frac{m_{\mathrm{n}}}{m_{\mathrm{s}}}\right) = 1\right\}, \tag{12}$$

and, as for the one-dimensional case, we take H^* as our choice for the definition of the *nucleation field* $h^n_{c3}(m_{\mathrm{n}}/m_{\mathrm{s}})$.

The value of σ is related, as we will show in Lemma 3 below, to the values of the one-dimensional eigenvalue problem

$$\tau(h, \omega) = \inf_{f \in H^1(\mathbb{R})} \frac{\int_{-\infty}^{\infty} (1/m)(f')^2 + h^2(\mu x - \omega)^2 f^2 + (a+1)\chi_+ f^2 \, \mathrm{d}x}{\int_{-\infty}^{\infty} f^2 \, \mathrm{d}x}, \tag{13}$$

where $h > 0, \omega \in \mathbb{R}$, and χ_+ is the characteristic function of $\{x \in \mathbb{R} : x > 0\}$. We analyze (13) in [14] while studying the stability of normal states for (7). In fact, in the one-dimensional setup, using the second variation of (7), we find that for $\tau > 1$ the normal states are stable local minimizers, whereas for $\tau < 1$ there are directions where the normal state is not a local minimizer. In particular, we show that for $h > 0$, $\omega \in \mathbb{R}$ and all other parameters fixed, there is an eigenfunction $f_\omega \in H^1(\mathbb{R})$ where the infimum in (13) is attained (cf. [14, Theorem 4.2]); and, we analyze the properties of τ and (h, ω) where $\tau = 1$.

Lemma 1 ([14]). *Let $m_{\mathrm{n}}, m_{\mathrm{s}}, a, \mu_{\mathrm{n}}, \mu_{\mathrm{s}}$ be fixed and $h > 0$, $\omega \in \mathbb{R}$.*

(1) *There exists a unique $h = h(\omega, m_{\mathrm{n}}/m_{\mathrm{s}})$ such that $\tau(h(\omega, m_{\mathrm{n}}/m_{\mathrm{s}}), \omega) = 1$;*
(2) *If $m_{\mathrm{n}}/m_{\mathrm{s}} \leq 1$ then $h(\omega, m_{\mathrm{n}}/m_{\mathrm{s}}) < 1$;*
(3) *If $m_{\mathrm{n}}/m_{\mathrm{s}} > 1$ then $\frac{1}{3} < h(\omega, m_{\mathrm{n}}/m_{\mathrm{s}}) < m_{\mathrm{n}}/m_{\mathrm{s}}$.*

The proof of the lemma above uses estimates, obtained in [21], for the following eigenvalue problems:

$$\lambda(z) = \inf_{u \in H^1_0([0,\infty))} \frac{\int_0^\infty \{|u'|^2 + (z+t)^2|u|^2\} \, \mathrm{d}t}{\int_0^\infty |u|^2 \, \mathrm{d}t}, \tag{14}$$

and

$$\beta_\gamma(z) = \inf_{u\in H^1([0,\infty))} \frac{\int_0^\infty \{|u'|^2 + (z+t)^2|u|^2\}\,\mathrm{d}t + \gamma|u(0)|}{\int_0^\infty |u|^2\,\mathrm{d}t}, \tag{15}$$

where $z, t \in \mathbb{R}$.

We then show that for our definition of the nucleation field there is a dependence on the ratio of m_n to m_s, in the fashion indicated by the physical literature.

Theorem 2 ([14]). *For $\omega \in \mathbb{R}$, let $h(\omega, m_\mathrm{n}/m_\mathrm{s})$ be defined as in Lemma 1.*

(1) *If $m_\mathrm{n}/m_\mathrm{s} \le 1$, then*

$$h^*\left(\frac{m_\mathrm{n}}{m_\mathrm{s}}\right) \equiv \sup_{\omega\in\mathbb{R}} h\left(\omega, \frac{m_\mathrm{n}}{m_\mathrm{s}}\right) = 1, \tag{16}$$

the value $h^*(m_\mathrm{n}/m_\mathrm{s})$ is not attained, and for every $h > h^*(m_\mathrm{n}/m_\mathrm{s})$ every normal state is a local minimizer, while for $h \le h^*(m_\mathrm{n}/m_\mathrm{s})$ there are normal states which are not local minimizers;

(2) *If $m_\mathrm{n}/m_\mathrm{s} > 1$, then there exists an $\omega^* < 0$ such that*

$$h\left(\omega^*, \frac{m_\mathrm{n}}{m_\mathrm{s}}\right) = h^*\left(\frac{m_\mathrm{n}}{m_\mathrm{s}}\right) \equiv \sup_{\omega\in\mathbb{R}} h\left(\omega, \frac{m_\mathrm{n}}{m_\mathrm{s}}\right) > 1. \tag{17}$$

For every $h \ge h^*(m_\mathrm{n}/m_\mathrm{s})$ every normal state is a local minimizer, while for $h < h^*(m_\mathrm{n}/m_\mathrm{s})$ there are normal states which are not local minimizers.

(3) *If $1 < m_1 < m_2$ then*

$$h^*(m_1) < h^*(m_2). \tag{18}$$

(4) *For β_0^* as in [21, Lemma 5.2], it holds*

$$\lim_{m_\mathrm{n}/m_\mathrm{s}\to\infty} h^*\left(\frac{m_\mathrm{n}}{m_\mathrm{s}}\right) = \frac{1}{\beta_0^*}. \tag{19}$$

In this context we define $h^n_{c3}(m_\mathrm{n}/m_\mathrm{s}) = h^*(m_\mathrm{n}/m_\mathrm{s})$.

We make in the following lemma the connection between the eigenvalue σ as defined in (9) in two dimensions, and the eigenvalue τ of the problem when reduced to one dimension, in a way equivalent to the one presented by Lu–Pan in [21], for the case when the normal material is replaced by vacuum.

Lemma 3. *Let $m_\mathrm{n}/m_\mathrm{s} > 0$ then:*

$$\sigma\left(h, \frac{m_\mathrm{n}}{m_\mathrm{s}}\right) = \inf_{\omega\in\mathbb{R}} \tau(h,\omega) \tag{20}$$

Proof. Following Lu–Pan [21], for any $h > 0$ we find

$$\sigma\left(h, \frac{m_\mathrm{n}}{m_\mathrm{s}}\right) \le \tau(h,\omega), \quad \omega \in R, \tag{21}$$

by using, in the definition of σ, the test function $\phi_n = \eta_n(y)\mathrm{e}^{ih\omega y} f_\omega(x)$, where f_ω is the eigenfunction associated with $\tau(h,\omega)$, and $\eta_n(t)$ is a smooth cut-off function with $0 \le \eta_n \le 1$, compact support in $\{|t| < n+1\}$, $\eta_n(t) = 1$ for $|t| < n$ and $\eta_n' \le 2$.

To show the opposite inequality we follow again Lu–Pan. Let $\phi \in H^1(R^2)$ with $(\mathrm{i}\partial_y\phi - h\mu x\phi) \in L^2(R^2)$, and denote by $\mathcal{F}\big(\phi(\cdot,y)\big) = f(\cdot,\omega)$ the Fourier transform in y, applying Parseval's identity we obtain:

$$\begin{aligned}\int_{-\infty}^{\infty}\left|\frac{1}{m}(\mathrm{i}\nabla+(0,h\mu x))\phi\right|^2+(a+1)\chi_+\psi^2\,\mathrm{d}y\\ &=\int_{-\infty}^{\infty}\left\{\frac{1}{m}(|\mathcal{F}\partial_x\phi|^2+|\mathcal{F}(\mathrm{i}\partial_y\phi+h\mu x\phi)|^2)+(a+1)\chi_+(x)|\mathcal{F}\phi|^2\right\}\mathrm{d}\omega\\ &=\int_{-\infty}^{\infty}\left\{\frac{1}{m}(|\partial_x f|^2+(h\mu x-\omega)^2|f(x,\omega)|^2)+(a+1)\chi_+(x)|f(x,\omega)|^2\right\}\mathrm{d}\omega\\ &=\int_{-\infty}^{\infty}\left\{\frac{1}{m}(|\partial_x f|^2+h^2(\mu x-\widetilde{\omega})^2|f(x,h\widetilde{\omega})|^2)+(a+1)\chi_+(x)|f(x,h\widetilde{\omega})|^2\right\}h\,\mathrm{d}\widetilde{\omega}\end{aligned}$$

where $\widetilde{\omega}=\omega/h$.

If we now use the variational definition of τ, we continue

$$\begin{aligned}&=\int_{-\infty}^{\infty}\int_{\infty}^{\infty}\left\{\frac{1}{m}(|\partial_x f|^2+h^2(\mu x-\widetilde{\omega})^2|f(x,h\widetilde{\omega})|^2)\right.\\ &\qquad\qquad\left.+(a+1)\chi_+(x)|f(x,h\widetilde{\omega})|^2\right\}h\,\mathrm{d}\widetilde{\omega}\,\mathrm{d}x\end{aligned}\tag{22}$$

$$\begin{aligned}&\geq\int_{-\infty}^{\infty}\left\{\tau(h,\widetilde{\omega})\int_{-\infty}^{\infty}|f|^2(x,h\widetilde{\omega})\,\mathrm{d}x\right\}h\,\mathrm{d}\widetilde{\omega}\\ &=\int_{-\infty}^{\infty}\int_{-\infty}^{\infty}\tau(h,\omega/h)|f|^2(x,\omega)\,\mathrm{d}x\,\mathrm{d}\omega\\ &\geq\left(\inf_{\omega\in\mathbb{R}}\tau(h,\omega)\right)\int_{-\infty}^{\infty}\int_{-\infty}^{\infty}|f|^2\,\mathrm{d}\omega\,\mathrm{d}x.\end{aligned}$$

Since ϕ is an arbitrary test function for which the Rayleigh quotient is finite in the variational problem determining σ, it follows that $\sigma(h,m_{\mathrm{n}}/m_{\mathrm{s}})\geq\inf_\omega\tau(\omega,h)$. □

The relation between σ and τ implies that the critical values we obtain for the one-dimensional problem are also the critical values in two dimensions, as we show in the following lemma.

Lemma 4. *We have* $H^*(m_{\mathrm{n}}/m_{\mathrm{s}})=h^*(m_{\mathrm{n}}/m_{\mathrm{s}})$.

PROOF. Note that from Lemma 3, we know that

$$H^*\left(\frac{m_{\mathrm{n}}}{m_{\mathrm{s}}}\right)=\inf\left\{h>0:\inf_{\omega\in\mathbb{R}}\tau(\omega,h)=1\right\}.\tag{23}$$

Additionally, a direct computation shows that $\tau(h,\omega)$ is continuous and strictly increasing in $h>0$, when other parameters are fixed.

We claim that $H^*(m_{\mathrm{n}}/m_{\mathrm{s}})\geq h^*(m_{\mathrm{n}}/m_{\mathrm{s}})$. In fact, if $h<h^*$ then there is an ω_0 such that $h<h(\omega_0,m_{\mathrm{n}}/m_{\mathrm{s}})$, which implies $\tau(h,\omega_0)<1$ and thus $\inf_\omega\tau(h,\omega)<1$.

On the other hand, we can show that $\inf_\omega\tau(h^*,\omega)=1$.

If $m_{\mathrm{n}}/m_{\mathrm{s}}>1$, from Theorem 2 we know that $h^*=h(\omega^*,m_{\mathrm{n}}/m_{\mathrm{s}})$ for some ω^*, so $\inf_\omega\tau(h^*,\omega)\leq 1$. If $\inf_\omega\tau(h^*,\omega)<1$, there is an $\omega_1\in\mathbb{R}$ such that $\tau(h^*,\omega_1)<1$. But, for this ω_1, we have that $h(\omega_1,m_{\mathrm{n}}/m_{\mathrm{s}})\leq h^*$ and $\tau(h(\omega_1,m_{\mathrm{n}}/m_{\mathrm{s}}),\omega_1)=1$. As before we use the strictly increasing property of τ for fixed ω to find:

$$1=\tau\left(h\left(\omega_1,\frac{m_{\mathrm{n}}}{m_{\mathrm{s}}}\right),\omega_1\right)\leq\tau(h^*,\omega_1)<1$$

a contradiction. Hence $\inf_\omega \tau(h^*,\omega) = 1$. This with the previous inequality implies that $H^*(m_n/m_s) = h^*$.

If $m_n/m_s \le 1$ from the inequality $\tau(h,\omega) \le h\lambda(h^{1/2}\omega)$, where λ is defined as in (14), and the fact that $\inf_z \lambda(z) = 1$ proven in [21], we have $\inf_\omega \tau(h,\omega) \le h$, and since by Lemma 2 we have $h^* = 1$, we conclude $\inf_\omega \tau(h^*,\omega) \le 1$, and as above we obtain $H^*(m_n/m_s) = h^*$. □

Note that the above lemma confirms the complete experimental picture also for this particular two-dimensional setup.

The stability of normal states in a similar context has been studied in the case of a bulk superconductor occupying the whole space $\mathbb{R}^2$. It is well known ([9, 21]) that the variational problem has an $L^2(\mathbb{R}^2)$ eigenfunction, associated with the eigenvalue $\sigma_{\mathbb{R}^2} = 1$. If the superconducting domain is a half-plane with the other half occupied by a vacuum, in [21] it is proven the nonexistence of an $L^2(\mathbb{R}^2_+)$ eigenfunction. Furthermore, in [9] it is shown that any bounded solution to the associated PDE corresponding to the smallest eigenvalue, must have the form $e^{i\omega y}f(x)$ where f is the first eigenfunction of: $f'' + h^2(x-\omega^*)^2 f = \beta_0^* f$, $f'(0) = 0$. In particular, the solutions decay only away from the interface, while all points along the interface are equally supportive of the nucleation process. In the case of wedges with corner angle $\theta \le \pi/2$, L^2 eigenfunctions exist and decay away from the corner ([4, 19, 23]). It is then natural to ask the question if the superconducting/normal infinity domains discussion falls in either case. As expected, see Corollary 5 below, it is straightforward to check that substituting a vacuum with a normal material has no effect from the point of view of the nonexistence of nontrivial L^2 eigenfunctions.

Corollary 5. *For any $m_n/m_s > 0$, the eigenvalue problem has no nontrivial $L^2(\mathbb{R}^2)$ eigenfunction. Moreover,*

(1) *For $m_n/m_s \le 1$ there is at least one bounded solution for each field strength $h < H^*(m_n/m_s)$.*
(2) *For $m_n/m_s > 1$ there is at least one bounded solution for each field strength $h \le H^*(m_n/m_s)$.*

PROOF. From (21) and the computational method in (22) we can show that (9) has no $L^2(\mathbb{R}^2)$ eigenfunctions following the argument of Lu and Pan in [21]. By contradiction, assume there is ϕ a (nontrivial) eigenfunction in $H^1(R^2)$ associated with σ, with $\mathcal{F}\phi(\cdot,y) \equiv f(\cdot,\omega)$ as in (22) we have that

$$\sigma\left(h, \frac{m_n}{m_s}\right)\int_{-\infty}^{\infty}\int_{-\infty}^{\infty}|f|^2\,dx\,d\omega \equiv \sigma\left(h, \frac{m_n}{m_s}\right)\int_{-\infty}^{\infty}\int_{-\infty}^{\infty}|\phi|^2\,dx\,dy$$
$$\ge \int_{-\infty}^{\infty}\int_{-\infty}^{\infty}\tau(h,\omega/h)|f|^2\,dx\,d\omega,$$

that is

$$\int_{-\infty}^{\infty}\int_{-\infty}^{\infty}\left(\sigma\left(h, \frac{m_n}{m_s}\right) - \tau(h,\omega/h)\right)|f|^2\,dx\,d\omega \ge 0. \tag{24}$$

Since $\sigma(h, m_n/m_s) \le \tau(h,\tilde{\omega})$ for any $\tilde{\omega} \in \mathbb{R}$, and σ does not depend on $\tilde{\omega}$, we conclude that either $|f|^2 = 0$ almost everywhere, and so $\phi = 0$ a contradiction, or there is some value of h such that $\tau(h,\tilde{\omega}) = \tau_0 > 0$ for any $\tilde{\omega} \in \mathbb{R}$. On the other hand, if f_0 is the eigenfunction of $\tau(h,0)$, for any $\widetilde{\omega}$ one has

$$\tau(h,\widetilde{\omega})\int f_0^2\,\mathrm{d}x \le \tau(h,0)\int f_0^2\,\mathrm{d}x + h^2\widetilde{\omega}\Big[\widetilde{\omega}\int \mu f_0^2\,\mathrm{d}x - 2\int \mu x f_0^2\,\mathrm{d}x\Big], \tag{25}$$

so that for $\widetilde{\omega}$ large, we would have $\tau_0 < \tau_0$, which is a contradiction. The rest of the corollary is proven by taking $\psi(x,y) = \mathrm{e}^{\mathrm{i}\omega y} f_{(h,\omega)}(x)$ where $f_{(h,\omega)}$ is the eigenfunction associated with $\tau(h(\omega, m_{\mathrm{n}}/m_{\mathrm{s}}),\omega) = 1$, and noticing that this function satisfies (11). □

Acknowledgments. The authors would like to thank the organizers of the workshop for providing the opportunity to attend a very interesting series of talks.

References

1. C. Amrouche, V. Girault, and J. Giroire, *Weighted Sobolev spaces for Laplace's equation in* $\mathbb{R}^n$, J. Math. Pures Appl. (9) **73** (1994), no. 6, 579–606.
2. P. Bauman, D. Phillips, and Q. Tang, *Stable nucleation for the Ginzburg–Landau system with an applied magnetic field*, Arch. Rational Mech. Anal. **142** (1998), no. 1, 1–43.
3. C. Bolley and B. Helffer, *Rigorous results on Ginzburg–Landau models in a film submitted to an exterior parallel magnetic field*. I, Nonlinear Stud. **3** (1996), no. 1, 1–29.
4. V. Bonnaillie, *On the fundamental state energy for a Schrödinger operator with magnetic field in domains with corners*, Asymptot. Anal. **41** (2005), no. 3-4, 215–258.
5. J. P. Burger, G. Deutscher, E. Guyon, and A. Martinet, *Proximity effects under magnetic fields*, Phys. Lett. **17** (1965), no. 3, 180–181.
6. J. P. Burger, G. Deutscher, J.-P. Hurault, and A. Martinet, *Surface effects in superconductor coated with a normal metal*, Proceedings of the Tenth International Conference on Low Temperature Physics (Moscow, 1966), Vol. II.A, VINITI, Moscow, 1967, pp. 186–189.
7. S. J. Chapman, Q. Du, and M. D. Gunzburger, *A Ginzburg–Landau type model for superconducting/normal junctions including Josephson junctions*, European J. Appl. Math. **6** (1995), no. 2, 97–114.
8. P.-G. de Gennes, *Superconductivity of metals and alloys*, translated by P. A. Pincus, Advanced Book Classics, Addison Wesley, Redwood City, CA, 1989.
9. M. del Pino, P. L. Felmer, and P. Sternberg, *Boundary concentration for eigenvalue problems related to the onset of superconductivity*, Comm. Math. Phys. **210** (2000), no. 2, 413–446.
10. G. Deutscher and P.-G. de Gennes, *Proximity effects*, Superconductivity. Part 2 (R. D. Parks, ed.), Marcel Dekker, Inc, New York, 1969, pp. 1005–1034.
11. Q. Du, *Numerical approximations of the Ginzburg-Landau models for superconductivity*, J. Math. Phys. **46** (2005), no. 9, 095109.
12. Q. Du, M. D. Gunzburger, and J. Peterson, *Computational simulation of type-*II *superconductivity including pinning phenomena*, Phys. Rev. B (3) **51** (1995), no. 22, 16194–16203.
13. T. Giorgi, *Superconductors surrounded by normal materials*, Proc. Roy. Soc. Edinburgh Sect. A **135** (2005), no. 2, 331–356.
14. T. Giorgi and H. Jadallah, *The onset of superconductivity at a superconducting/normal interface*, European J. Appl. Math. **17** (2006), no. 6, 633–650.
15. T. Giorgi and D. Phillips, *The breakdown of superconductivity due to strong fields for the Ginzburg–Landau model*, SIAM J. Math. Anal. **30** (1999), no. 2, 341–359.
16. T. Giorgi and R. G. Smits, *Remarks on the existence of global minimizers for the Ginzburg–Landau energy functional*, Nonlinear Anal. **53** (2003), no. 2, 147–155.
17. E. Hill, J. Rubinstein, and P. Sternberg, *A modified Ginzburg–Landau model for Josephson junctions in a ring*, Quart. Appl. Math. **60** (2002), no. 3, 485–503.
18. J.-P. Hurault, *Surface nucleation in a superconductor coated with a normal metal*, Phys. Lett. **20** (1966), no. 6, 587–588.
19. H. Jadallah, *The onset of superconductivity in a domain with a corner*, J. Math. Phys. **42** (2001), no. 9, 4101–4121.
20. A. Kachmar, *On the stability of normal states for a generalized Ginzburg–Landau model*, available at `arXiv:math-ph/0609026`.

21. K. Lu and X.-B. Pan, *Gauge invariant eigenvalue problems in $\mathbb{R}^2$ and in $\mathbb{R}^2_+$*, Trans. Amer. Math. Soc. **352** (2000), no. 3, 1247–1276.
22. ———, *Surface nucleation of superconductivity in 3-dimensions*, J. Differential Equations **168** (2000), no. 2, 386–452.
23. X.-B. Pan, *Upper critical field for superconductors with edges and corners*, Calc. Var. Partial Differential Equations **14** (2002), no. 4, 447–482.
24. J. Rubinstein and P. Sternberg, *Limits of a Ginzburg–Landau model with codimension-one defects*, J. Math. Phys. **44** (2003), no. 3, 1240–1251.
25. D. Saint-James, G. Sarma, and E. J. Thomas, *Type-*II *superconductivity*, Pergamon, New York, 1969.
26. Y. Yang, *Existence, regularity, and asymptotic behavior of the solutions to the Ginzburg–Landau equations on $\mathbb{R}^3$*, Comm. Math. Phys. **123** (1989), 147–161.
27. ———, *The Ginzburg–Landau equations for superconducting films and the Meissner effect*, J. Math. Phys. **31** (1990), no. 5, 1284–1289.
28. ———, *The existence of Ginzburg–Landau solutions on the plane by a direct variational method*, Ann. Inst. H. Poincaré Anal. Non Linéaire **11** (1994), no. 5, 517–536.

DEPARTMENT OF MATHEMATICAL SCIENCES, NEW MEXICO STATE UNIVERSITY, LAS CRUCES, NM 88003, USA
E-mail address: tgiorgi@nmsu.edu

DEPARTMENT OF MATHEMATICS, ILLINOIS STATE UNIVERSITY, NORMAL, IL 61790, USA
E-mail address: htree2000@yahoo.com

Centre de Recherches Mathématiques
CRM Proceedings and Lecture Notes
Volume **44**, 2008

Global Questions for Map Evolution Equations

Meijiao Guan, Stephen Gustafson, Kyungkeun Kang, and Tai-Peng Tsai

Abstract. Just as the harmonic map equation is a geometric analogue of the classical Laplace equation for harmonic functions, so the classical linear evolution PDEs, the heat, wave, and Schrödinger equations, have geometric "map" analogues: the harmonic map heat-flow, wave map, and Schrödinger map equations. These equations are nonlinear when the target space geometry is nontrivial. Quite remarkably, these equations are all of physical (as well as mathematical) interest, at least when the target space is a 2-sphere, arising variously in the study of ferromagnets (and anti-ferromagnets), liquid crystals, and general relativity. In this article we review some recent results for map evolution equations (focusing on the Landau–Lifshitz family of equations, which includes as special cases the heat-flow and Schrödinger map) concerning the basic global questions: singularity formation vs. global regularity, and long-time asymptotics.

1. Introduction

Let us begin with the harmonic map equation. From the outset, in order to streamline the presentation and make the analysis more concrete, we fix a specific choice of domain and target manifold for our maps:

$$\mathbf{u}\colon \mathbb{R}^n \to \mathbb{S}^2,$$

mostly $n = 2$. We realize $\mathbb{S}^2$ as the unit sphere in $\mathbb{R}^3$:

$$\mathbb{S}^2 := \{\mathbf{u} = (u_1, u_2, u_2) \in \mathbb{R}^3 \mid |\mathbf{u}| = 1\} \subset \mathbb{R}^3.$$

(*Notation.* 3-vectors will be bold-faced throughout.) Harmonic maps are critical points of the Dirichlet energy functional

$$\mathcal{E}(\mathbf{u}) := \frac{1}{2}\int_{\mathbb{R}^n} |\nabla \mathbf{u}|^2 \,\mathrm{d}x = \frac{1}{2}\int_{\mathbb{R}^n} \sum_{j=1}^{n}\sum_{k=1}^{3} \left|\frac{\partial u_k}{\partial x_j}\right|^2 \mathrm{d}x,$$

and so (if regular) solve the corresponding Euler–Lagrange equation

$$0 = -\mathcal{E}'(u) = P^{\mathbf{u}}\Delta\mathbf{u} = \Delta\mathbf{u} + |\nabla\mathbf{u}|^2 u \qquad \text{(HM)}$$

where $P^{\mathbf{u}}$ denotes the orthogonal projection from $\mathbb{R}^3$ onto the tangent plane

$$T_{\mathbf{u}}\mathbb{S}^2 := \{\boldsymbol{\xi} \in \mathbb{R}^3 \mid \boldsymbol{\xi}\cdot\mathbf{u} = 0\}$$

2000 *Mathematics Subject Classification.* 35Q55.

This is the final form of the paper.

to $\mathbb{S}^2$ at $\mathbf{u}$. Equation (HM) is the equation for harmonic maps between $\mathbb{R}^n$ and $\mathbb{S}^2$. It generalizes Laplace's equation to maps.

1.1. Map evolution equations. Now we let our maps vary with time as well, so that for each time $t \geq 0$,

$$u(\cdot, t) \colon \mathbb{R}^n \to \mathbb{S}^2,$$

or equivalently

$$\mathbf{u}(x,t) = \big(u_1(x,t), u_2(x,t), u_3(x,t)\big) \in \mathbb{R}^3 \text{ with pointwise constraint } |\mathbf{u}(x,t)| \equiv 1.$$

Harmonic map heat-flow. Harmonic maps (between general Riemannian manifolds) have for many years been of interest to differential geometers, and in order to study them, [8] introduced the gradient-flow equations for the energy $\mathcal{E}$, the *harmonic map heat flow* equations $\partial \mathbf{u}/\partial t = -\mathcal{E}'(\mathbf{u})$, which in our setting read

$$\text{(HMHF)} \qquad \frac{\partial \mathbf{u}}{\partial t} = \Delta \mathbf{u} + |\nabla \mathbf{u}|^2 \mathbf{u}$$

The harmonic map heat flow generalizes the linear heat equation to maps.

Landau–Lifshitz equations. Physically, equation (HMHF) is the special case $b = 0$ of the *Landau–Lifshitz* (sometimes *Landau–Lifshitz–Gilbert*) equations modeling dynamics in ferromagnets:

$$\text{(LL)} \qquad \frac{\partial \mathbf{u}}{\partial t} = a(\Delta \mathbf{u} + |\nabla \mathbf{u}|^2 \mathbf{u}) + b\mathbf{u} \times \Delta \mathbf{u}$$

with $a \geq 0$, $b \in \mathbb{R}$, and where $\times$ denotes the usual cross-product in $\mathbb{R}^3$. In fact equation (LL) itself is a special case of a more general equation incorporating additional physical effects such as anisotropy, and demagnetization (see, e.g., [17, 19]).

Schrödinger maps. The opposite limiting case ($a = 0$, i.e., no dissipation) of (LL), can be written

$$\text{(SM)} \qquad \frac{\partial \mathbf{u}}{\partial t} = \mathbf{u} \times \Delta \mathbf{u} = -J^{\mathbf{u}} \mathcal{E}'(\mathbf{u})$$

where the operator

$$J^{\mathbf{u}} := \mathbf{u} \times \, : T_{\mathbf{u}}\mathbb{S}^2 \to T_{\mathbf{u}}\mathbb{S}^2$$

TABLE 1

Linear $u\colon \mathbb{R}^n \to \mathbb{R}$ $(\mathbb{C})$	**Geometric** $u\colon \mathbb{R}^n \to \mathbb{S}^2 \subset \mathbb{R}^3$	**Main Application**
Laplace $\Delta u = 0$	harmonic map $\Delta u = -\lvert\nabla u\rvert^2 u$	geometry
heat $\frac{\partial u}{\partial t} = \Delta u$	h.m. heat-flow $\frac{\partial u}{\partial t} = \Delta u + \lvert\nabla u\rvert^2 u$	geometry, ferromagnets
wave $\frac{\partial^2 u}{\partial t^2} = \Delta u$	wave map $\frac{\partial^2 u}{\partial t^2} = \Delta u + \left(\lvert\nabla u\rvert^2 - \left\lvert\frac{\partial u}{\partial t}\right\rvert^2\right) u$	relativity
Schrödinger $\frac{\partial u}{\partial t} = i\Delta u$	Schrödinger map $\frac{\partial u}{\partial t} = u \times \Delta u$	ferromagnets

gives a rotation through $\pi/2$ on the tangent plane $T_{\mathbf{u}}\mathbb{S}^2$, and so endows $\mathbb{S}^2$ with a complex structure. Thus equation (SM) can immediately be written for general maps from Riemannian manifolds into Kähler manifolds (see, e.g., [6, 10, 28]).

Since it generalizes the linear Schrödinger equation to maps, (SM) is known as the *Schrödinger map* (sometimes *Schrödinger flow*) equation.

Wave maps. Finally, the *wave map* equation is $P^{\mathbf{u}}(\partial^2\mathbf{u}/\partial t^2 + \Delta\mathbf{u}) = 0$, which in our setting is

$$\text{(WM)} \qquad \left(\frac{\partial^2}{\partial t^2} - \Delta\right)\mathbf{u} + \left(\left|\frac{\partial \mathbf{u}}{\partial t}\right|^2 - |\nabla \mathbf{u}|^2\right)\mathbf{u} = 0,$$

generalizes the linear wave equation to maps. It is has been studied as a toy model in particle physics ("nonlinear σ-model"), but its main interest, aside from the inherent mathematical one, is in general relativity, where it is studied as a (comparatively simple) model for understanding singularity formation (for some background, see, e.g., [23, 24]).

1.2. The energy landscape and equivariant symmetry. The energy $\mathcal{E}(\mathbf{u})$ plays a central role in all of our analysis. We begin by observing that the energy behaves well under the various dynamics introduced above.

Energy identity. Formally taking the dot product of (LL) with $\mathcal{E}'(\mathbf{u}) = -\Delta\mathbf{u} - |\nabla\mathbf{u}|^2\mathbf{u} \in T_{\mathbf{u}}\mathbb{S}^2$ and integrating in space and time yields the basic energy identity

$$\text{(1.1)} \qquad \mathcal{E}\big(\mathbf{u}(t)\big) + a\int_0^t \int_{\mathbb{R}^n} |\Delta\mathbf{u} + |\nabla\mathbf{u}|^2\mathbf{u}|^2 \, dx\, dt = \mathcal{E}\big(\mathbf{u}(0)\big).$$

For (SM) ($a = 0$) this means energy conservation, while for $a > 0$ (including the (HMHF) case $b = 0$), energy is nonincreasing. A conserved Hamiltonian functional for (WM) is obtained by adding $\frac{1}{2}\int_{\mathbb{R}^n} |\partial\mathbf{u}/\partial t|^2 \, dx$ to $\mathcal{E}(\mathbf{u})$.

Two space dimensions is energy critical. The energy scales as

$$\mathcal{E}\big(\mathbf{u}(\cdot)\big) = s^{2-n}\mathcal{E}\big(\mathbf{u}(\cdot/s)\big)$$

for $s > 0$, which makes the space dimension $n = 2$ "energy critical." This has important consequences (see below) and in particular leads to the intuition that $n = 2$ should be a borderline case for the formation of singularities for our map dynamics. So $n = 2$ turns out to be particularly interesting mathematically (and of course $n = 2$ and $n = 3$ are physically the most interesting space dimensions). For these reasons, we specialize to $n = 2$ from here on in.

Equivariant symmetry. Since the analysis of our flow equations is a big challenge, a good starting point is to assume some symmetry. Fix an integer $m \in \mathbb{Z}$. By an *m-equivariant map* $\mathbf{u}\colon \mathbb{R}^2 \to \mathbb{S}^2 \subset \mathbb{R}^3$, we mean a map of the form

$$\mathbf{u}(r,\theta) = e^{m\theta R}\,\mathbf{v}(r)$$

where (r,θ) are polar coordinates on $\mathbb{R}^2$, $\mathbf{v}\colon [0,\infty) \to \mathbb{S}^2$, and R is the matrix generating rotations around the u_3-axis:

$$R = \begin{bmatrix} 0 & -1 & 0 \\ 1 & 0 & 0 \\ 0 & 0 & 0 \end{bmatrix}, \qquad e^{\alpha R} = \begin{bmatrix} \cos\alpha & -\sin\alpha & 0 \\ \sin\alpha & \cos\alpha & 0 \\ 0 & 0 & 1 \end{bmatrix}.$$

Radial maps arise as the case $m = 0$, and we may always assume $m \geq 0$ (a trivial transformation flips the sign of m). m-equivariance is preserved by all of the evolution equations considered above. A *subclass* of m-equivariant maps are those of the form

$$\mathbf{u}(r,\theta) = \big(\cos m\theta \sin\phi(r), \sin m\theta \sin\phi(r), \cos\phi(r)\big) \tag{1.2}$$

which depend on a single radial function $\phi(r)$. This subclass is preserved by (HMHF) and (WM), and is much used in the corresponding literature, since the map equation reduces to a scalar PDE for $\phi(r,t)$ (which allows ready use of the maximum principle in the (HMHF) case, for example). This subclass is notably *not* preserved by (LL) or (SM), just as the wave and heat equations preserve real functions, while the Schrödinger equation (or a heat-Schrödinger mix) does not. We will work in the m-equivariant class for most of what follows.

Topological lower bound on energy. There is a well-known energy lower bound

$$\mathcal{E}(\mathbf{u}) \geq 4\pi|\deg(\mathbf{u})|$$

where $\deg(\mathbf{u})$ is the *degree* of the map $\mathbf{u}$, considered (compactifying the domain $\mathbb{R}^2$ via stereographic projection) as a map from $\mathbb{S}^2$ to itself (defined, for example, by integrating the pullback by $\mathbf{u}$ of the volume form on $\mathbb{S}^2$).

This bound is particularly easy to understand when $\mathbf{u}$ is an m-equivariant map, so that

$$\mathcal{E}(\mathbf{u}) = \pi\int_0^\infty \left(\left|\frac{\partial\mathbf{v}}{\partial r}\right|^2 + \frac{m^2}{r^2}(v_1^2+v_2^2)\right) r\,\mathrm{d}r.$$

If $\mathcal{E}(\mathbf{u}) < \infty$, then $\mathbf{v}(r)$ is continuous, and the limits $\lim_{r\to 0}\mathbf{v}(r)$ and $\lim_{r\to\infty}\mathbf{v}(r)$ exist (see [12]), and so we must have $\mathbf{v}(0), \mathbf{v}(\infty) = \pm\widehat{\mathrm{k}}$, where $\widehat{\mathrm{k}} = (0,0,1)$. Without loss of generality we fix $\mathbf{v}(0) = -\widehat{\mathrm{k}}$. The two cases $\mathbf{v}(\infty) = \pm\widehat{\mathrm{k}}$ then correspond to different topological classes of maps. We denote by Σ_m the class of m-equivariant maps with $\mathbf{v}(\infty) = \widehat{\mathrm{k}}$:

$$\Sigma_m = \big\{\mathbf{u} : \mathbb{R}^2 \to \mathbb{S}^2 \mid \mathbf{u} = \mathrm{e}^{m\theta R}\mathbf{v}(r), \mathcal{E}(\mathbf{u}) < \infty, \mathbf{v}(0) = -\widehat{\mathrm{k}}, \mathbf{v}(\infty) = \widehat{\mathrm{k}}\big\}.$$

For $\mathbf{u} \in \Sigma_m$, the energy $\mathcal{E}(\mathbf{u})$ can be rewritten by "completing the square":

$$\mathcal{E}(\mathbf{u}) = \pi\int_0^\infty \left(\left|\frac{\partial\mathbf{v}}{\partial r}\right|^2 + \frac{m^2}{r^2}|J^{\mathbf{v}}R\mathbf{v}|^2\right) r\,\mathrm{d}r = \pi\int_0^\infty \left|\frac{\partial\mathbf{v}}{\partial r} - \frac{|m|}{r}J^{\mathbf{v}}R\mathbf{v}\right|^2 r\,\mathrm{d}r + \mathcal{E}_{\min}$$

(recall $J^{\mathbf{v}} := \mathbf{v}\times$) with

$$\mathcal{E}_{\min} = 2\pi\int_0^\infty \mathbf{v}_r \cdot \frac{|m|}{r}J^{\mathbf{v}}R\mathbf{v}\, r\,\mathrm{d}r = 2\pi|m|\int_0^\infty (v_3)_r\,\mathrm{d}r = 4\pi|m|.$$

Thus we arrive at

$$\mathbf{u} \in \Sigma_m \implies \mathcal{E}(\mathbf{u}) \geq 4\pi|m|.$$

Harmonic maps. This topological lower bound is clearly saturated if and only if

$$\frac{\partial\mathbf{v}}{\partial r} = \frac{|m|}{r}J^{\mathbf{v}}R\mathbf{v}, \tag{1.3}$$

and the minimal energy is attained precisely an explicit two-parameter family of harmonic maps:

$$\mathcal{O}_m := \big\{\mathrm{e}^{(m\theta+\alpha)R}\mathbf{h}(r/s) \mid s > 0,\ \alpha\in\mathbb{R}\big\} \tag{1.4}$$

where

$$\mathbf{h}(r) = \begin{pmatrix} h_1(r) \\ 0 \\ h_3(r) \end{pmatrix}, \quad h_1(r) = \frac{2}{r^{|m|} + r^{-|m|}}, \quad h_3(r) = \frac{r^{|m|} - r^{-|m|}}{r^{|m|} + r^{-|m|}}. \tag{1.5}$$

The rotation parameter α is determined only up to shifts of 2π (i.e., really $\alpha \in \mathbb{S}^1$). Note that $\mathcal{O}_m$ is just the orbit of the harmonic map $e^{m\theta R}\mathbf{h}(r)$ under the symmetries of the energy $\mathcal{E}$ which preserve equivariance: scaling and rotation. Of course, these harmonic maps are each static solutions of all of the map evolution equations introduced above.

This phenomenon — attainment of the minimal energy at solutions of a *first-order* PDE, is not special to equivariant maps, it occurs in general, amounting (after stereographically projecting $\mathbb{S}^2$ to the complex plane $\mathbb{C}$) to the Cauchy–Riemann equations for meromorphic (or anti-meromorphic) functions. This is another reason that $n = 2$ is so interesting: the harmonic map problem has a beautiful structure, and a wealth of explicit solutions.

1.3. A little recent history. Here we describe some of the important results for the various map dynamics described above, continuing to focus on maps from $\mathbb{R}^2$ to $\mathbb{S}^2$. We do not attempt to be exhaustive, but rather to point out a few highlights relevant for our discussion.

Harmonic map heat-flow. Of the map evolution problems we are considering, (HMHF) has been studied the longest, and is certainly the best understood. The energy space theory goes back (at least for compact 2-dimensional domains) to [25], where global weak solutions are constructed, with at most finitely many singular space-time points where nonconstant harmonic maps "separate." The small energy solutions are global. Working in the subclass (1.2) with $m = 1$, and on a disk, [4] showed that, indeed, finite time blow-up *does* occur in some solutions. An interesting question, which was studied via formal asymptotics in [29], and addressed rigorously in [11] and in the next section, is the relation between the possibility of singularity formation, and the degree m.

Landau–Lifshitz equation. Once the Schrödinger-type term ($b \neq 0$) in included in (LL), our understanding diminishes considerably. Though the problem is still dissipative, maximum principle-type arguments are not readily applied, and even partial regularity results become more difficult and weaker (see, e.g., [16] and references therein). Singularity formation is an open question, partly because the class (1.2) is no longer preserved. Indeed, the (HMHF) blow-up may not provide a reliable guide for the (LL) problem.

Schrödinger maps. In the absence of dissipation ($a = 0$), the analysis becomes still more difficult. Even the *local* theory is just beginning to be understood. In fact, despite a great deal of recent work on the local well-posedness problem in two space dimensions ([1, 7, 14, 21, 26]; see also [15, 22] for the "modified Schrödinger map" case), there is no general local result for energy space data. For the class of data we consider in the next section, m-equivariant solutions with energy near the minimal energy $2\pi|m|$, an energy-space local well-posedness result is given in [13]. It is worth remarking that the existence time furnished by this theorem depends not on the energy (reflecting the energy-space critical nature of the equation in dimension $n = 2$), but rather on more refined information about the initial data: the "length scale" of the $\dot{H}^1$-nearest harmonic map (see [13] for details).

Very few global results are known. We single out the result of [5] showing that *small energy, equivariant* solutions are globally regular. The global results of [12, 13] we describe in the next section can be thought of as analogues of the [5] result for large energy, where the problem is considerably enriched by the presence of the harmonic map family.

Wave maps. Wave maps have received more attention, for a longer time than have Schrödinger maps. There is a large literature, especially concerning local questions, which we will not attempt to summarize here (see, e.g., [24] for some background). Because of the close connection with the $\mathbb{R}^2 \to \mathbb{S}^2$ Schrödinger map problem we are focusing on, we mention only that the possibility of finite-time blow-up for the energy-space critical ($n = 2$) wave maps was established only quite recently, first in [23] for higher degree equivariant maps, and then in [18] for degree $m = 1$. The former result is quite analogous to the Schrödinger map result of [13] which we will describe in the next section, though this is in fact a *no blow-up* result. The essential difference is that the wave map problem allows specification of initial *momentum*, which makes singularity formation easier to force.

2. Global results for Landau – Lifshitz and Schrödinger maps

Here we state some recent results concerning the question of global regularity vs. singularity formation for the Landau – Lifshitz (LL) family of equations

$$\text{(LL)} \qquad \frac{\partial \mathbf{u}}{\partial t} = a(\Delta \mathbf{u} + |\nabla \mathbf{u}|^2 \mathbf{u}) + b\mathbf{u} \times \Delta \mathbf{u}, \quad \mathbf{u}(x, 0) = \mathbf{u}_0(x)$$

($a \geq 0$, $b \in \mathbb{R}$) which of course includes as special cases the harmonic map heat-flow ($a = 1$, $b = 0$) and the Schrödinger map ($a = 0$, $b = 1$). This is mostly work from the papers [12, 13] which address the Schrödinger case, and the paper [11] which addresses the heat-flow case.

The results concern m-equivariant maps with energy near the minimal energy $\mathcal{E}_{\min} = 4\pi|m|$ (the harmonic map energy), and so the standing assumption on the initial data is

$$\mathbf{u}_0 \in \Sigma_m, \quad \mathcal{E}(\mathbf{u}_0) = 4\pi|m| + \delta_0^2, \quad \delta_0 \ll 1.$$

Let

$$\mathbf{u}(t) \in C([0, T); \Sigma_m)$$

(Σ_m topologized with the energy ($\dot{H}^1$) norm) be the solution of (LL) corresponding to the initial data $\mathbf{u}_0$ (which is a priori just a local-in-time solution — see [13] for local well-posedness for this class of data).

Theorem 2.1 ([12] **"orbital stability" of harmonic maps).** *For δ_0 sufficiently small, there exist $s(t) \in \mathcal{C}\big([0, T); (0, \infty)\big)$ and $\alpha(t) \in \mathcal{C}([0, T); \mathbb{R})$ so that*

$$(2.1) \qquad \big\|\mathbf{u}(x, t) - e^{(m\theta + \alpha(t))R}\mathbf{h}\big(r/s(t)\big)\big\|_{\dot{H}_1(\mathbb{R}^2)} \lesssim \delta_0, \quad \forall t \in [0, T).$$

Moreover, suppose $T < \infty$. Then T is the maximal existence time ($\mathbf{u}(t)$ doesn't extend past T as a solution continuous into Σ_m) if and only if

$$(2.2) \qquad \liminf_{t \to T^-} s(t) = 0.$$

This theorem can be viewed, on one hand, as an *orbital stability* result for the family $\mathcal{O}_m$ of harmonic maps (at least up to the possible blow-up time), and on the other hand as a characterization of blow-up for energy near $\mathcal{E}_{\min}$: solutions blow-up if and only if the "length-scale" $s(t)$ goes to zero.

The first statement of the theorem is really a kind of convexity result for the energy functional — the only property of the dynamics that is needed is the nonincrease of the energy, as follows from the energy identity (1.1). The details can be found in [12]

The second statement comes from the energy-space local existence theory — see [13].

Of course this theorem leaves open the question of whether or not finite-time singularities can form. The next result shows that when the degree is sufficiently high, singularities will *not* form, and moreover, solutions converge to specific harmonic maps as $t \to \infty$.

Theorem 2.2 ([11, 13] **global regularity and asymptotic stability for high degree).** *For* (LL) *with* $a > 0$, *assume* $|m| \geq 3$. *For* (SM) ($a = 0$), *assume* $|m| \geq 4$. *As before,* δ_0 *is sufficiently small. Then*

(1) *there is no finite-time blow-up: the solution can be extended to* $\mathbf{u} \in C([0, \infty); \Sigma_m)$.

(2) *For any* $r \in (2, \infty]$, $p \in [2, \infty)$ *with* $1/r + 1/p = \frac{1}{2}$, *we have*

$$\left\| \nabla \left[\mathbf{u}(x,t) - e^{(m\theta + \alpha(t))R} \mathbf{h}(r/s(t)) \right] \right\|_{L_t^r L_x^p(\mathbb{R}^2 \times [0,\infty))} \lesssim C_p \delta_0$$

(*if* $a > 0$ *we may include* $(r, p) = (2, \infty)$)

(3) *furthermore, there exist* $s_+ > 0$ *and* α_+ *with*

$$s(t) \to s_+, \quad \alpha(t) \to \alpha_+, \quad \text{as } t \to \infty.$$

The space-time estimates above imply asymptotic *convergence* of the solutions to the family of harmonic maps (in a space-time norm ("dispersive") sense, which is the best we can expect for the Schrödinger case $a = 0$). Hence we say the harmonic maps are *asymptotically stable* under the Landau – Lifshitz flow for $|m| \geq 3$ ($|m| \geq 4$ for (SM)).

The question of blow-up for lower degree maps is unresolved, except in the pure heat-flow case. There we have

Theorem 2.3 ([11] **heatflow blow-up for $m = 1$).** *Let* $m = 1$. *For any* $\delta > 0$, *there exists* $u_0 \in \Sigma_1$ *with* $0 < \mathcal{E}(u_0) - 4\pi \leq \delta^2$ *such that the corresponding solution of the harmonic map heat flow blows up in finite time, in the sense that (for example)* $\|\nabla \mathbf{u}(\cdot, t)\|_{L_x^\infty} \to \infty$.

This result is an adaptation of the blow-up proof of [4] for a disk domain, to the case of $\mathbb{R}^2$. In particular, it must be verified that the construction (based on a subsolution argument) can be achieved for data with energy arbitrarily close to the harmonic map energy. The reader is referred to [11] for details.

We will present a few of the central ideas of our approach — and in particular of the proof of Theorem 2.2 — in the next two sections.

3. The approach: two geometric coordinate systems

The framework is to write our maps $\mathbf{u}(x,t)$ using two essentially different "coordinate systems" in the energy space of maps. The first decomposes a map into a "nearby" harmonic map (finite-dimensional part), and a deviation from the harmonic map family (infinite-dimensional part). This system is used to track the time-varying parameters of the "nearest" harmonic map. The second coordinate system — the "generalized Hasimoto transform" of [5] — involves a kind of projection which removes the harmonic map component, and produces an equation for which we can do the hard estimates showing the perturbation away from the harmonic maps "decays" with time.

3.1. Splitting and orthogonality. We split $\mathbf{u}(x,t) = e^{m\theta R}\mathbf{v}(r,t) \in \Sigma_m$ as a harmonic map with time-varying parameters, plus a perturbation:

$$\mathbf{v}(r,t) = e^{\alpha(t)R}[\mathbf{h}(\rho) + \boldsymbol{\xi}(\rho,t)], \quad \rho := \frac{r}{s(t)}$$

The choice of the parameter paths $s(t)$ and $\alpha(t)$ is important, and we will address it later.

We further split the perturbation into tangent and normal components (to the sphere at the harmonic map) using the explicit orthonormal basis of $T_{\mathbf{h}(\rho)}\mathbb{S}^2$

$$\hat{\jmath} := \begin{pmatrix} 0 \\ 1 \\ 0 \end{pmatrix}, \quad \text{and} \quad J^{\mathbf{h}(\rho)}\hat{\jmath} = \begin{pmatrix} -h_3(\rho) \\ 0 \\ h_1(\rho) \end{pmatrix},$$

$$\xi(\rho,t) = z_1(\rho,t)\hat{\jmath} + z_2(\rho,t)J^{\mathbf{h}(\rho)}\hat{\jmath} + \gamma(\rho,t)\mathbf{h}(\rho)$$

where the pointwise constraint $|u(x,t)| \equiv 1$ forces $\gamma = O(|z|^2)$ for $|\xi|$ small. In this way, the complex function

$$z(\rho,t) = z_1(\rho,t) + iz_2(\rho,t),$$

together with a choice of the parameters $s(t)$ and $\alpha(t)$, gives a full description of the original solution $\mathbf{u}(x,t)$ (provided $|\boldsymbol{\xi}| \le 1$).

It can then be shown that if $\mathbf{u}(x,t)$ satisfies (LL), then $z(\rho,t)$ satisfies a nonlinear equation of (in general) mixed heat-Schrödinger type, of the form

$$s^2\frac{\partial z}{\partial t} = -(a+ib)Hz + (s^2\dot{\alpha} - \mathrm{i}m\, s\dot{s})h_1 + F(\rho,t), \quad H := -\partial_\rho^2 - \frac{1}{\rho}\partial_\rho + \frac{m^2}{\rho^2}(1-2h_1^2),$$

where F denotes terms nonlinear in z, $\dot{s}$, and $\dot{\alpha}$.

We next address the question of how to choose $s(t)$ and $\alpha(t)$. Supposing for a moment that $s(t) \equiv 1$, $\alpha(t) \equiv 0$, the linearized equation for $z(\rho,t)$ is

$$\partial_t z = -(a+ib)Hz.$$

and

$$H = L_0^* L_0, \quad L_0 := \partial_\rho + \frac{m}{\rho}h_3 = h_1\partial_\rho\frac{1}{h_1}$$

(the adjoint L_0^* is taken in the $L^2(\rho\, d\rho)$ inner product). So in particular, $\ker H = \operatorname{span}\{h_1\}$, and the linearized equation admits the constant (in time) solution $z(\rho,t) \equiv h_1(\rho)$. Since we would like $z(\rho,t)$ to have some decay in time, we must choose $s(t)$ and $\alpha(t)$ in such a way as to avoid such constant solutions. And since H is self-adjoint in L^2, the natural choice is to work in the subspace of functions z satisfying

$$(z,h_1)_{L^2} = \int_0^\infty z(\rho)h_1(\rho)\rho\, d\rho \equiv 0, \tag{3.1}$$

which is invariant under the linearized equation. We will see below that this choice is good for several other reasons. Imposition of this condition will determine the dynamics of $s(t)$ and $\alpha(t)$. But there is an important drawback to this approach:

neither z, nor h_1, lies in L^2 in general. The natural (energy) space for z is given by the norm (see [12])

$$\|z\|_X^2 := \int_0^\infty \left\{|z_\rho|^2 + \frac{|z|^2}{\rho^2}\right\}\rho\,\mathrm{d}\rho$$

and so the best we can do is

$$|(z, h_1)_{L^2}| = \left|\left(\frac{z}{\rho}, \rho h_1\right)_{L^2}\right| \le \|z\|_X \|\rho h_1\|_{L^2}.$$

So to make sense of our orthogonality condition, we require

$$\rho h_1(\rho) = \frac{2\rho}{\rho^m + \rho^{-m}} \in L^2(\rho, \mathrm{d}\rho),$$

which only holds if $m \ge 3$. This is the main reason we cannot handle the small $|m|$ cases in Theorem 2.2. (The further restriction $|m| > 3$ for (SM) is needed for different reasons, to be explained shortly.)

Differentiating the orthogonality condition with respect to t, and using the equation for z, we arrive at a system of ODEs for $s(t)$ and $\alpha(t)$, coupled to $z(\rho, t)$. A crucial aspect of our choice of orthogonality condition is that the linear terms drop out, and $\dot{s}$ and $\dot{\alpha}$ are *quadratic* in z:

$$|s\dot{s}| + |s^2\dot{\alpha}| \lesssim \left|\big(h_1, F(\rho, t)\big)\right| \tag{3.2}$$

The objective then is to obtain estimates on $z(\rho, t)$ which are L^2 in time, which will show $\dot{s}$ and $\dot{\alpha}$ are integrable in time, and so $s(t)$ and $\alpha(t)$ converge to limits as $t \to \infty$.

3.2. Generalized Hasimoto transform. The equation for $z(\rho, t)$, however, is not suitable for obtaining estimates, for at least two reasons: (a) the orthogonality condition has to be imposed to avoid nondecaying solutions, and (b) there are nonlinear terms containing derivatives of z. Fortunately, there is a neat way around these problems: the *generalized Hasimoto transform* of [5].

The Landau–Lifshitz equation (SM), written in terms of $\mathbf{v}(r, t)$, can be factored as

$$\frac{\partial \mathbf{v}}{\partial t} = (a + bJ^{\mathbf{v}})[D_r^{\mathbf{v}} + \frac{1}{r} - \frac{m}{r}v_3]\mathbf{W}$$

where

$$\mathbf{W}(r) := \mathbf{v}_r(r) - \frac{m}{r}J^{\mathbf{v}}R\mathbf{v}(r) \in T_{\mathbf{v}(r)}\mathbb{S}^2$$

and

$$D_r^{\mathbf{v}} := P^{\mathbf{v}(r)}\partial_r$$

denotes the *covariant derivative* (with respect to r, along $\mathbf{v}$).

Let $\mathbf{e}(r) \in T_{\mathbf{v}(r)}\mathbb{S}^2$ be a unit-length tangent field satisfying the "gauge condition" $D_r^{\mathbf{v}}\mathbf{e} \equiv 0$. Expressing $\mathbf{W}$ in the orthonormal frame $\{\mathbf{e}, J^{\mathbf{v}}\mathbf{e}\}$,

$$\mathbf{W} = q_1\mathbf{e} + q_2J^{\mathbf{v}}\mathbf{e},$$

it is not difficult to arrive at the following equation for the complex function $q(r, t) := q_1(r, t) + \mathrm{i}q_2(r, t)$:

$$q_t = (a + \mathrm{i}b)\left(\Delta_r - \frac{1}{r^2}\big((1 - mv_3)^2 + mr(v_3)_r\big)\right)q - \mathrm{i}Sq \tag{3.3}$$

where the function $S(r,t)$ arises as $D_t^{\mathbf{v}}\mathbf{e} = SJ^{\mathbf{v}}\mathbf{e}$. The curvature relation

$$[D_r, D_t]\mathbf{e} = -\operatorname{Re}\left[\left(\partial_r + \frac{1}{r} - \frac{m}{r}v_3\right)q\overline{\left(q + \frac{m}{r}\nu\right)}\right]J^{\mathbf{v}}\mathbf{e},$$

where $P^{\mathbf{v}(r)}\widehat{\mathrm{k}} = \widehat{\mathrm{k}} - v_3\mathbf{v} = \nu_1\mathbf{e} + \nu_2 J^{\mathbf{v}}\mathbf{e}$, then leads to

$$S(r,t) = -\frac{1}{2}Q(r,t) + \int_r^\infty \frac{1}{\tau}Q(\tau,t)\,\mathrm{d}\tau, \quad Q := |q|^2 + \frac{2m}{r}\operatorname{Re}(\bar{\nu}q).$$

So our equation resembles a cubic nonlinear heat-Schrödinger equation, keeping in mind (a) there are nonlocal nonlinear terms, and (b) it is not self-contained: the unknown map $\mathbf{v}(r,t)$ itself appears in several places (including through ν).

A key point is that

$$1 \gg \mathcal{E}(\mathbf{u}) - 4\pi m = \tfrac{1}{2}\|\mathbf{W}\|_{L^2}^2 = \pi\|q\|_{L^2(r\,\mathrm{d}r)}^2.$$

The transformation has "killed" the harmonic map component, leaving us with a *small L^2-data problem* for for the q equation (3.3), even though the map $\mathbf{u}$ is not a small-energy map). What's more, this equation is amenable to estimates.

3.3. Relating the two coordinate systems. Of course, it is useless (and impossible) to obtain estimates for $q(r,t)$, unless we can control z (and hence $\mathbf{v}$) in terms of q. This is certainly *only* possible if we have a supplementary condition such as the orthogonality condition (3.1), since $q \equiv 0$ just means that $\mathbf{v}(r) = e^{\alpha R}\mathbf{h}(r/s)$ for some s, α).

We have: provided $|m| \geq 3$ and (3.1) holds, and $\|z\|_X \ll 1$ is sufficiently small,

(1) $\|z_\rho\|_{L^p} + \|z/\rho\|_{L^p} \lesssim s^{1-2/p}\|q\|_{L^p}$ $2 \leq p \leq \infty$

(2) if further $|m| \geq 4$, then $\|z_\rho/\rho\|_{L^2} + \|z/\rho^2\|_{L^2} \lesssim s\|q/r\|_{L^2}$.

The proofs can be found in [13] (the $p = \infty$ case of the first statement was not considered there, but it can be shown also to hold).

The second statement is the source of the extra restriction $|m| \geq 4$ in the (SM) case. The point is that we need $\dot{s}$ and $\dot{\alpha}$ to be L_t^1, and hence by (3.2) we need an estimate on z which is L_t^2. As we will see in the next section, we can estimate q in $L_t^2 L_x^\infty$ if $a > 0$ (and so we may use the $p = \infty$ case of the first statement above), but we don't have this estimate when $a = 0$, and so must rely on $q/|x| \in L_t^2 L_x^2$, forcing us to use the second statement above.

4. Linear evolution estimates with critical-decay potentials

So the remaining task is to obtain estimates for $q(r,t)$ (including L^2-in-time estimates). The key here is to understand the linear part, obtained by substituting $\mathbf{h}(r)$ for $\mathbf{v}(r,t)$ in the q equation (3.3), and dropping nonlinear terms, to arrive at

$$q_t = (a + ib)\left[q_{rr} - \frac{1}{r}q_r - \frac{1}{r^2}(1 + m^2 - 2mh_3)q\right].$$

Thus we need space-time estimates for the linear evolution operator

$$e^{-(a+ib)tH}, \quad H = -\Delta + V(|x|), \quad V(r) = \frac{1 + m^2 - 2mh_3(r)}{r^2}$$

where $a \geq 0$, $b \in \mathbb{R}$. This turns out to be an interesting mathematical problem, precisely because the $1/r^2$ behavior of the potential $V(r)$ at the origin and at infinity places it just beyond the reach (on the borderline, in fact) of the typical perturbative arguments used to obtain estimates for such operators. Indeed, the question of estimates for such operators has quite recently been addressed: see [30] for the purely diffusive case ($b = 0$), and [2, 3] for the purely conservative case ($a = 0$). In both cases, the relevant results hold for space dimensions ≥ 3, because of the lack of Hardy inequality in dimension 2. Our job, then is to recover 2-dimensional versions of these estimates, exploiting our symmetry (equivariance) assumption. We summarize the resulting estimates for L^2 initial data:

Theorem 4.1. *Let the exponent pairs (r, p) and $(\tilde{r}, \tilde{p})$ both satisfy the "admissibility" condition $1/r + 1/p = \frac{1}{2}$. The space-time estimates*

$$\|e^{-t(a+ib)H}\varphi\|_{L_t^r L_x^p} + \|\int_0^t e^{-(t-s)(a+ib)H} f(\cdot, s)\, ds\|_{L_t^r L_x^p} \lesssim \|\varphi\|_{L^2} + \|f\|_{L_t^{\tilde{r}'} L_x^{\tilde{p}'}}$$

hold in the following cases:

• **pure heat ($a > 0$, $b = 0$)** *for general ϕ, f, excluding the endpoint case $r = \tilde{r} = 2$ (so $p = \tilde{p} = \infty$), which is false in general — see the example of* [27]

• **mixed case ($a > 0$, $b \in \mathbb{R}$)** *for $\phi(\cdot)$ and $f(\cdot, t)$* radial, *but* including *the endpoint case $r = \tilde{r} = 2$, provided $m \geq 2$.*

• **pure Schrödinger case ($a = 0$, $b \neq 0$)** *holds for $\phi(\cdot)$ and $f(\cdot, t)$* radial, *but* not including *the endpoint case $r = \tilde{r} = 2$ (which is open), again provided $m \geq 2$.*

In addition, for all cases, *if $m \geq 2$ and ϕ and f are radial, we also have weighted L^2 estimates. Denoting $\|g\|_{L^{2,\pm 1}} := \||x|^{\pm 1} g\|_{L^2}$:*

$$\|e^{-t(a+ib)H}\varphi\|_{L_t^2 L_x^{2,-1}} \lesssim \|\varphi\|_{L^2}$$

$$\left\|\int_0^t e^{-(t-s)(a+ib)H} f(\cdot, s)\, ds\right\|_{L_t^2 L_x^{2,-1} \cap L_t^r L_x^p} \lesssim \min(\|f\|_{L_t^{\tilde{r}'} L_x^{\tilde{p}'}}, \|f\|_{L_t^2 L_x^{2,1}})$$

(excluding $r = \tilde{r} = 2$ only if $a = 0$).

Remarks on the proofs of these estimates. The goal is to estimate $w(x, t)$, a solution of the linear inhomogeneous initial value problem

$$w_t + (a + ib)Hu = f, \quad w(x, 0) = \varphi(x).$$

• **L^2 contraction for any $a \geq 0$.** We begin with this trivial observation, using $V \geq 0$. Multiply the equation for w by $\overline{w}$, take the real part, integrate in space and time, and use the Hölder inequality to produce the basic L^2 estimate:

$$\|w\|_{L_t^\infty L_x^2} \lesssim \|\varphi\|_{L^2} + \|f\|_{L_t^1 L_x^2}.$$

• **Estimates for $a > 0$, radial data, and $m \geq 2$.** These estimates are the easiest, coming as they do directly from "energy estimates." First, to compensate for the lack of Hardy inequality in dimension 2, change the function:

$$\widetilde{w}(x, t) := e^{ii\theta} w(r, t)$$

so that

$$|w_r|^2 + |\frac{w}{r}|^2 = |\nabla \widetilde{w}|^2,$$

where now $\widetilde{w}(x, t)$ solves

$$\widetilde{w}_t + (a + ib)(-\Delta + \widetilde{V})\widetilde{w}, \quad \widetilde{w}(x, 0) = e^{i\theta}\varphi(r),$$

and the key fact is that the "new" potential satisfies

$$r^2 \widetilde{V}(r) = m^2 - 2mh_3(r) \geq m^2 - 2m \geq 0 \quad \text{for } m \geq 2.$$

To get an L^∞_x estimate, we employ the simple embedding inequality

$$\|w\|_{L^\infty} \lesssim \|w_r\|_{L^2}^{1/2} \|\frac{w}{r}\|_{L^2}^{1/2}$$

for radial functions, established in [12]. So now, as above, multiply the equation by $\overline{\widetilde{w}}$, take the real part, integrate over space and time, and use $a > 0$ and $\widetilde{V} \geq 0$ to arrive at

$$\|w\|^2_{L^\infty_t L^2_x \cap L^2_t L^{2,-1}_x \cap L^\infty_t L^2_x} \leq \|\nabla \widetilde{w}\|^2_{L^2_t L^2_x} + \|\widetilde{w}\|^2_{L^\infty_t L^2_x} \lesssim \|\varphi\|^2_{L^2} + \|wf\|_{L^1_t L^1_x}$$

which, by Hölder and interpolation yields all the desired estimates in this case (radial data, $a > 0$, $m \geq 2$).

• **Estimates for $b = 0$.** Not surprisingly, we obtain the finest estimates in the "pure heat" case. The first step is to establish the time decay estimates

$$\|e^{-tH}\varphi\|_{L^p_x} \lesssim t^{-(1/a-1/p)/2}\|\varphi\|_{L^a}$$

for $1 \leq a \leq p \leq \infty$. This is done via the maximum principle, comparing with the the heat flow $e^{t\Delta}$ for which these estimates follow immediately from the explicit fundamental solution. A necessary preliminary step is to establish boundedness of solutions via an embedding inequality $\|w\|_{L^\infty} \lesssim \|Hw\|_{L^2} + \|w\|_{L^2}$. The second step is to follow [9] in applying the Marcinkiewicz interpolation theorem to the time decay estimates to generate the homogeneous space-time estimates

$$\|e^{-tH}\varphi\|_{L^r_t L^p} \leq C\|\varphi\|_{L^a}$$

for $1/r = 1/a - 1/p'$, $r \geq a > 1$. Finally, the corresponding nonhomogeneous estimates then follow from these, together with the the Hardy – Littlewood inequality.

• **Estimates for $a = 0$.** The "pure Schrödinger" case is the most delicate. Here we adopt the approach from [2, 3], working with radial functions to compensate for the lack of Hardy inequality in 2 dimensions. The idea is to first obtain the weighted L^2 estimates (in this context a "Kato smoothing" estimate) through a resolvent estimate which itself comes directly from estimates on the relevant equation, and by exploiting the positivity and repulsivity of the potential. The various other estimates can then be obtained by perturbative arguments from the reference operator $-\Delta + 1/r^2$, provided $m \geq 2$. We refer to [13] for details.

5. Conclusions and future directions

The map equations discussed in this article are of both physical and geometric interest, and yet it is only very recently that the global behavior of solutions is starting to be understood, and that only in very limited settings. Much work remains to be done. We single out a few pressing directions of inquiry:

• **Move beyond the equivariant setting.** To now, global results (e.g., blow-up or asymptotic behavior) for energy-critical map problems are confined to equivariant maps. It is important to remove this restrictive symmetry assumption, especially for stability and asymptotic questions. Among other things, this requires (1) understanding the energy landscape near the (now much larger) harmonic map family outside the equivariant class (note that some of this analysis described above is essentially one-dimensional in nature, and may not survive the removal of the symmetry assumption); (2) adapting the "Hasimoto transform" used above, which introduces "loss-of-regularity" problems.

• **Singularity formation in $m = 1$ Schrödinger maps.** This is the most obvious unanswered question. Blow-up arguments for the pure heat-flow with degree $m = 1$ rely heavily on the maximum principle, unavailable for the Landau–Lifshitz problem (with $b \neq 0$). On the other hand, the analysis described above breaks down (in several ways) for $m = 1$, as indeed it should if blow-up is possible.

• **The full physical model.** The Landau–Lifshitz equation described above is a "bare-bones" model, incorporating only the "exchange energy," and (for $a > 0$) dissipation. The question of how physically important effects such as anisotropy, and demagnetization (a nonlocal term) affect solutions has hardly been touched.

References

1. I. Bejenaru, *On Schrödinger maps*, available at `arXiv:math/0604255`.
2. N. Burq, F. Planchon, J. Stalker, and S. Tahvilder-Zadeh, *Strichartz estimates for the wave and Schrödinger equations with the inverse-square potential*, J. Funct. Anal. **203** (2003), no. 2, 519–549.
3. ———, *Strichartz estimates for the wave and Schrödinger equations potentials of critical decay*, Indiana Univ. Math. J. **53** (2004), no. 6, 1665–1680.
4. K.-C. Chang, W. Y. Ding, and R. Ye, *Finite-time blow-up of the heat flow of harmonic maps from surfaces*, J. Differential Geom. **36** (1992), no. 2, 507–515.
5. N.-H. Chang, J. Shatah, and K. Uhlenbeck, *Schrödinger maps*, Comm. Pure Appl. Math. **53** (2000), no. 5, 590–602.
6. W. Y. Ding, *On the Schrödinger flows*, Proceedings of the International Congress of Mathematicians. Vol.II (Beijing, 2002) (T. Li, ed.), Higher Ed. Press, Beijing, 2002, pp. 283–291, available at `arXiv:math.0304263`.
7. W. Y. Ding and Y. D. Wang, *Local Schrödinger flow into Kähler manifolds*, Sci. China Ser. A **44** (2001), no. 11, 1446–1464.
8. J. Eells and J. H. Sampson, *Harmonic mappings of Riemannian manifolds*, Amer. J. Math. **86** (1964), 109–160.
9. Y. Giga, *Solutions for semilinear parabolic equations in L^p and regularity of weak solutions of the Navier–Stokes system*, J. Differential Equations **62** (1986), no. 2, 186–212.
10. M. Grillakis and V. Stefanopoulos, *Lagrangian formulation, energy estimates, and the Schrödinger map problem*, Comm. Partial Differential Equations **27** (2002), no. 9-10, 1845–1877.
11. M. Guan, S. Gustafson, and T.-P. Tsai, *Global existence and blow-up for harmonic map heat flow*, preprint.
12. S. Gustafson, K. Kang, and T.-P. Tsai, *Schrödinger flow near harmonic maps*, Comm. Pure Appl. Math. **60** (2007), no. 4, 463-499.
13. ———, *Asymptotic stability of harmonic maps under theSchrödinger flow*, available at `arXiv:math/0609591`.
14. A. D. Ionescu and C. E. Kenig, *Low-regularity Schrödinger maps*, Differential Integral Equations **19** (2006), no. 11, 1271–1300, available at `arXiv:math/0605210`.
15. J. Kato and H. Koch, *Uniqueness of the modified Schrödinger map in $H^{3/4+\epsilon}(\mathbb{R}^2)$*, Comm. Partial Differential Equations **32** (2007), no. 1-3, 415–429, available at `arXiv:math/0508423`.
16. J. Ko, *The construction of a partially regular solution to the Landau–Lifshitz–Gilbert equation in $\mathbb{R}^2$*, Nonlinearity **18** (2005), no. 6, 2681–2714.
17. A. M. Kosevich, B. A. Ivanov, and A. S. Kovalev, *Magnetic solitons*, Phys. Rep. **194** (1990), no. 3-4, 117–238.
18. J. Krieger, W. Schlag, and D. Tataru, *Renormalization and blow-up for charge one equivariant critical wave maps*, available at `arXiv:math/0610248`.
19. M. Kružík and A. Prohl, *Recent developments in the modeling, analysis, and numerics in ferromagnetism*, SIAM Rev. **48** (2006), no. 3, 439–483.
20. L. D. Landau and E. M. Lifshitz, *On the theory of the dispersion of magnetic permeability in ferromagnetic bodies*, Physikalische Zeitschrift der Sowjetunion **8** (1935), 153–169; reprinted in *Collected papers of L. D. Landau*, Pergamon Press, New York, NY, 1965, pp. 101–114.
21. H. McGahagan, *An approximation scheme for Schrödinger maps*, Comm. Partial Differential Equations **32** (2007), no. 1-3, 375–400.

22. A. Nahmod, A. Stefanov, and K. Uhlenbeck, *On Schrödinger maps*, Comm. Pure Appl. Math. **56** (2003), no. 1, 114–151; Erratum, **57** (2004), no. 6, 833–839.
23. I. Rodnianski and J. Sterbenz, *On the formation of singularities in the critical* O(3) σ*-model*, available at arXiv:math/0605023.
24. J. Shatah and M. Struwe, *Geometeric wave equations*, Courant Lect. Notes Math., vol. 2, Courant Inst. Math. Sci., New York, NY, 1998.
25. M. Struwe, *On the evolution of harmonic mappings of Riemannian surfaces*, Comment. Math. Helv. **60** (1985), no. 4, 558–581.
26. P.-L. Sulem, C. Sulem, and C. Bardos, *On the continuous limit for a system of classical spins*, Comm. Math. Phys. **107** (1986), no. 3, 431–454.
27. T. Tao, *Spherically averaged endpoint Strichartz estimates for the two-dimensional Schrödinger equation*, Comm. Partial Differential Equations **25** (2000), no. 7-8, 1471–1485.
28. C.-L. Terng and K. Uhlenbeck, *Schrödinger flows on Grassmannians*, Integrable Systems, Geometry, and Topology (C.-L. Terng, ed.), AMS/IP Stud. Adv. Math., vol. 36, Amer. Math. Soc., Providence RI, 2006, pp. 235–256, available at arXiv:math/9901086.
29. J. B. van den Berg, J. Hulshof, and J. R. King, *Formal asymptotics of bubbling in the harmonic map heat flow*, SIAM J. Appl. Math. **63**, no. 5, 1682–1717.
30. J. L. Vazquez and E. Zuazua, *The Hardy inequality and the asymptotic behavior of the heat equation with an inverse-square potential*, J. Funct. Anal. **173** (2000), no. 1, 103–153.

Department of Mathematics, University of British Columbia, Vancouver, BC V6T 1Z2, Canada
E-mail address, M. Guan: guanmj@math.ubc.ca
E-mail address, S. Gustafson: gustaf@math.ubc.ca

Sungkyunkwan University and Institute of Basic Science, Suwon 440-746, Republic of Korea
E-mail address: kkang@skku.edu

Department of Mathematics, University of British Columbia, Vancouver, BC V6T 1Z2, Canada
E-mail address: ttsai@math.ubc.ca

Centre de Recherches Mathématiques
CRM Proceedings and Lecture Notes
Volume **44**, 2008

Pohožaev-Type Identities for an Elliptic Equation

Radu Ignat

ABSTRACT. We present some Pohožaev identities for the equation $-\Delta u = |u|^{p-1}u - \lambda u$ and as an application, we prove some nonexistence results.

1. Introduction

In this paper we present some Pohožaev-type identities for the following nonlinear elliptic equation:

$$-\Delta_g u = |u|^{p-1}u - \lambda u \quad \text{on } M. \tag{1}$$

Here, $p > 1$, $\lambda \in \mathbb{R}$ and M is a ball in $\mathbb{R}^n$ or on the unit sphere S^n, $n \geq 3$, equipped with the standard metric g and Δ_g stands for the Laplace–Beltrami operator on (M, g). The goal is to prove nonexistence results for (1) in different ranges of λ.

Motivated by the study of Brezis and Nirenberg [5], we first consider the Dirichlet problem associated to (1) in the unit ball $B_1 \subset \mathbb{R}^n$, i.e.,

$$\begin{cases} -\Delta u = |u|^{p-1}u - \lambda u, \ u \not\equiv 0 & \text{in } B_1, \\ u = 0 & \text{on } \partial B_1. \end{cases} \tag{2}$$

We prove the following identity:

Lemma 1. *Let $\psi\colon [0,\infty) \to \mathbb{R}$ be a smooth function (with $\psi'(0) = \psi''(0) = 0$ if $n \geq 4$). If u is a solution of (2), then*

$$\begin{aligned} \int_{B_1} \Big\{ r\psi'''(r) + 3\psi''(r) - \frac{(n-1)(n-3)}{r}\psi'(r) - 4\lambda\big(r\psi'(r) + \psi(r)\big) \Big\} \frac{u^2}{2} \,\mathrm{d}x \\ = \psi(1) \int_{\partial B_1} \left| \frac{\partial u}{\partial r} \right|^2 \mathrm{d}\mathcal{H}^{n-1}(x) + 2 \int_{B_1} r\psi'(r) \left(|\nabla u|^2 - \left| \frac{\partial u}{\partial r} \right|^2 \right) \mathrm{d}x \\ - \int_{B_1} \Big\{ \frac{p+3}{p+1} r\psi'(r) - \Big(\frac{p-1}{p+1} n - 2 \Big) \psi(r) \Big\} |u|^{p+1} \,\mathrm{d}x, \end{aligned} \tag{3}$$

where $r = |x|$ and $\partial u(x)/\partial r = (x/|x|) \cdot \nabla u(x)$ stands for the radial derivative of u.

As a consequence, we deduce the nonexistence result of Brezis and Nirenberg [5] for positive solutions of (2) in the supercritical case $p \geq (n+2)/(n-2)$:

2000 *Mathematics Subject Classification.* Primary 35J60; Secondary 35B33.
This is the final form of the paper.

Theorem 2 (Brezis and Nirenberg [5]). *Let $p \geq (n+2)/(n-2)$ and $\lambda_1 := \lambda_1(-\Delta; B_1)$ be the first eigenvalue of the Laplace operator with Dirichlet boundary condition in B_1. If one of the following two conditions is satisfied*

(i) $n = 3$ *and* $\lambda \notin (-\lambda_1, -\lambda_1/4)$,
(ii) $n \geq 4$ *and* $\lambda \notin (-\lambda_1, 0)$,

then there is no positive solution of (2).

Remark 1. (a) The set of positive and nodal *radial* solutions (regular or singular) of (2) is described by Benguria, Dolbeault and Esteban [2].
(b) The question if there is no (nodal) solution of (2) for $n = 3$ and $\lambda \in (-\lambda_1/4, 0)$ is still open.

Next we study the Dirichlet problem associated to (1) on a geodesic ball D_{θ^*} centered at the North pole in S^3 of radius $\theta^* \in (0, \pi)$:

$$\begin{cases} -\Delta_g u = |u|^{p-1}u - \lambda u, \ u \not\equiv 0 & \text{in } D_{\theta^*}, \\ u = 0 & \text{on } \partial D_{\theta^*}. \end{cases} \tag{4}$$

We want to obtain a similar identity to (3) for any solution u of (4). Using the stereographic projection $\Phi_Q \colon S^3 \setminus \{Q\} \to \mathbb{R}^3$ with vertex at the South pole Q in S^3, the equation (4) writes as

$$-\frac{1}{\rho^3} \operatorname{div}(\rho \nabla U) = |U|^{p-1}U - \lambda U \quad \text{in } B_{R^*} \subset \mathbb{R}^3$$

where $\rho(x) = 2/(1+|x|^2)$, $U(x) = u\big(\Phi_Q^{-1}(x)\big)$ for every $x \in B_{R^*}$ and $R^* = \tan(\theta^*/2)$. The transformation

$$v(x) = U(x)\sqrt{\rho(x)}$$

turns (4) into

$$\begin{cases} -\Delta v = \rho(x)^{(5-p)/2}|v|^{p-1}v + \dfrac{3-4\lambda}{4}\rho^2 v, \ v \not\equiv 0 & \text{in } B_{R^*}, \\ v = 0 & \text{on } \partial B_{R^*}. \end{cases} \tag{5}$$

We prove the following identity for a solution v of (5):

Lemma 3. *Let $\psi \colon [0, \infty) \to \mathbb{R}$ be a smooth function. If v is a solution of* (5), *then*

$$\begin{aligned} \int_{B_{R^*}} \Big\{ r\psi''' + 3\psi'' + \frac{3-4\lambda}{2}[r(1+r^2)\psi' + (1-r^2)\psi]\rho^3(x) \Big\} \frac{v^2}{2} \, \mathrm{d}x \\ = R^*\psi(R^*) \int_{\partial B_{R^*}} \left| \frac{\partial v}{\partial r} \right|^2 \mathrm{d}\mathcal{H}^2(x) + 2\int_{B_{R^*}} r\psi' \left(|\nabla v|^2 - \left| \frac{\partial v}{\partial r} \right|^2 \right) \mathrm{d}x \\ - \int_{B_{R^*}} \Big\{ (1 + \frac{2}{p+1}) r\psi' - \frac{p-5}{p+1} \cdot \frac{1-r^2}{1+r^2}\psi \Big\} |v|^{p+1} \rho(x)^{(5-p)/2} \, \mathrm{d}x. \end{aligned} \tag{6}$$

From here, we deduce the result of Bandle and Benguria [1], i.e., to determine the range of values of the parameters θ^* and λ for which there exists no positive solution of (4) in the critical case $p = 5$.

Theorem 4 (Bandle and Benguria [1]). *Let $p = 5$ and $D_{\theta^*} \subset S^3$ be a geodesic ball of radius θ^*. Set $\lambda_1^* = (\pi^2 - (\theta^*)^2)/(\theta^*)^2$ be the first eigenvalue*

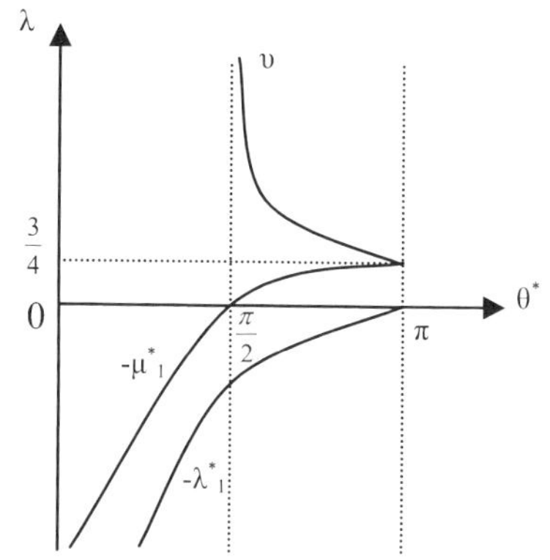

FIGURE 1. Range of values of λ for nonexistence of positive radial solutions

of the Laplace – Beltrami operator with Dirichlet boundary condition in D_{θ^*} and $\mu_1^* = (\pi^2 - 4(\theta^*)^2)/(4(\theta^*)^2)$. If one of the following conditions is satisfied

(i) $\lambda \leq \frac{3}{4}$ *and* $\lambda \notin (-\lambda_1^*, -\mu_1^*)$,
(ii) $\lambda > \frac{3}{4}$ *and* $\theta^* \in (0, \pi/2]$,

then there is no positive solution of (4). Moreover, there exists a curve in the strip $(\theta^*, \lambda) \subset (\pi/2, \pi) \times (\frac{3}{4}, \infty)$, denoted by $\nu(\theta^*) = \lambda$ such that $\nu(\theta^*) \to \frac{3}{4}$ as $\theta^* \to \pi$ and for $\lambda \in \left(\frac{3}{4}, \nu(\theta^*)\right)$ there is no radial *solution of (4) (see Figure 1).*

Remark 2. (a) In [6], Brezis and Peletier proved that for any $\theta^* \in (\pi/2, \pi)$, there exist positive radial solutions of (4) for λ sufficiently large; therefore, $\nu(\theta^*) \to +\infty$ as $\theta^* \to \pi/2$.

(b) The question if there is no solution of (4) in the strip $(\theta^*, \lambda) \subset (\pi/2, \pi) \times (\frac{3}{4}, \infty)$ below the curve ν is still open (see discussion in Section 4).

Finally, we deal with positive solutions of (1) in the case of the whole unit sphere S^n, $n \geq 3$:

$$
\begin{cases} -\Delta_g u = u^p - \lambda u \\ u > 0 \end{cases} \quad \text{on } S^n. \tag{7}
$$

For $\lambda \leq 0$, there is no solution of (7) (it directly follows by integration of (7) on S^n). Therefore, we consider the range $\lambda > 0$. The goal is to present a simplified proof of the following Pohožaev-type identity due to Gidas and Spruck [9]:

Lemma 5 (Gidas and Spruck [9]). *Let $n \geq 3$ and u be a solution of (7). Set*

$$
w = u^{-2/(n-2)} \tag{8}
$$

and

$$
J(x) = \frac{1}{w^{n-1}} \sum_{i,j=1}^{n} \left(\nabla_i \tilde{w}_j - \frac{1}{n} \Delta_g w \, \delta_{ij} \right)^2 \geq 0,
$$

where $\nabla_i \tilde{w}_j$ *denotes the* j *component of the covariant derivative* ∇_i *of the vector field* $\left((1/\rho^2)\partial_j w\right)_{1\le j\le n}$. *For any* $\gamma \in \mathbb{R}$ *we have*

$$
\begin{aligned}
(9)\quad & \frac{(n-2)^2}{2}\int_{S^n} J(x)u^{-\gamma} + \gamma(1-\gamma)\int_{S^n} u^{-\gamma-(2n-2)/(n-2)}|\mathrm{d}u|^4 \\
&+ \frac{2(n-1)}{n}\left(\frac{n+2}{n-2} - p - \gamma\frac{n+2}{2(n-1)}\right)\int_{S^n} u^{p-\gamma-n/(n-2)}|\mathrm{d}u|^2 \\
&+ 2\left[n-1-\lambda\left(\frac{4(n-1)}{n(n-2)} - \gamma\frac{n+2}{2n}\right)\right]\int_{S^n} u^{-\gamma-2/(n-2)}|\mathrm{d}u|^2 = 0.
\end{aligned}
$$

From here, it follows the following uniqueness result for solutions of (7):

Corollary 6. *Let* $n \ge 3$. *Assume that one of the following conditions holds:*

(i) $1 < p < (n+2)/(n-2)$ *and*

$$
(10)\qquad \begin{cases} 0 < \lambda < \min\{n/(p-1), a_n\} & \text{if } n < 8, \\ 0 < \lambda \le n/(p-1) & \text{if } n \ge 8, \end{cases}
$$

where

$$
a_n = \frac{2n(n-1)(n-2)}{-n^2+8n-4};
$$

(ii) $p = (n+2)/(n-2)$ *and* $0 < \lambda < n(n-2)/4$.

Then the only solution of (7) *is the trivial constant solution* $\lambda^{1/(p-1)}$.

This result was extended by M.-F. Bidaut-Véron et L. Véron [3] using the Bochner–Lichnerowicz–Weitzenböck formula [10]:

Theorem 7 (Bidaut-Véron and Véron [3]). *Assume that*

$$
(11)\qquad 1 < p \le \frac{n+2}{n-2} \quad \text{and} \quad 0 < \lambda \le \frac{n}{p-1}
$$

where at least one of the two inequalities (11) *is strict. Then the only solution of* (7) *is the constant* $\lambda^{1/(p-1)}$.

Recently, Brezis and Li [4] proved Theorem 7 in the case of $\lambda \le n(n-2)/4$ using the theory of moving planes; they also showed that for subcritical exponent p, there exist nonconstant solutions of (7) if $\lambda > n/(p-1)$ with $|\lambda - n/(p-1)|$ small. For the critical exponent $p = (n+2)/(n-2)$, Corollary 6 and Theorem 7 are also sharp since there is a well-known branch of nonconstant solutions if $\lambda = n(n-2)/4$ (see [7]).

The outline of the paper is the following: we start with some preliminaries on the geometry of the unit sphere S^n that we use in the proof of Lemma 5. In Section 3, we prove Lemma 1 and Theorem 2. In Section 4, we show Lemma 3 and Theorem 4. Finally, we give a simplified proof of the Gidas–Spruck result.

2. Preliminaries

In this section we introduce some notations that we use throughout of the paper. Let $P = (0,\ldots,0,1) \in \mathbb{R}^{n+1}$ and $Q = (0,\ldots,0,-1) \in \mathbb{R}^{n+1}$ be the North and the

South pole of S^n and set $\Omega_P = S^n \setminus \{P\}$ and $\Omega_Q = S^n \setminus \{Q\}$. The stereographic projection $\Phi_P \colon \Omega_P \to \mathbb{R}^n$ of pole P (resp. $\Phi_Q \colon \Omega_Q \to \mathbb{R}^n$ of pole Q) is defined as

$$\Phi_P(y) = \left(\frac{y_1}{1-y_{n+1}}, \dots, \frac{y_n}{1-y_{n+1}}\right), \quad \forall y = (y_1, \dots, y_{n+1}) \in \Omega_P$$

$$\left(\text{resp.} \quad \Phi_Q(y) = \left(\frac{y_1}{1+y_{n+1}}, \dots, \frac{y_n}{1+y_{n+1}}\right), \quad \forall y = (y_1, \dots, y_{n+1}) \in \Omega_Q\right).$$

We easily check that Φ_P (resp. Φ_Q) is a homeomorphism of Ω_P (resp. Ω_Q) into $\mathbb{R}^n$ and the inverse function $\Phi_P^{-1} \colon \mathbb{R}^n \to \mathbb{R}^{n+1}$ (resp. $\Phi_Q^{-1} \colon \mathbb{R}^n \to \mathbb{R}^{n+1}$) is given by

$$\Phi_P^{-1}(x) = \rho(x)\left(x_1, \dots, x_n, \frac{|x|^2-1}{2}\right), \quad \forall x = (x_1, \dots, x_n) \in \mathbb{R}^n$$

$$\left(\text{resp.} \quad \Phi_Q^{-1}(x) = \rho(x)\left(x_1, \dots, x_n, \frac{1-|x|^2}{2}\right), \quad \forall x = (x_1, \dots, x_n) \in \mathbb{R}^n\right),$$

where

$$\rho(x) = \frac{2}{1+|x|^2}$$

and $|x|^2 = \sum_{i=1}^n x_i^2$ for any $x \in \mathbb{R}^n$. In the following we omit the argument of maps. In the local charts (Ω_P, Φ_P) and (Ω_Q, Φ_Q), the standard metric g on S^n writes as

$$g_{ij} = \rho^2 \delta_{ij}, \quad 1 \le i, j \le n$$

where δ_{ij} denotes the Kronecker's symbol. For a function $u \colon S^n \to \mathbb{R}$, we use the following notations:

$$u_i = \partial_i u, \ \tilde{u}_i = \frac{1}{\rho^2} u_i, \quad 1 \le i \le n$$

$$|\mathrm{d}u|^2 = \sum_{i=1}^n u_i \tilde{u}_i,$$

$$\Delta_g u = \frac{1}{\rho^n} \sum_{i=1}^n \partial_i(\rho^n \tilde{u}_i).$$

For $1 \le i \le n$, let ∇_i be the covariant derivative. The Cristoffel symbols are given by

$$\Gamma_{ij}^k = \begin{cases} x_k\rho & \text{if } i = j \ne k, \\ -x_i\rho & \text{if } i = j = k, \\ -x_j\rho & \text{if } i \ne j,\ i = k, \\ -x_i\rho & \text{if } i \ne j,\ j = k. \end{cases}$$

If V_i are the components of a vector field V, we associate the vector field $\widetilde{V}$ of components $\widetilde{V}_i = (1/\rho^2)V_i$. Standard computations yield that

$$(\nabla_i V)_j = \partial_i(V_j) + \sum_{k=1}^n \Gamma_{ik}^j V_k, \quad 1 \le i, j \le n,$$

$$\sum_{i=1}^n (\nabla_i V)_i = \frac{1}{\rho^n} \sum_{i=1}^n \partial_i(\rho^n V_i), \tag{12}$$

$$\sum_{j=1}^n (\nabla_j \nabla_i V - \nabla_i \nabla_j V)_j = (n-1)\rho^2 V_i, \quad 1 \le i \le n. \tag{13}$$

One can check that

$$\nabla_i \tilde{u}_j = \nabla_j \tilde{u}_i, \quad 1 \le i, j \le n, \tag{14}$$

$$\Delta_g u = \sum_{i=1}^n \nabla_i \tilde{u}_i, \tag{15}$$

$$\partial_i(|\mathrm{d}u|^2) = 2\sum_{j=1}^n u_j \nabla_j \tilde{u}_i, \quad 1 \le i \le n, \tag{16}$$

$$\sum_{i,j=1}^n (\partial_i \tilde{u}_j \nabla_j \tilde{u}_i + \tilde{u}_j \nabla_j \nabla_i \tilde{u}_i) = \sum_{i,j=1}^n \big(\nabla_i \tilde{u}_j \nabla_j \tilde{u}_i + \tilde{u}_j \partial_j(\nabla_i \tilde{u}_i)\big), \tag{17}$$

where $\nabla_i \tilde{u}_j$ denotes the j component of the covariant derivative ∇_i of the vector field with the components $\tilde{u}_j$.

3. The case of the unit ball in $\mathbb{R}^n$. Proofs of Lemma 1 and Theorem 2

We start by proving identity (3):

Proof of Lemma 1. Let u be a solution of (2). Following the ideas of Brezis and Nirenberg [5], we first multiply (2) by $\psi(r)x \cdot \nabla u(x) \in C^1(B_1)$ (without any condition on ψ) and integrating by parts, we obtain:

$$\begin{aligned} -\int_{B_1} [r\psi' - (n-2)\psi]|\nabla u|^2 \,\mathrm{d}x + \psi(1)\int_{\partial B_1} \left|\frac{\partial u}{\partial r}\right|^2 \mathrm{d}\mathcal{H}^{n-1}(x) \\ + 2\int_{B_1} r\psi'\left(|\nabla u|^2 - \left|\frac{\partial u}{\partial r}\right|^2\right) \mathrm{d}x = \int_{B_1} (r\psi' + n\psi)\left(\frac{2}{p+1}|u|^{p+1} - \lambda u^2\right) \mathrm{d}x. \end{aligned} \tag{18}$$

Next we multiply (2) by $[r\psi' - (n-2)\psi]u \in C^1(B_1)$ (since $\psi'(0) = 0$ if $n \ge 4$) and it results by integration by parts:

$$\begin{aligned} \int_{B_1} [r\psi' - (n-2)\psi]|\nabla u|^2 \,\mathrm{d}x - \int_{B_1} \left[r\psi''' + 3\psi'' - \frac{(n-1)(n-3)}{r}\psi'\right] \frac{u^2}{2} \,\mathrm{d}x \\ = \int_{B_1} [r\psi' - (n-2)\psi](|u|^{p+1} - \lambda u^2) \,\mathrm{d}x. \end{aligned} \tag{19}$$

(Here, we used that $\psi''(0) = 0$ if $n \ge 4$.) Combining (18) and (19), the conclusion follows immediately. □

Using identity (3), we show Theorem 2:

Proof of Theorem 2. Let u be a positive solution of (2). Choosing $\psi = 1$, (3) becomes the standard Pohožaev identity [11]:

$$-2\lambda \int_{B_1} u^2 \,\mathrm{d}x - \left(\frac{p-1}{p+1}n - 2\right)\int_{B_1} u^{p+1} \,\mathrm{d}x = \int_{\partial B_1} \left|\frac{\partial u}{\partial r}\right|^2 \mathrm{d}\mathcal{H}^{n-1}(x).$$

Therefore, no solution exists for (2) if $\lambda \ge 0$ and $p \ge (n+2)/(n-2)$. Set φ be a positive eigenfunction associated to the first eigenvalue λ_1 in B_1. First, we prove the nonexistence result for $\lambda \le -\lambda_1$. Indeed, multiplying (2) by φ, we deduce:

$$\lambda_1 \int_{B_1} u\varphi = -\int_{B_1} u\Delta\varphi = -\int_{B_1} \Delta u\, \varphi = \int_{B_1} (u^p - \lambda u)\varphi.$$

That is,

$$\int_{B_1} [u^p - (\lambda + \lambda_1)u]\varphi = 0;$$

therefore, we get a contradiction with $u > 0$. Now we treat the remaining case: $n = 3$ and $\lambda \in [-\lambda_1/4, 0)$ where $\lambda_1 = \pi^2$. By the symmetry result of Gidas, Ni and Nirenberg [8] applied for positive solutions of (2), we know that u is radial. Then (3) becomes:

$$(20)\quad \int_{B_1} \{r\psi''' + 3\psi'' - 4\lambda(r\psi' + \psi)\}\frac{u^2}{2}\,\mathrm{d}x$$
$$= \psi(1)\int_{\partial B_1}\left|\frac{\partial u}{\partial r}\right|^2 \mathrm{d}\mathcal{H}^2(x) - \int_{B_1}\left\{\frac{p+3}{p+1}r\psi' - \frac{p-5}{p+1}\psi\right\}u^{p+1}\,\mathrm{d}x,$$

for any smooth function $\psi\colon [0,\infty) \to \mathbb{R}$. Following the argument in [5], we choose the smooth function

$$\psi(r) = \frac{\sin(2\sqrt{|\lambda|}\,r)}{r} > 0, \quad \forall r \in [0,1).$$

Then

$$r\psi''' + 3\psi'' - 4\lambda(r\psi' + \psi) = 0,$$

and

$$r\psi' = \frac{2\sqrt{|\lambda|}\,r\cos(2\sqrt{|\lambda|}\,r) - \sin(2\sqrt{|\lambda|}\,r)}{r} < 0,$$

$\forall r \in (0,1]$. Since $p \geq 5$, (20) leads to a contradiction. □

Now we discuss the case $n = 3$ and $p = 5$, i.e.,

$$(21)\quad \begin{cases} -\Delta u = u^5 - \lambda u,\ u \not\equiv 0 & \text{in } B_1 \subset \mathbb{R}^3, \\ u = 0 & \text{on } \partial B_1. \end{cases}$$

Our aim is to present a list of properties for a solution u. We start by proving another identity "à la Pohožaev" related to (3):

Lemma 8. *Let $\psi\colon [0,\infty) \to \mathbb{R}$ be a smooth function. If u is a solution of (21), then*

$$(22)\quad -\int_{B_1}\left\{r\psi'''(r) + 5\psi''(r) + \frac{4}{r}\psi'(r) + 4\lambda\big(r\psi'(r) + \psi(r)\big)\right\}\frac{u^2}{2}\,\mathrm{d}x$$
$$= \psi(1)\int_{\partial B_1}\left|\frac{\partial u}{\partial r}\right|^2 \mathrm{d}\mathcal{H}^2(x) - 2\int_{B_1} r\psi'(r)\left(\left|\frac{\partial u}{\partial r}\right|^2 + \lambda u^2 - \frac{1}{3}u^6\right)\mathrm{d}x.$$

PROOF. First, identity (3) writes in the case $n = 3$ and $p = 5$ as

$$(23)\quad \int_{B_1}\{r\psi'''(r) + 3\psi''(r) - 4\lambda\big(r\psi'(r) + \psi(r)\big)\}\frac{u^2}{2}\,\mathrm{d}x$$
$$= \psi(1)\int_{\partial B_1}\left|\frac{\partial u}{\partial r}\right|^2 \mathrm{d}\mathcal{H}^2(x) + \int_0^1 r\psi'(r)\int_{\partial B_r}\left(2|\nabla_{S^2}u|^2 - \frac{4}{3}u^6\right)\mathrm{d}\mathcal{H}^2\,\mathrm{d}r,$$

where

$$\nabla_{S^2}u = \frac{\partial u}{\partial\theta}\frac{\vec{\theta}}{r} + \frac{\partial u}{\partial\varphi}\frac{\vec{\varphi}}{r\sin\theta}, \quad \nabla u = \frac{\partial u}{\partial r}\vec{r} + \nabla_{S^2}u$$

are written in the spherical coordinates $(r,\theta,\varphi) \in (0,1)\times(0,\pi)\times(0,2\pi)$. We compute the last term in (23): multiplying (21) by u and integrating by parts in the variables θ and φ on ∂B_r, $r\in(0,1)$, we obtain:

$$(24)\quad \int_0^1 r\psi'(r)\int_{\partial B_r}\left(2|\nabla_{S^2}u|^2-\frac{4}{3}u^6\right)\mathrm{d}\mathcal{H}^2\mathrm{d}r \\ = \int_{B_1} r\psi'(r)\left[\frac{2}{3}u^6-2\lambda u^2+\frac{2}{r^2}\frac{\partial}{\partial r}\left(r^2\frac{\partial u}{\partial r}\right)u\right]\mathrm{d}x.$$

If we integrate by parts the last term in (24) with respect to r, we get that

$$(25)\quad \int_{B_1}\frac{2\psi'(r)}{r}\frac{\partial}{\partial r}\left(r^2\frac{\partial u}{\partial r}\right)u\,\mathrm{d}x \\ = \int_{B_1}\left\{r\psi'''(r)+4\psi''(r)+\frac{2}{r}\psi'(r)\right\}u^2\,\mathrm{d}x-2\int_{B_1}r\psi'(r)\left|\frac{\partial u}{\partial r}\right|^2\mathrm{d}x.$$

Combining (23), (24) and (25), we conclude with (22). □

By Lemmas 1 and 8, we obtain the following properties:

Proposition 9. *If u is a solution of* (21), *then:*

(i)
$$\int_{\partial B_1}\left|\frac{\partial u}{\partial r}\right|^2\mathrm{d}\mathcal{H}^2(x)=2\lambda\int_{B_1}u^2\,\mathrm{d}x.$$

(ii) *If $\psi\colon[0,\infty)\to\mathbb{R}$ is a smooth function, then*
$$\int_{B_1}\left\{r\psi'''(r)+4\psi''(r)+\frac{2}{r}\psi'(r)\right\}u^2\,\mathrm{d}x=2\int_{B_1}r\psi'(r)(|\nabla u|^2+\lambda u^2-u^6t)\,\mathrm{d}x.$$

(iii) *If $\lambda<0$ and $\psi(r)=(1/r)\sin(2\sqrt{|\lambda|}\,r)$, $\forall r\in[0,1)$, then*
$$\psi(1)\int_{\partial B_1}\left|\frac{\partial u}{\partial r}\right|^2\mathrm{d}\mathcal{H}^2(x)+2\int_{B_1}r\psi'(r)\left(|\nabla_{S^2}u|^2-\frac{2}{3}u^6\right)\mathrm{d}x=0.$$

(iv) *If $\lambda<0$ and $\psi(r)=(1/r)\int_0^r(1/t)\sinh(2\sqrt{|\lambda|}\,t)\,\mathrm{d}t$, $\forall r\in[0,1)$, then*
$$\psi(1)\int_{\partial B_1}|\frac{\partial u}{\partial r}|^2\,\mathrm{d}\mathcal{H}^2(x)=2\int_{B_1}r\psi'(r)\left(\left|\frac{\partial u}{\partial r}\right|^2+\lambda u^2-\frac{1}{3}u^6\right)\mathrm{d}x.$$

PROOF. The first point (i) follows from (3) by taking $\psi\equiv 1$. The identity in (ii) comes by subtracting (23) from (22). Notice that the following ODE:
$$r\psi'''+4\psi''+\frac{2}{r}\psi'=0$$
has the solution
$$\psi(r)=c_1+\frac{c_2}{r}+c_3\ln r.$$
(If we approximate $\psi=1/r$ by smooth functions, we obtain by (ii) the obvious relation $\int_{B_1}(|\nabla u|^2+\lambda u^2-u^6)\,\mathrm{d}x=0$.) Point (iii) follows from (23) since ψ is the solution of the ODE:
$$(r\psi)'''-4\lambda(r\psi)'=0.$$
Similarly, (iv) comes from (22) since $g=(r\psi)'$ satisfies the ODE:
$$-g''-\frac{2}{r}g'=4\lambda g.$$
□

4. The case of caps in S^3. Proofs of Lemma 3 and Theorem 4

First we present the proof of identity (6):

PROOF OF LEMMA 3. Let v be a solution of (5). Following the same argument as in the proof of Lemma 1, we multiply (5) by $\psi(r)x\cdot\nabla v(x)\in C^1(B_{R^*})$ and integrating by parts, we obtain:

$$(26)\quad -\int_{B_{R^*}}(r\psi'-\psi)|\nabla v|^2\,\mathrm{d}x+R^*\psi(R^*)\int_{\partial B_{R^*}}\left|\frac{\partial v}{\partial r}\right|^2\mathrm{d}\mathcal{H}^2 +2\int_{B_{R^*}}r\psi'\left(|\nabla v|^2-\left|\frac{\partial v}{\partial r}\right|^2\right)\mathrm{d}x$$
$$=\frac{3-4\lambda}{8}\int_{B_{R^*}}\left[r(1+r^2)\psi'+(3-r^2)\psi\right]\rho^3(x)v^2\,\mathrm{d}x +\frac{2}{p+1}\int_{B_{R^*}}\left\{r\psi'+\left(3+\frac{(p-5)r^2}{1+r^2}\right)\psi\right\}|v|^{p+1}\rho(x)^{(5-p)/2}\,\mathrm{d}x.$$

Then, multiplying (5) by $(r\psi'-\psi)v\in C^1(B_{R^*})$, we obtain:

$$(27)\quad \int_{B_{R^*}}(r\psi'-\psi)|\nabla v|^2\,\mathrm{d}x-\int_{B_{R^*}}(r\psi'''+3\psi'')\frac{v^2}{2}\,\mathrm{d}x =\int_{B_{R^*}}(r\psi'-\psi)\left(\rho(x)^{(5-p)/2}|v|^{p+1}+\frac{3-4\lambda}{4}\rho^2v^2\right)\mathrm{d}x.$$

By summation of (26) and (27), we get the identity (6). □

As an application, we give the proof of Theorem 4:

PROOF OF THEOREM 4. Let u be a positive solution of (4). As in the proof of Theorem 2, multiplying (4) with a positive eigenfunction associated to λ_1^*, we obtain that no positive solution of (4) exists if $\lambda\le-\lambda_1^*$. Set $v(x)=u\big(\Phi_Q^{-1}(x)\big)\sqrt{\rho(x)}$ be the corresponding solution of (5). We distinguish the following two cases:

(i) $\lambda\in[-\mu_1^*,\frac{3}{4}]$. Since $\lambda\le\frac{3}{4}$, by the symmetry result in [8] applied for positive solutions of (5), we deduce that v is radial. Therefore, (6) becomes:

$$(28)\quad \int_{B_{R^*}}\left\{r\psi'''+3\psi''+\frac{3-4\lambda}{2}[r(1+r^2)\psi'+(1-r^2)\psi]\rho^3(x)\right\}\frac{v^2}{2}\,\mathrm{d}x =R^*\psi(R^*)\int_{\partial B_{R^*}}\left|\frac{\partial v}{\partial r}\right|^2\mathrm{d}\mathcal{H}^2(x)-\frac{4}{3}\int_{B_{R^*}}r\psi'v^6\mathrm{d}x.$$

Set

$$(29)\qquad w=\sqrt{4(1-\lambda)}$$

and we use the change of variable

$$\theta=2\arctan r\quad\text{for } r\in[0,R^*].$$

Remark that $w\ge1$ and the assumption $\lambda\ge-\mu_1^*$ turns into $w\theta^*\le\pi$. Like in the proof of Theorem 2, we choose the smooth function

$$\psi(r)=\frac{\sin(w\theta)}{\sin\theta}>0,\quad\forall\theta\in(0,\theta^*).$$

Then

$$r\psi''' + 3\psi'' + \frac{3-4\lambda}{2}[r(1+r^2)\psi' + (1-r^2)\psi]\rho^3(r) = 0. \tag{30}$$

Moreover,

$$\psi'(r) = 2\frac{w\cos(w\theta)\sin\theta - \cos\theta\sin(w\theta)}{(1+\tan^2(\theta/2))\sin^2\theta} \leq 0, \quad \forall\theta \in (0, \theta^*).$$

Indeed, if we denote

$$F(\theta) = w\cos(w\theta)\sin\theta - \cos\theta\sin(w\theta),$$

then $F(0) = 0$ and $F'(\theta) = (4\lambda - 3)\sin\theta\sin(w\theta) \leq 0$ for every $\theta \in (0, \theta^*)$; therefore, we conclude that $F(\theta) \leq 0$ on $(0, \theta^*)$. Using (28), we obtain a contradiction.

(ii) $\lambda > \frac{3}{4}$ and $\theta^* \in (0, \frac{\pi}{2}]$, i.e., $R^* \leq 1$. If we take $\psi = 1$, (6) writes as:

$$\frac{3-4\lambda}{4}\int_{B_{R^*}} (1-r^2)\rho^3(x)v^2\, dx = R^* \int_{\partial B_{R^*}} \left|\frac{\partial v}{\partial r}\right|^2 d\mathcal{H}^2(x).$$

Therefore, the nonexistence result also follows in this situation.

Using a similar argument as Bandle and Benguria [1], we prove the nonexistence of radial (nodal) solutions u below a curve $\lambda = \nu(\theta^*)$ in the region $(\lambda, \theta^*) \in (\frac{3}{4}, \infty) \times (\frac{\pi}{2}, \pi)$. We distinguish three cases:

(I) $\lambda = 1$. We consider $m(\theta) := a\theta - \theta^2$ and

$$\psi(r) = \frac{m(\theta)}{\sin\theta} \tag{31}$$

where $a \in (\pi/2, \pi)$ is to be chosen in such a way that

$$\psi(r) \geq 0 \quad \text{and} \quad \psi'(r) \leq 0 \tag{32}$$

for every $0 < \theta < a$. This is equivalent with $G(\theta) > 0$, $\forall\theta \in (0, a)$ where

$$G(\theta) = \cos\theta m(\theta) - \sin\theta m'(\theta). \tag{33}$$

Since $G(0) = 0$, we ask that $G'(\theta) = \sin\theta(\theta^2 - a\theta + 2) \geq 0$ for every $\theta \in (0, a)$; for example, $a = 2\sqrt{2} = 2.828\ldots$. Bandle and Benguria [1] numerically found a better value $a = 3.042$. Since (30) holds, by (28), we get a contradiction. Therefore, for the largest a we set $\nu(a) = 1$.

(II) $\lambda \in (\frac{3}{4}, 1)$. Let $w \in (0, 1)$ be given by (29). We consider

$$m(\theta) := \sin(w\theta) - a\cos(w\theta) + a$$

and ψ be defined as in (31) where a is to be chosen in such a way that (32) holds for the largest range of θ. Denote it by $\nu^{-1}(\lambda)$. Then (30) is satisfied and by (28), we deduce that no radial (nodal) solution of (4) exists if $\theta^* < \nu^{-1}(\lambda)$. Let us check that

$$\nu^{-1}(\lambda) > \frac{\pi}{2}.$$

For that, we take a negative a in the interval

$$a \in \left(\frac{\sin(w\pi/2)}{\cos(w\pi/2) - 1/(1-w^2)}, -\frac{\cos(w\pi/2)}{\sin(w\pi/2)}\right). \tag{34}$$

Notice that a is well defined since $w^2 + \cos(w\pi/2) \leq 1$ for every $w \in (0, 1)$. We want to prove that (32) holds in $(0, \theta_a)$ for some $\theta_a > \pi/2$. We have that

$$m'(\theta) = w\big(\cos(w\theta) + a\sin(w\theta)\big) \quad \text{and} \quad m''(\theta) = w^2\big(a\cos(w\theta) - \sin(w\theta)\big).$$

Therefore, $m''(\theta) \le 0$ for $\theta \in [0, \pi/2]$, i.e., m is concave. Since $m(0) = 0$ and $m(\pi/2) > 0$ (by (34)), we get that $m \ge 0$ in $[0, \theta_a)$ for $\theta_a > \pi/2$ and close to $\pi/2$. The same argument yields that

$$(m + m'')(\theta) = (1 - w^2)\big(\sin(w\theta) - a\cos(w\theta)\big) + a$$

is concave in $[0, \pi/2]$. The choice (34) leads to $(m+m'')(\pi/2) > 0$. Since $(m+m'')(0) = aw^2 < 0$, we deduce that $m + m''$ changes sign just once in $(0, \pi/2)$. Define G as in (33); then

$$G'(\theta) = -\sin\theta(m + m'')(\theta).$$

Hence, G' also changes sign once in $(0, \pi/2)$ and $G' \ge 0$ for θ close to 0. Since $G(0) = 0$ and $G(\pi/2) = -m'(\pi/2) > 0$ (by (34)), we obtain

$$G(\theta) \ge \min\{G(0), G(\pi/2)\} \ge 0, \quad \forall\theta \in (0, \theta_a),$$

i.e., (32) is satisfied in $(0, \theta_a)$. Finally, we check that

$$\nu^{-1}(\lambda) \to \pi \quad \text{as} \quad \lambda \downarrow \tfrac{3}{4}.$$

Indeed, let $\varepsilon > 0$ be very small. We consider λ be close to $\frac{3}{4}$ such that $1 - w^2 = O(\varepsilon^2)$. Choose $a < 0$ with $|a| = O(\varepsilon)$. Then $m''(\theta) \ge 0$ in an interval $(0, \theta_\varepsilon)$ with $\theta_\varepsilon \to \pi$ as $\varepsilon \to 0$. Therefore m is concave in $(0, \theta_\varepsilon)$, and eventually by shrinking that interval, we can assume that $m(\theta_\varepsilon) > 0$, and thus, m is positive in $(0, \theta_\varepsilon)$. Now notice that $m + m'' < 0$ in $(0, \pi)$ and hence, G is positive in $(0, \pi)$. We conclude that (32) holds in the interval $(0, \theta_\varepsilon)$ that tends to $(0, \pi)$ as $\varepsilon \to 0$.

(III) $\lambda \in (1, \infty)$. Set $w = \sqrt{-4(1 - \lambda)}$. Consider

$$m(\theta) := \sinh(w\theta) - a\cosh(w\theta) + a$$

and ψ be as in (31) where a is to be chosen in such a way that (32) holds for the largest range of θ. Denote it by $\nu^{-1}(\lambda)$. The same argument as before gives that $\nu^{-1}(\lambda) > \pi/2$, i.e., ν is well-defined; for that, it suffices to choose a positive a in the interval

$$a \in \left(\frac{\cosh(w\pi/2)}{\sinh(w\pi/2)}, \frac{\sinh(w\pi/2)}{\cosh(w\pi/2) - 1/(1 + w^2)}\right).$$

Since (30) is satisfied, we conclude by (28) that no radial (nodal) solution of (4) exists if $\theta^* < \nu^{-1}(\lambda)$. □

Notice that for $\theta^* \in (\pi/2, \pi)$, we don't know in general if a positive solution of (4) is radial; moreover, for λ large enough, nonradial solutions do exist as announced by Bandle and Wei. As mentioned in Remark 2, it would be interesting to see if no solution of (4) exists below the curve ν in the strip $(\theta^*, \lambda) \subset (\pi/2, \pi) \times (\frac{3}{4}, \infty)$. We believe that the answer to this question is related to the open question raised in Remark 1.

5. The simplified proof of the Gidas – Spruck result

In the following we present the proof of Lemma 5:

PROOF OF LEMMA 5. The relation (8) between w and u leads to

$$w_i = -\frac{2}{n-2}u^{-n/(n-2)}u_i \quad \text{and} \quad |dw|^2 = \frac{4}{(n-2)^2}u^{-2n/(n-2)}|du|^2. \tag{35}$$

By (7), w satisfies

$$-\frac{1}{n}w\Delta_g w + \frac{1}{2}|\mathrm{d}w|^2 = \frac{2}{n(n-2)}\left(\lambda w^2 - w^{((n-2)/2)((n+2)/(n-2)-p)}\right) \tag{36}$$

which writes in terms of u as

$$\Delta_g w = \frac{2n}{(n-2)^2}u^{-2(n-1)/(n-2)}|\mathrm{d}u|^2 + \frac{2}{n-2}\left(u^{p-n/(n-2)} - \lambda u^{-2/(n-2)}\right). \tag{37}$$

We will use the vector field defined in [9] that has the components

$$V_i = \frac{1}{w^{n-1}}\left(\frac{1}{2}\partial_i(|\mathrm{d}w|^2) - \frac{1}{n}w_i\Delta_g w\right). \tag{38}$$

Using the equations (35) and (37), the expression of V_i in function of u writes as

$$V_i = \frac{2}{(n-2)^2}\Big[\partial_i(u^{-2/(n-2)}|\mathrm{d}u|^2) + (n-2)\partial_i\left(u^{-2/(n-2)}\right)|\mathrm{d}u|^2 + \frac{2}{n}\left(u^{p-2/(n-2)} - \lambda u^{(n-4)/(n-2)}\right)u_i\Big]. \tag{39}$$

Notice that by (14) and (15), we have that

$$J(x) = \frac{1}{w^{n-1}}\left(\sum_{i,j=1}^{n}\nabla_i\tilde{w}_j\nabla_j\tilde{w}_i - \frac{1}{n}(\Delta_g w)^2\right).$$

Now we compute the co-differential of the vector $\tilde{V} = (1/\rho^2)V$:

$$\begin{aligned}
\sum_{i=1}^{n}\nabla_i\widetilde{V}_i
&= \sum_{i=1}^{n}\nabla_i\left[\frac{1}{w^{n-1}}\left(\frac{1}{2\rho^2}\partial_i(|\mathrm{d}w|^2) - \frac{1}{n}\tilde{w}_i\Delta_g w\right)\right]\\
&\overset{(15),(16)}{=} \frac{1}{w^{n-1}}\left[\sum_{i,j=1}^{n}(\tilde{w}_j\nabla_i\nabla_j\tilde{w}_i + \partial_i(\tilde{w}_j)\nabla_j\tilde{w}_i) - \frac{1}{n}(\Delta_g w)^2 - \frac{1}{n}\sum_{i=1}^{n}\tilde{w}_i\partial_i(\Delta_g w)\right]\\
&\qquad + \sum_{i=1}^{n}\partial_i\left(\frac{1}{w^{n-1}}\right)\left[\frac{1}{2\rho^2}\partial_i(|\mathrm{d}w|^2) - \frac{1}{n}\tilde{w}_i\Delta_g w\right]\\
&\overset{(13),(17)}{=} J(x) + \frac{n-1}{w^{n-1}}\rho^2\sum_{j=1}^{n}\tilde{w}_j^2 + \frac{n-1}{nw^{n-1}}\sum_{i=1}^{n}\tilde{w}_i\partial_i(\Delta_g w)\\
&\qquad - \sum_{i=1}^{n}\frac{(n-1)w_i}{w^n}\left[\frac{1}{2\rho^2}\partial_i(|\mathrm{d}w|^2) - \frac{1}{n}\tilde{w}_i\Delta_g w\right]\\
&\overset{(36)}{=} J(x) + \frac{n-1}{w^{n-1}}|\mathrm{d}w|^2\\
&\qquad + \frac{2(n-1)}{n(n-2)}\left[\left(\frac{n+2}{2} - p\frac{n-2}{2}\right)w^{-n/2-p(n-2)/2} - 2\lambda w^{1-n}\right]|\mathrm{d}w|^2.
\end{aligned}$$

In terms of u, the co-differential of $\tilde{V}$ becomes

$$(40)\quad \sum_{i=1}^{n} \nabla_i \widetilde{V}_i \overset{(35)}{=} J(x) + \frac{4(n-1)}{(n-2)^2} u^{-2/(n-2)} |du|^2 \\ + \frac{8(n-1)}{n(n-2)^3} \left[\left(\frac{n+2}{2} - p\frac{n-2}{2} \right) u^{p-n/(n-2)} - 2\lambda u^{-2/(n-2)} \right] |du|^2.$$

Now let $\gamma \in \mathbb{R}$. Integration by parts yields

$$(41)\qquad 0 = \int_{S^n} \sum_{i=1}^{n} \nabla_i (u^{-\gamma} \widetilde{[V]})_i = \int_{S^n} u^{-\gamma} \sum_{i=1}^{n} \nabla_i \widetilde{V}_i - \gamma \int_{S^n} u^{-\gamma-1} \sum_{i=1}^{n} u_i \widetilde{V}_i.$$

By (39), integrating by parts, we deduce:

$$\frac{(n-2)^2}{2} \int_{S^n} u^{-\gamma-1} \sum_{i=1}^{n} u_i \widetilde{V}_i \\ = - \int_{S^n} \sum_{i=1}^{n} \nabla_i (u^{-\gamma-1} \tilde{u}_i) u^{-2/(n-2)} |du|^2 - 2 \int_{S^n} u^{-\gamma-2(n-1)/(n-2)} |du|^4 \\ + \frac{2}{n} \int_{S^n} \left(u^{p-\gamma-n/(n-2)} - \lambda u^{-\gamma-2/(n-2)} \right) |du|^2 \\ \overset{(7)}{=} (\gamma - 1) \int_{S^n} u^{-\gamma-2(n-1)/(n-2)} |du|^4 \\ + \frac{n+2}{n} \int_{S^n} \left(u^{p-\gamma-n/(n-2)} - \lambda u^{-\gamma-2/(n-2)} \right) |du|^2.$$

The conclusion follows by (40) and (41). □

Corollary 6 is a trivial consequence of Lemma 5:

PROOF OF COROLLARY 6. Suppose that hypothesis (i) holds. Set

$$\gamma_0 = \frac{2(n-1)}{n+2} \left(\frac{n+2}{n-2} - p \right).$$

If $n \geq 8$, then $0 < \gamma_0 < 1$ and

$$(42)\qquad \lambda \left(\frac{4(n-1)}{n(n-2)} - \gamma_0 \frac{n+2}{2n} \right) \leq n - 1$$

provided that $\lambda \leq n/(p-1)$. Applying (9) for γ_0, we conclude that the second term in (9) must vanish, that means u is constant. If $3 \leq n < 8$, an easy computation shows that

$$a_n \geq \frac{n}{p-1} \iff p \geq \frac{n(n+2)}{2(n-1)(n-2)} \iff \gamma_0 \leq 1.$$

Therefore, if $a_n \geq n/(p-1)$, we choose $\gamma = \gamma_0$ in (9) and we deduce that the last term in (9) must be zero, i.e., u is constant provided that $\lambda < n/(p-1)$. Otherwise, (9) for $\gamma = 1$ also yields that the last term vanishes and the conclusion follows. Now suppose that (ii) holds. Then for $\gamma = 0$, (9) shows that the last term is zero, that is u must be constant provided that $\lambda < n/(p-1)$. □

Remark 3. When $\{n \geq 8, p \in (1, (n+2)/(n-2))\}$ and $\{3 \leq n < 8, p \in \big(n(n+2)/\big(2(n-1)(n-2)\big), (n+2)/(n-2)\big)\}$, Corollary 6 is sharp.

References

1. C. Bandle and R. D. Benguria, *The Brezis – Nirenberg problem on* $\mathbb{S}^3$, J. Differential Equations **178** (2002), no. 1, 264 – 279.
2. R. D. Benguria, J. Dolbeault, and M. J. Esteban, *Classification of the solutions of semilinear elliptic problems in a ball*, J. Differential Equations **167** (2000), no. 2, 438 – 466.
3. M.-F. Bidaut-Véron and L. Véron, *Nonlinear elliptic equations on compact Riemannian manifolds and asymptotics of Emden equations*, Invent. Math. **106** (1991), no. 3, 489 – 539.
4. H. Brezis and Y. Li, *Some nonlinear elliptic equations have only constant solutions*, J. Partial Differential Equations **19** (2006), no. 3, 208 – 217.
5. H. Brezis and L. Nirenberg, *Positive solutions of nonlinear elliptic equations involving critical Sobolev exponents*, Comm. Pure Appl. Math. **36** (1983), no. 4, 437 – 477.
6. H. Brezis and L. A. Peletier, *Elliptic equations with critical exponent on spherical caps of* S^3, J. Anal. Math. **98** (2006), 279 – 316.
7. M. G. Crandall and P. H. Rabinowitz, *Bifurcation from simple eigenvalues*, J. Functional Analysis **8** (1971), 321 – 340.
8. B. Gidas, W. M. Ni, and L. Nirenberg, *Symmetry and related properties via the maximum principle*, Comm. Math. Phys. **68** (1979), no. 3, 209 – 243.
9. B. Gidas and J. Spruck, *Global and local behavior of positive solutions of nonlinear elliptic equations*, Comm. Pure Appl. Math. **34** (1981), no. 4, 525 – 598.
10. A. Lichnerowicz, *Géométrie des groupes de transformations*, Travaux et recherches mathématiques, vol. 3, Dunod, Paris, 1958.
11. S. I. Pohožaev, *On the eigenfunctions of the equation* $\Delta u + \lambda f(u) = 0$, Dokl. Akad. Nauk SSSR **165** (1965), 36 – 39 (Russian); English transl., Soviet Math. Dokl. **6** (1965), 1408 – 1411.

Laboratoire Jacques-Louis Lions, Université Pierre et Marie Curie, 4 place Jussieu, BC 187, 75252 Paris Cedex 05, France

E-mail address: ignat@ann.jussieu.fr

Centre de Recherches Mathématiques
CRM Proceedings and Lecture Notes
Volume **44**, 2008

Some Remarks on Monge – Ampère Functions

Robert L. Jerrard

ABSTRACT. The space of Monge – Ampère functions is a rather large function space with the property that, if u is a Monge – Ampère function, then the determinant of the Hessian can be identified with a well-defined Radon measure, denoted $\operatorname{Det} D^2u$. Moreover, the map $u \mapsto \operatorname{Det} D^2u$ is continuous in a natural weak topology on the space of Monge – Ampère functions. These properties make Monge – Ampère functions potentially useful for certain applications in the calculus of variations. We attempt to give a reasonably elementary treatment of a portion of the relevant theory.

1. Introduction

This note presents some properties of Monge – Ampère functions, a class of functions introduced by J. H. G. Fu in [4, 5] and extended by the author in [13], and one that we believe is potentially useful for certain problems in the calculus of variations. This possible applicability arises from the fact that every locally Monge – Ampère function u has the property that the determinant of the Hessian matrix D^2u, as well as all other minors of the Hessian, can be represented by Radon measures in the product space $\Omega \times \mathbb{R}^n$. Moreover, the determinant measure and measures associated with other minors are all continuous with respect to a natural notion of weak convergence. It is the largest known function space with these properties. For example, Monge – Ampère functions need not be convex, nor generated in any obvious way by convex functions, so that one cannot define $\operatorname{Det} D^2u$ using any kind of monotonicity properties of the gradient map. Also, Monge – Ampère functions need not belong to $W^{2,p}_{\mathrm{loc}}$ for any $p \geq 1$, and so are not regular enough, in general, that one can make sense of $\det D^2u$ as a distribution.

The defining attribute of a Monge – Ampère function $u\colon \Omega \subset \mathbb{R}^n \to \mathbb{R}$ is that one can associate with it an integral current that can be thought of as corresponding to the graph in the product space $\Omega \times \mathbb{R}^n$ of the gradient Du. The basic property of Monge – Ampère functions is that, when this current exists, it is unique. This was proved by Fu [4], and recently extended in [13]. Measures corresponding to the determinant and other minors of the Hessian are defined in terms of the action of

---

2000 *Mathematics Subject Classification.* Primary 49Q15; Secondary 53C24.

The author was partially supported by the National Science and Engineering Research Council of Canada under operating Grant 261955.

This is the final form of the paper.

the associated integral current on certain n-forms. The uniqueness property of this current mentioned above means that the measures defined in this fashion are in some sense canonical, and implies the good weak continuity properties mentioned above.

Here we try to present some properties of Monge–Ampère functions in a way that makes them reasonably accessible to readers without an intimate familiarity with geometric measure theory. We start by summarizing some basic facts. The definitions inevitably are stated in terms of integral currents, but some consequences, such as the weak continuity results mentioned above, can be understood without reference to this machinery. These results complement those found in recent papers of Iwaniec [12], Jerrard and Jung [14], and Fonseca and Maly [3], which discuss weak continuity properties of $u \mapsto \operatorname{Det} D^2 u$ in the different framework of distributional determinants. Incidentally, all these authors were apparently unaware of Fu's earlier work — this was certainly true of the authors of [14] at the time that it was written.

We go on to prove that piecewise linear functions are Monge–Ampère. This is known to experts and perhaps obvious to at least some nonexperts, but we think that it is useful to present an elementary proof that records some facts for which we do not know any good references, and that illustrates the abstract theory in a concrete setting.

We next discuss regularity properties of Monge–Ampère functions. Technical conditions in Fu's work forced him to restrict his attention to Monge–Ampère functions that are locally Lipschitz. The author's generalization [13] of Fu's uniqueness theorem removes this restriction and thereby raises the question of what regularity properties are naturally enjoyed by Monge–Ampère functions. Here we do not give any positive results, but we construct examples of Monge–Ampère functions that fail to be $C^{0,\alpha}$ for α sufficiently close to 1. As far as we know this result is new. The construction depends only on results about piecewise constant functions from the previous section and is entirely elementary. A related issue is to understand the regularity properties that follow from control over the L^1 norm of $\det D^2 u$ alone (or the mass of $\operatorname{Det} D^2 u$, when it is only a measure), that is, without assuming anything about other minors of the Hessian. This is well understood when u is convex, and we recall some of the these results, as we believe that the comparison is instructive.

Next we discuss some results that characterize $\operatorname{Det} D^2 u$ as a measure in terms of the topology of level sets of the functions $y \mapsto u(y) - \xi \cdot y$, for $\xi \in \mathbb{R}^n$. Rather general results in this direction were proved in a somewhat different context by Fu [6, 7], Zähle [20], Rataj and Zähle [19] among others. A nice feature of these results is that they clarify the sense in which the determinant measure for a Monge–Ampère function generalizes the classical construction for convex functions. We discuss these results, and we work out in detail a simple example. We have again tried to keep our argument as elementary as possible, though perhaps with less success than elsewhere in this paper. Nonetheless, the results of this section indicate ways in which concrete geometric information is encoded in the integral currents that appear in the definition of Monge–Ampère functions.

The final section collects some open problems.

The current associated with a Monge–Ampère function can be seen as a Cartesian current associated with the gradient $\mathrm{D}u$, and satisfying an additional Lagrangian condition — this is what is responsible for the uniqueness. Thus a lot of related material can be found in the work of Giaquinta, Modica and Souček [9, 10] on Cartesian currents.

1.1. Some preliminaries. We often implicitly sum over repeated indices.

We will write points in $\Omega \times \mathbb{R}^n$ in the form (x, ξ). Thus an n-form in $\Omega \times \mathbb{R}^n$ can be written as

$$\phi = \sum_{|\alpha|+|\beta|=n} \phi^{\alpha\beta}(x, \xi)\, \mathrm{d}x^\alpha \wedge \mathrm{d}\xi^\beta. \tag{1.1}$$

Here α, β are multiindices, so that for example $\alpha = (\alpha_1, \ldots, \alpha_j) \in \mathbb{Z}^j$ with $1 \leq \alpha_1 < \cdots < \alpha_j \leq n$. We write $|\alpha|$ to denote the length of the multiindex, so that $|\alpha| = j$ in the above example. Also, for example $\mathrm{d}x^\alpha = \mathrm{d}x_{\alpha_1} \wedge \cdots \wedge \mathrm{d}x_{\alpha_j}$.

Given a multiindex α of length j, we write $\bar{\alpha}$ to denote the complementary multiindex, of length $n - j$, such that $(\alpha, \bar{\alpha})$ is a permutation of $(1, \ldots, n)$, and we write $\sigma(\alpha, \bar{\alpha})$ to denote the sign of this permutation. One can check that if u is a smooth function $u\colon \Omega \to \mathbb{R}$ and $G_{\mathrm{D}u}$ denotes the graph of $\mathrm{D}u$, then for ϕ as in (1.1),

$$\int_{G_{\mathrm{D}u}} \phi = \sum_{|\alpha|+|\beta|=n} \int_\Omega \phi^{\alpha\beta}\big(x, \mathrm{D}u(x)\big)\sigma(\alpha, \bar{\alpha}) M^{\bar{\alpha}\beta}(\mathrm{D}^2 u)\, \mathrm{d}x \tag{1.2}$$

where, if $A = (a_{ij})_{i,j=1}^n$ is an $n \times n$ matrix,

$$M^{\bar{\alpha}\beta}(A) := \det A^{\bar{\alpha}\beta},$$

and $A^{\bar{\alpha}\beta}$ is the $|\bar{\alpha}| \times |\beta|$ (square) matrix whose i, j entry is $a_{\bar{\alpha}_i, \beta_j}$. We use the convention $M^{00}(A) = 1$. This formula is the starting point of our subject. It can be verified by writing say $W(x) = \big(x, \mathrm{D}u(x)\big)$, so that $G_{\mathrm{D}u}$ is just the image $W(\Omega)$. Then

$$\int_{G_{\mathrm{D}u}} \phi = \int_{W(\Omega)} \phi = \int_\Omega W^\# \phi,$$

and upon expanding the right-hand side, one arrives at (1.2); see [9, Section 2.2.1] for example for a detailed exposition.

Although it is not needed in much of what follows, we recall that an n-dimensional current in $\Omega \times \mathbb{R}^n$ is by definition a linear functional that acts on compactly supported n-forms in $\Omega \times \mathbb{R}^n$, i.e., on objects of the form (1.1). An example of such a current is the functional $\phi \mapsto \int_{G_{\mathrm{D}u}} \phi$, if u is a given sufficiently smooth function. We will only consider currents that have no boundary in $\Omega \times \mathbb{R}^n$. We say that such a current is locally integral it has locally finite mass and can be represented in terms of an integral over what Federer [2] terms a countably $(\mathcal{H}^n, n)$-rectifiable set, equipped with an integer-valued weight function. This is a kind of regularity condition, allowing us to think of such currents as generalized n-dimensional submanifolds.

In general we mostly follow notation from [2]. We reluctantly follow conventions of geometric measure theory and write $\|T\|$ to denote the total variation measure associated with a current T of locally finite mass. Similarly, we write $\|\mu\|$ for the total variation measure associated with a vector-valued Radon measure μ. If X

is a Banach space, we will write $\|\cdot\|_X$ to indicate the norm in X, always with a subscript to avoid any possible ambiguity.

1.2. Definition of Monge – Ampère functions. The definition of a Monge–Ampère function u is stated in terms of a locally integral n-current $[\mathrm{d}u]$ that is required to satisfy a number of conditions. Informally, these state that $[\mathrm{d}u]$ can be thought of as a generalized submanifold corresponding to the graph of the gradient of u, and that it has some weak regularity and finiteness properties. Before stating these conditions, we emphasize that a basic example of a locally Monge – Ampère function is a C^2 function u; then one can verify that $[\mathrm{d}u](\phi) := \int_{G_{\mathrm{D}u}} \phi$ has the required properties. Indeed, this remains true for $u \in W^{2,n}_{\mathrm{loc}}(\Omega)$.

The definition we give below is only used in its full generality in Section 4. We do not recall all the notation, but following the definition we attempt to explain a couple of points. Also, the meaning of conditions such as (1.3) can be extracted from the discussion in Section 4, see for example (4.15). In Section 2, we consider piecewise linear functions, and there the conditions of the definition take on a very concrete form, see (2.4), (2.5), (2.6).

If Ω is an open subset of $\mathbb{R}^n$, then $u \in W^{1,1}_{\mathrm{loc}}(\Omega)$ is said to be locally a Monge – Ampère function in Ω if there exists an n-dimensional locally integral current $[\mathrm{d}u]$ in $\Omega \times \mathbb{R}^n$ such that

$$\partial[\mathrm{d}u] = 0 \quad \text{in } \Omega \times \mathbb{R}^n \tag{1.3}$$

$$[\mathrm{d}u] \text{ is Lagrangian} \tag{1.4}$$

which one can think of as meaning "weakly curl-free"; see below for more details;

$$\mathbf{M}\big([\mathrm{d}u] \llcorner (K \times \mathbb{R}^n)\big) < \infty \quad \text{whenever } K \subset \Omega \text{ is compact.} \tag{1.5}$$

and

$$[\mathrm{d}u](\phi\, \mathrm{d}x^1 \wedge \cdots \wedge \mathrm{d}x^n) = \int \phi\big(x, \mathrm{D}u(x)\big)\, \mathrm{d}x \tag{1.6}$$

for every $\phi \in C^\infty_c(R^n \times \mathbb{R}^n)$.

We say that u is Monge – Ampère if $\mathbf{M}([\mathrm{d}u]) < \infty$. We write $u \in \mathrm{MA}_{\mathrm{loc}}(\Omega)$ if u is locally Monge – Ampère in Ω, and $u \in \mathrm{MA}(\Omega)$ if u is Monge – Ampère.

Note that (1.6) states that for a "purely horizontal" n-forms $\phi\, \mathrm{d}x^1 \wedge \cdots \wedge \mathrm{d}x^n$ (with no $\mathrm{d}\xi^j$ s), $[\mathrm{d}u](\phi)$ exactly agree with integration over the graph of $\mathrm{D}u$ as displayed in (1.2) (recalling our convention that $M^{00}(\mathrm{D}^2 u) := 1$).

Observe also that if $u \in W^{2,n}$ then, as remarked earlier, u is Monge – Ampère with $[\mathrm{d}u](\phi) := \int_{G_{\mathrm{D}u}} \phi$ as given in (1.2), and then the definition of $\mathbf{M}(\cdot)$ implies that

$$\begin{aligned} \mathbf{M}\big([\mathrm{d}u] \llcorner (K \times \mathbb{R}^n)\big) &= \int_K \Big(\sum_{|\alpha|+|\beta|=n} \big[M^{\bar\alpha\beta}(\mathrm{D}^2 u(x))\big]^2 \Big)^{1/2} \mathrm{d}x \\ &\approx \sum_{|\alpha|+|\beta|=n} \|M^{\bar\alpha\beta}(\mathrm{D}^2 u)\|_{L^1(K)}. \end{aligned} \tag{1.7}$$

Thus control over the mass of $[\mathrm{d}u]$ assumed in (1.5) provides control over the L^1 norms of all minors of $\mathrm{D}^2 u$. A suitable generalization of this fact remains true for all Monge – Ampère functions.

We now give the definition of the Lagrangian condition appearing in (1.4). In fact we give two equivalent definitions. Before going into the details, we remark that one can check from (1.11) that if u is smooth, then $\phi \mapsto \int G_{\mathrm{D}u}(\phi)$ is Lagrangian, and moreover, this holds exactly because $\mathrm{D}^2 u = (\mathrm{D}^2 u)^T$. This motivates the interpretation of (1.4) as weakly curl-free.

An integral n-current in $\mathbb{R}^n \times \mathbb{R}^n$ is Lagrangian if $\mathcal{H}^n$ almost every approximate tangent n-plane P satisfies

$$\langle \omega, \tau_1 \wedge \tau_2 \rangle = 0 \text{ for any two vectors } \tau_1, \tau_2 \text{ tangent to } P \tag{1.8}$$

for $\omega = \sum_i \mathrm{d}x^i \wedge \mathrm{d}\xi^i$. An n-plane P with this property is said to be Lagrangian. It is not hard to check that P is Lagrangian if and only if for every basis $\{\tau_i\}_{i=1}^n$ of the tangent space to P,

$$\langle \omega \wedge \eta, \tau_1 \wedge \cdots \wedge \tau_n \rangle = 0 \quad \text{for every } (n-2)\text{-covector } \eta. \tag{1.9}$$

This equivalence is based on the identity

$$\langle \omega \wedge \eta, \tau_1 \wedge \ldots, \wedge \tau_n \rangle = \sum_{\alpha \in I(2,n)} \operatorname{sgn}(\alpha, \bar{\alpha}) \langle \omega, \tau_{\alpha_1} \wedge \tau_{\alpha_2} \rangle \langle \eta, \tau_{\bar{\alpha}_1} \wedge \cdots \wedge \tau_{\bar{\alpha}_{n-2}} \rangle \tag{1.10}$$

From (1.9) one can check that an integral n-current T is Lagrangian if and only if

$$T(\omega \wedge \phi) = 0 \text{ for every smooth compactly supported } (n-2)\text{-form } \phi. \tag{1.11}$$

The advantage of this formulation is that it makes it clear that this property is preserved under weak convergence.

1.3. Basic properties of Monge–Ampère functions. The *raison d'être* of the space of Monge–Ampère functions is provided by the following

Theorem 1.1 ([4, 13]). *Suppose that $u \in \mathrm{MA}_{\mathrm{loc}}(\Omega)$. Then there is* exactly *one current in $\Omega \times \mathbb{R}^n$ satisfying* (1.3), (1.4), (1.5), (1.6).

Fu's original definition was actually slightly different from that given above: in place of (1.5) he assumed the stronger condition

$$(K \times \mathbb{R}^n) \cap \operatorname{supp}[\mathrm{d}u] \text{ is compact whenever } K \subset \Omega \text{ is compact.} \tag{1.12}$$

Thus in his version of Theorem 1.1, this hypothesis appeared in place of (1.5). As mentioned above, functions satisfying (1.12) are necessarily locally Lipschitz, so this condition is genuinely restrictive. The theorem in the form stated above is proved in [13]. Fu's terminology also differed somewhat — he defined Monge–Ampère functions on $\mathbb{R}^n$, and his definition corresponded to what we are here calling a local Monge–Ampère function.

For general $u \in \mathrm{MA}_{\mathrm{loc}}(\Omega)$, not necessarily smooth, we use the current $[\mathrm{d}u]$ to *define* all minors of $\mathrm{D}^2 u$ as measures in the product space $\Omega \times \mathbb{R}^n$. That is, we can define a measure $M^{\bar{\alpha}\beta}(\mathrm{D}^2 u)(\mathrm{d}x, \mathrm{d}\xi)$ by requiring that

$$\int_{\Omega \times \mathbb{R}^n} \phi(x,\xi) \sigma(\alpha, \bar{\alpha}) M^{\bar{\alpha}\beta}(\mathrm{D}^2 u)(\mathrm{d}x, \mathrm{d}\xi) := [\mathrm{d}u](\phi \, \mathrm{d}x^\alpha \wedge \mathrm{d}\xi^\beta) \tag{1.13}$$

for all $\phi \in C_c^\infty(\Omega \times \mathbb{R}^n)$. This definition is motivated by (1.2). Of particular interest is the special case

$$\int_{\Omega \times \mathbb{R}^n} \phi(x,\xi) \operatorname{Det} \mathrm{D}^2 u(\mathrm{d}x, \mathrm{d}\xi) := [\mathrm{d}u](\phi \, \mathrm{d}\xi^1 \wedge \cdots \wedge \mathrm{d}\xi^n). \tag{1.14}$$

The definitions imply that

$$\sum_{|\alpha|+|\beta|=n} \|M^{\bar{\alpha}\beta}(\mathrm{D}^2 u)\|(K \times \mathbb{R}^n) \approx \mathbf{M}([\mathrm{d}u] \llcorner (K \times \mathbb{R}^n))$$

so that, as in the smooth case (1.7), the mass of measures associated with minors of D^2u is controlled by the mass of $[du]$ (and conversely).

The following is a straightforward consequence of Theorem 1.1, see [4,9,10,13].

Corollary 1.1. *If u_k is a sequence of Monge–Ampère functions on a domain $\Omega \subset \mathbb{R}^n$ and if*

$$\mathbf{M}([du_k]) \le C \quad \text{and} \quad u_k \to u \text{ in } L^1_{\rm loc}$$

then $u \in \mathrm{MA}(\Omega)$ and $\mathbf{M}([du]) \le \limsup_k \mathbf{M}([du_k])$. Moreover, for every α, β such that $|\alpha| + |\beta| = n$,

$$M^{\bar\alpha\beta}(D^2u_k) \rightharpoonup M^{\bar\alpha\beta}(D^2u) \quad \text{weakly as measures.}$$

The corollary implies that if $\{u_k\}$ is a family of C^∞ functions such that $\|M^{\bar\alpha\beta}(D^2u_k)\|_{L^1} \le C$ for every α, β, then any $L^1_{\rm loc}$ limit of a convergent subsequence is Monge–Ampère. (Such a sequence is precompact in $W^{1,p}_{\rm loc}$ for $1 < p < n/n-1$, so we could equally well assume convergence in these spaces.)

2. Piecewise linear functions

We say that a function u is a piecewise linear function on a bounded set if the domain $\Omega \subset \mathbb{R}^n$ of u is a *finite* union $\Omega = \bigcup_{i=1}^N F_i^n$, where $\{F_i^n\}$ is a collection of closed bounded polygons with open, pairwise disjoint interiors; and if $u \in W^{1,\infty}(\Omega)$, with Du equal to a constant, say $Du = p_i$, on the interior of each F_i^n. In this section we prove that every such function is Monge–Ampère. This will be the basis for some examples we construct in later sections. As remarked earlier, this is certainly known to experts, but we hope that the proof gives some insight into the rather abstract general theory. In the piecewise linear setting, the geometric measure theory conditions of the definition of Monge–Ampère functions become much more concrete. In particular, in this section, all currents that appear are just finite linear combinations of (oriented) polygons.

We will write $\{F_i^k\}_{i=1}^{N(k)}$ to denote the set of k-faces of the polygons $\{F_i^n\}$, where a k-face is defined as a k-dimensional polygon that is the intersection of 2 or more $(k+1)$-faces, or equivalently of $n-k$ or more polygons F_i^n.

We fix an orientation for each F_i^k; we will below specify this further as convenient. We will not distinguish in our notation between the oriented polygon F_i^k and the integral current corresponding to integration over F_i^k.

We will say that a piecewise linear function is *generic* if each $(n-k)$-face is the intersection of exactly $k+1$ polygons. The main result of this section is

Lemma 2.1. *Suppose that u is a piecewise linear function on $\Omega \subset \mathbb{R}^n$. Then u is Monge–Ampère, and for $k = 0, \dots, n$ there exist polygonal k-currents $\{P_i^k\}_{i=1}^{N(n-k)}$ such that*

$$[du] = \sum_{k=0}^{n} \sum_{i=1}^{N(n-k)} F_i^{n-k} \times P_i^k. \tag{2.1}$$

If u is a generic piecewise linear function, then each P_i^k is a k-dimensional simplex with vertices $p_{\alpha_0}, \dots, p_{\alpha_k}$, where $\alpha_0, \dots, \alpha_k$ are such that F_i^{n-k} is a shared $(n-k)$-face of $F_{\alpha_0}^n, \dots F_{\alpha_k}^n$. In particular, if u is generic then

$$\mathbf{M}([du]) = \mathcal{L}^n(\Omega) + \sum_{k=1}^{n} \sum_{i=1}^{N(n-k)} \mathcal{H}^{n-k}(F_i^{n-k}) J_k(F_i^{n-k}) \tag{2.2}$$

where $J_k(F_i^{n-k})$ *denotes the* k*-dimensional jump of* $\mathrm{D}u$ *along* F_i^{n-k}*, defined by*

$$J_k(F_i^{n-k}) = \mathcal{H}^k\big(\mathrm{co}(p_{\alpha_0}, \dots p_{\alpha_k})\big), \quad F_i^{n-k} \text{ is a } (n-k)\text{-face of } F^n_{\alpha_0}, \dots, F^n_{\alpha_k},$$

and $\mathrm{co}(\cdots)$ *denotes the convex hull.*

If u is not generic, it follows from (2.1) that

$$\mathbf{M}([\mathrm{d}u]) = \sum_{k=0}^{n} \sum_{i=1}^{N(n-k)} \mathcal{H}^{n-k}(F_i^{n-k})\mathbf{M}(P_i^k)$$

but we do not know any easy formula for $\mathbf{M}(P_i^k)$ except in the generic case, where $\mathbf{M}(P_i^k)$ is exactly what we have called $J_k(F_i^{n-k})$.

We point out that by inspection of the proof below one can see the way in which, in this simple context at least, the Lagrangian condition enforces the uniqueness of $[\mathrm{d}u]$. The same mechanism is the main point in the more difficult proof of the basic uniqueness result, Theorem 1.1.

PROOF. For every $k \geq 1$ and every i, the boundary of F_i^k is a union of $k-1$ faces, so there are constants $\sigma_{i,j}^k \in \{0, \pm 1\}$ such that

$$\partial F_i^k = \sum_{j=1}^{N(k-1)} \sigma_{i,j}^k F_j^{k-1}. \tag{2.3}$$

Note that $\sigma_{i,j}^k \neq 0$ if and only if F_j^{k-1} is a $(k-1)$-face of F_i^k.

To prove the lemma, we will show that there exists a current of the form (2.1) with all the properties required of $[\mathrm{d}u]$. We first translate these properties into conditions on the currents $\{P_i^k\}$. By using (2.3), one checks that for a current T of the form (2.1),

$$\partial T = \sum_{k=0}^{n} \sum_{j=1}^{N(n-k)} F_j^{n-k} \times \left((-1)^{n-k} \partial P_j^k + \sum_{i=1}^{N(n-k+1)} \sigma_{i,j}^{n-k+1} P_i^{k-1} \right)$$

Thus $\partial T = 0$ if and only if

$$\partial P_j^k = (-1)^{n-k+1} \sum_{i=1}^{N(n-k+1)} \sigma_{i,j}^{n-k+1} P_i^{k-1} \tag{2.4}$$

for all k, j. One can further verify from the definition (1.8) that a current of the form (2.1) is Lagrangian if and only if

$$\mathrm{T}\, F_i^{n-k} = (\mathrm{T}\, P_i^k)^{\perp} \quad \text{for all } i, k \text{ such that } P_i^k \neq 0 \tag{2.5}$$

where $\mathrm{T}\, F_i^{n-k}, \mathrm{T}\, P_i^k$ denote the tangent spaces to F_i^{n-k}, P_i^k respectively. (Strictly speaking these tangent spaces belong to different copies of $\mathbb{R}^n$; for simplicity, throughout the proof we identify these in the canonical way.) Finally, it follows from the definitions that a current T of the form (2.1) satisfies (1.6) if and only if

$$P_i^0 = [p_i], \quad i = 1, \dots, N(n), \tag{2.6}$$

where we recall that p_i is the constant value of $\mathrm{D}u$ on F_i^n. (Here we are implicitly taking the polygons F_i^n to be oriented in the standard fashion.)

So we must show that we can select polygons $\{P_i^k\}$ with finite mass that satisfy (2.4), (2.5), and (2.6). Our proof will also show that

$$P_i^k \subset \text{the plane generated by } \{p_{\alpha_j} : F_i^{n-k} \text{ is a face of } F_{\alpha_j}^n\} \tag{2.7}$$

for every $k \in \{0, \dots, n\}$ and $i \in \{0, \dots, N(n-k)\}$. Indeed, we will see that this is more or less equivalent in this setting to the Lagrangian condition (2.5). Here the plane generated by a finite set $\{v_0, \dots, v_N\}$ is just the set $\{v_0 + \sum_{i=1}^N a_i(v_i - v_0) : a_i \in \mathbb{R}\}$. From (2.6) we note that (2.7) is trivially true when $k = 0$.

Since $\{P_i^0\}_{i=0}^{N(n)}$ are already determined by (2.6), we can use (2.4) and (2.5) to determine $\{P_i^1\}_{i=1}^{N(n-1)}$. To do this, fix $j \in \{1, \dots, N(n-1)\}$, and note that the $(n-1)$-face F_j^{n-1} is the intersection of exactly two simplices, say $F_{i_1}^n, F_{i_2}^n$. That is, $\sigma_{i,j}^n \neq 0$ if and only if $i = i_1$ or i_2. Moreover, the set $F_{i_1}^n \cap F_{i_2}^n$ inherits opposite orientations from $F_{i_1}^n$ and $F_{i_2}^n$; exactly one of these must agree with the orientation of F_j^{n-1}. Thus (2.4) reduces (in view of (2.6)) to

$$\partial P_j^1 = \pm([p_{i_1}] - [p_{i_2}]).$$

Thus (2.4) is satisfied if we specify that $P_j^1 = \pm[p_{i_1}, p_{i_2}]$. Moreover, because u is continuous across F_j^{n-1}, it follows that $(p_{i_1} - p_{i_2}) \cdot \tau = 0$ for every vector τ tangent to F_j^{n-1}, which shows that our choice of P_j^1 satisfies (2.5). Note also that (2.7) holds.

Now for $k \geq 2$ we assume by induction that we have found $\{P_i^l\}_{i=1}^{N(n-l)}$ with the required properties, for $l = 0, \dots, k-1$, and we show that the system of equations (2.4) with constraints (2.5) can be solved, and that the solutions satisfy (2.7). We first check that currents P_j^k solving (2.4) exist. To do this we need to check that the right-hand side of (2.4) is a boundary. Using the induction hypothesis,

$$\partial\Big(\sum_{i=1}^{N(n-k+1)} \sigma_{i,j}^{n-k+1} P_i^{k-1}\Big) = \sum_{i_1=1}^{N(n-k+1)} \sum_{i_2=1}^{N(n-k+2)} \sigma_{i_1,j}^{n-k+1} \sigma_{i_2,i_1}^{n-k+2} P_{i_2}^{k-2}. \tag{2.8}$$

However,

$$0 = \partial^2 F_i^k = \partial \sum_{j_1} \sigma_{i,j_1}^k F_{j_1}^{k-1} = \sum_{j_1,j_2} \sigma_{i,j_1}^k \sigma_{j_1,j_2}^{k-1} F_{j_2}^{k-1}$$

Thus

$$\sum_{j_1} \sigma_{i,j_1}^k \sigma_{j_1,j_2}^{k-1} = 0$$

for every i, j_2, which implies that the right-hand side of (2.8) vanishes, as desired.

It follows that P_j^k solving (2.4) exist for every $j \in \{1, \dots, N(n-k)\}$. We now show that we can find such a current satisfying in addition the Lagrangian condition. To see this, observe that by induction, (2.7) holds for every P_i^{k-1}, so that the right-hand side of (2.4) is supported in the

$$\text{plane generated by } \{p_{\alpha_l} : F_j^{n-k} \text{ is a face of } F_{\alpha_l}^n\}.$$

Therefore we can find P_j^k solving (2.4) and supported in the same plane. We may assume that P_j^k is nonzero, as otherwise (2.5) is vacuous. Then clearly $\mathrm{T}\, P_j^k$ is contained in the tangent space to this plane, which is

$$\operatorname{span}\{p_{\alpha_l} - p_{\alpha_l'} : F_j^{n-k} \text{ is a face of both } F_{\alpha_l}^n \text{ and } F_{\alpha_l'}^n\}.$$

And since u is continuous, if τ is any vector tangent to F_j^{n-k}, then $(p_{\alpha_l} - p_{\alpha_l'})\cdot\tau = 0$ when F_j^{n-k} is a face of both $F_{\alpha_l}^n$ and $F_{\alpha_l'}^n$, exactly as in the $k=1$ case. Thus $\mathrm{T}\,P_j^k \subset (\mathrm{T}\,F_j^{n-k})^\perp$. The opposite inclusion follows from noting that $\mathrm{T}\,P_j^k$ is k-dimensional, since P_i^k is nonzero, and $\mathrm{T}\,F_j^{n-k}$ is $n-k$-dimensional.

In the generic case we can write down an explicit formula for $[du]$. This necessitates a bit more notation. Suppose that $F_{i_0}^n, \dots, F_{i_k}^n$ are n-faces that share a common $(n-k)$-face. This common face will be denoted $F_{i_0\dots i_k}^{n-k}$, the orientation of which is taken to depend on the order in which the indices $i_0 \dots i_k$ are listed, in such a way that

$$\partial F_{i_0\dots i_k}^{n-k} = (-1)^n \sum_{i_{k+1}} F_{i_0\dots i_k i_{k+1}}^{n-k-1}. \tag{2.9}$$

where we use the convention that $F_{i_0\dots i_j}^{n-j} = 0$ if $F_{i_0}^n, \dots, F_{i_j}^n$ do not share a common $(n-j)$-face. One can check that if σ is any permutation of $k+1$ elements, then

$$F_{i_0\dots i_k}^{n-k} = \operatorname{sign}(\sigma) F_{\sigma(i_0)\dots\sigma(i_k)}^{n-k}. \tag{2.10}$$

Next, for $p_0, \dots, p_k \in \mathbb{R}^n$ with $0 \le k \le n$, we write $[p_0, \dots, p_k]$ to denote the k-dimensional simplex with vertices at the given points. We stipulate that such a simplex is oriented as as described in Federer [2, 4.1.11] which implies in particular that a 0-simplex $[p]$ is taken with the standard orientation, and that

$$\partial[p_0, \dots, p_k] = \sum_{i=0}^{k} (-1)^i [p_0, \dots, p_{i-1}, p_{i+1}, \dots, p_k]. \tag{2.11}$$

We claim that if u is a generic piecewise linear function, then

$$[du] = \sum_{k=0}^{n} \sum_{i_0 < \dots < i_k} F_{i_0\dots i_k}^{n-k} \times [p_{i_0}, \dots, p_{i_k}]. \tag{2.12}$$

Our earlier arguments show that (1.6) and (1.4), which as above reduce to (2.6) and (2.5), are satisfied. So we only need to check that the current on the right-hand side above has vanishing boundary. To do this, apply the operator ∂ to the right-hand side of (2.12), and then collect all terms containing some fixed $F_{i_0\dots i_{k+1}}^{n-k-1}$. Then the terms of the form $\partial F_{\dots}^{n-k} \times [\cdots]$ and those of the form $(-1)^{n-k} F_{\dots}^{n-k} \times \partial[\cdots]$ can be see from (2.9), (2.10), and (2.11) to cancel.

The remaining claim (2.2) follows from the explicit formula (2.12), since it is clear that the terms on the right-hand side of (2.12) are pairwise disjoint, and that

$$\mathbf{M}(F_{i_0\dots i_k}^{n-k} \times [p_{i_0}, \dots, p_{i_k}]) = \mathcal{H}^{n-k}(F_{i_0\dots i_k}^{n-k}) \mathcal{H}^k([p_{i_0}, \dots, p_{i_k}]),$$

and that $\mathcal{H}^k([p_{i_0}, \dots, p_{i_k}]) = J_k(F_{i_0\dots i_k}^{n-k})$ as defined in the statement of the theorem. □

3. Regularity

Regularity properties of Monge–Ampère functions are not well-understood. It is rather easy to see that if u is a Monge–Ampère function, then the second derivatives of u are measures[1] and it is also easy to construct examples of Monge–Ampère functions — for example, the piecewise affine functions of the last section — that do not belong to $W^{2,p}$ for any $p \geq 1$. Beyond these statements, not much is known.

In particular, when $n \geq 3$ it is not known whether Monge–Ampère functions on $\mathbb{R}^n$ are necessarily continuous, although this seems overwhelmingly likely For $n \leq 2$, continuity follows from the fact that the second derivatives are measures.

In this section we show that Monge–Ampère functions can fail to be $C^{0,\alpha}$ when $\alpha > 2/(n+1)$. We also give a lemma, which is no doubt classical, showing that if u is a function that is convex on its support, then

$$[u]_{C^{0,1/n}} \leq C(\operatorname{diam}\operatorname{supp} u)\|\det \mathrm{D}^2 u\|_{L^1}^{1/n},$$

and we show that no estimate of this sort can hold for $\alpha > 1/n$.

Lemma 3.1. *If Ω is a bounded, open subset of $\mathbb{R}^n$, $n \geq 2$, then there exists $u \in \mathrm{MA}(\Omega)$ with compact support in Ω and such that*

$$u \notin C^{0,\alpha}(\Omega) \quad \textit{for any } \alpha > \frac{2}{n+1}.$$

PROOF. For concreteness we take $\Omega := \{x \in \mathbb{R}^n : |x| < 2\}$.

(1) We first show that for $\varepsilon \in (0,1]$ and $q > 0$, there exists a function $u_{\varepsilon,q} \in \mathrm{MA}(\Omega)$ such that

$$\mathbf{M}([\mathrm{d}u_{\varepsilon,q}]) - \mathcal{L}^n(\Omega) \leq C\varepsilon^{n-1-q}, \quad [u_{\varepsilon,q}]_{C^{0,\alpha}} \geq c\varepsilon^{2-\gamma(2+q)} \tag{3.1}$$

where the constants C, c are independent of ε, q. Indeed, we define

$$u_{\varepsilon,q}(x) = (\min\{\varepsilon x_1, \ldots, \varepsilon x_n, \varepsilon^{-q}(\varepsilon - \sum x_i)\})^+$$

where $a^+ = \max\{a, 0\}$. Note that $u_{\varepsilon,q}$ is supported in

$$S_\varepsilon : \left\{(x_1, \ldots, x_n) : x_i \geq 0 \ \forall i, \ \sum x_i \leq \varepsilon\right\}, \tag{3.2}$$

and in particular has compact support in Ω.

We first estimate $\mathbf{M}([\mathrm{d}u_{\varepsilon,q}])$ using (2.2) from the previous section. Note that $u_{\varepsilon,q}$ is what we have called a generic piecewise linear function, so this result is applicable. Let us write $p_0 := -\varepsilon^{-q}(1, \ldots, 1)$, $p_i = \varepsilon e_i$ for $i = 1, \ldots, n$, and $p_{n+1} = 0$, where e_i denotes the standard unit vector. Then the values assumed by $\mathrm{D}u_{\varepsilon,q}$ are $\{p_0, \ldots, p_n, p_{n+1}\}$. If $k \geq 1$, then any simplex $[p_{\alpha_0}, \ldots, p_{\alpha_k}]$ has its k-dimensional volume bounded by $\prod_{j=1}^k |p_{\alpha_j} - p_{\alpha_0}|$. By permuting the p_{α_i}s, we can assume that $\alpha_0 \neq 0$, which implies that $|p_{\alpha_j} - p_{\alpha_0}| \leq C\varepsilon$ except for at most 1 choice of j, and $|p_{\alpha_j} - p_{\alpha_0}| \leq C\varepsilon^{-q}$ for all j. Thus

$$\mathcal{H}^k([p_{\alpha_0}, \ldots, p_{\alpha_k}]) \leq C\varepsilon^{k-1-q}.$$

---

[1]Note that in (1.13) the we have already defined the 1×1 minors $M^{ij}(\mathrm{D}^2 u)$ as measures on $\Omega \times \mathbb{R}^n$. The point here is that if we let $u_{x_i x_j}$ denote the marginal on Ω of $M^{ij}(\mathrm{D}^2 u)$, i.e., $u_{x_i x_j}(A) := M^{ij}(\mathrm{D}^2 u)(A \times \mathbb{R}^n)$ whenever the latter is well-defined, then one can verify that $(u_{x_i x_j})_{j=1}^n$ are the distributional derivatives of u_{x_i}, using the condition $\partial[\mathrm{d}u] = 0$. For a full proof see Fu [4].

Also, if F_i^{n-k} is an $n-k$ face of the triangulation associated with $u_{\varepsilon,q}$, then F^{n-k} is contained in a ball of size $C\varepsilon$, and so $\mathcal{H}^{n-k}(F_i^{n-k}) \leq C\varepsilon^{n-k}$. From these considerations and (2.2) we conclude that

$$\mathbf{M}([\mathrm{d}u_{\varepsilon,q}]) \leq C\varepsilon^{n-1-q} + \mathcal{L}^n(\Omega).$$

Let $x_{\varepsilon,q}$ be the point at which $\varepsilon x_1 = \cdots = \varepsilon x_n = \varepsilon^{-q}(\varepsilon - \sum x_i)$; this is the unique point at which $u_{\varepsilon,q}$ attains its maximum. Also, let $y_{\varepsilon,q} = (\varepsilon/n)(1,\ldots,1)$ One can check that

$$x_{\varepsilon,q} = \frac{\varepsilon}{n+\varepsilon^{1+q}}(1,\ldots,1) = y_{\varepsilon,q} - \frac{\varepsilon^{q+2}}{n^2}(1,\ldots,1) + O(\varepsilon^{2q+3})$$

and that

$$u_{\varepsilon,q}(x_{\varepsilon,q}) = \frac{\varepsilon^2}{n+\varepsilon^{1+q}} = \frac{\varepsilon^2}{n} + O(\varepsilon^{q+3}).$$

Thus for any $\gamma \in (0,1]$, since $u_{\varepsilon,q}(y_{\varepsilon,q}) = 0$,

$$[u_{\varepsilon,q}]_{C^{0,\gamma}} \geq \frac{|u_{\varepsilon,q}(x_{\varepsilon,q}) - u_{\varepsilon,q}(y_{\varepsilon,q})|}{|x_{\varepsilon,q} - y_{\varepsilon,q}|^\gamma} \approx \varepsilon^{2-\gamma(2+q)}.$$

Thus we have proved (3.1).

(2) Now for $j \geq 1$, let $u_j(x) = u_{\varepsilon_j,q_j}(x - 2^{-j}e_1)$ where $\varepsilon_j = 2^{-j^2}$ and $q_j = n-1-1/j$. Here e_1 is the standard unit vector $(1,0,\ldots,0)$.

Then define $U_k(x) = \sum_{j=1}^k u_j(x)$, and finally let $u = \lim_{k\to\infty} U_k$. (It is easy to see that $U_k(x)$ converges in L^1 for example as $k \to \infty$.)

Observe from (3.2) that the supports of $u_j, u_{j'}$ are separated by a positive distance if $j \neq j'$, so that U_k is a generic piecewise linear function for every k. This disjoint support property and (2.2) further imply that

$$\mathbf{M}([\mathrm{d}U_k]) - \mathcal{L}^n(\Omega) = \sum_{j=1}^k \left[\mathbf{M}([\mathrm{d}u_j] - \mathcal{L}^n(\Omega))\right]$$

It then follows from (3.1) that

$$\mathbf{M}([\mathrm{d}U_k]) \leq \mathcal{L}^n(\Omega) + C\sum_{j=1}^k (2^{-j^2})^{1/j} \leq C$$

where the constant is independent of k.

In particular, since $U_k \to u$ in L^1 and $\mathbf{M}([dU_k]) \leq C$, it follows from Corollary 1.1 that u is a Monge–Ampère function.

Now define $x_j = x_{\varepsilon_j,q_j} + 2^{-j}e_1$, $y_j = y_{\varepsilon_j,q_j} + 2^{-j}e_1$, (using notation from step (1) above) so that

$$|x_j - y_j| \leq C\varepsilon_j^{q_j+2}, \quad |u_j(x_j) - u_j(y_j)| \geq c\varepsilon_j^2.$$

Note also that $u(x_j) = u_j(x_j)$ and $u(y_j) = u_j(y_j)$. Then for any $\gamma > 2/(n+1)$,

$$[u]_{C^{0,\gamma}} \geq \frac{|u(x_j) - u(y_j)|}{|x_j - y_j|^\gamma} \geq c\varepsilon_j^{2-\gamma(q_j+2)} \to \infty$$

as $j \to \infty$. Thus $u \notin C^{0,\gamma}(\Omega)$ for $\gamma > 2/(n+1)$. □

The above lemma naturally raises the question of what sort of continuity properties follow from the condition $\mathbf{M}([du]) < \infty$. This is related to (but probably harder than) the question of what continuity properties are implied by L^1 bounds on $\det D^2u$. In the next lemma we recall that this easier question is more or less completely understood in the convex case. This is presumably classical, although we do not know a reference for it. In this lemma, unlike in the rest of this paper, $\operatorname{Det} D^2u$ is understood in the sense of Alexandrov as a measure on Ω, so that for a Borel set $A \subset \Omega$,

$$\operatorname{Det} D^2u(A) = \mathcal{L}^n(\{p : p \in \partial u(x) \text{ for some } x \in A\}),$$

where $\partial u(x)$ denotes the subgradient of u at x. In particular, if $u \in W^{2,n}(\Omega)$ then $\operatorname{Det} D^2u(A) = \int_A \det D^2u(x)\, dx$.

Lemma 3.2. *Suppose that $\Omega \subset \mathbb{R}^n$ is a bounded convex open set and that $u\colon \bar\Omega \to \mathbb{R}$ is a continuous convex function with $u = 0$ on $\partial\Omega$. If in addition* $\operatorname{Det} D^2u(\Omega) < \infty$, *then $u \in C^{0,1/n}(\bar\Omega)$, and*

$$[u]_{C^{0,1/n}} \leq C[\operatorname{Det} D^2u(\Omega)]^{1/n} \tag{3.3}$$

where the constant C depends only on the diameter of Ω.

Conversely, there exists a convex $\Omega \subset \mathbb{R}^n$ and a function $u : \Omega \to \mathbb{R}$ which is convex in Ω and vanishing on $\partial\Omega$, such that $\operatorname{Det} D^2u(\Omega) < \infty$ *and*

$$u \notin C^{0,\gamma}(\bar\Omega) \quad \text{for any } \gamma > \frac{1}{n}. \tag{3.4}$$

The (perhaps limited) relevance of this lemma to our main concerns is as follows: first, if $u \in \mathrm{MA}_{\mathrm{loc}}(\Omega)$ and u vanishes on $\partial\Omega$, then (3.3) suggests that one might hope that $u \in C^{0,1/n}$ and that $[u]_{C^{0,1/n}} \leq C(\Omega)|\operatorname{Det} D^2u|(\Omega \times \mathbb{R}^n)$, where $\operatorname{Det} D^2u$ is now understood in the sense of Monge – Ampère functions, as a measure in $\Omega \times \mathbb{R}^n$ as in (1.14), and $|\operatorname{Det} D^2u|$ denotes the associated total variation measure.

Second, (3.4) shows that, if it is true that Monge – Ampère functions are $C^{0,\alpha}$ for some $\alpha > 1/n$, then any proof of this fact must exploit not only control over $|\operatorname{Det} D^2u|$, but must also use some other information encoded in $[du]$, such as bounds on second derivatives of u or on $k \times k$ minors of D^2u for some $k < n$.

We now give the

Proof of Lemma 3.2. (1) We first prove (3.3), using classical ideas of Alexandrov to argue that if $[u]_{C^{0,1/n}}$ is large then necessarily the image of the subgradient map has large measure.

We may assume after rescaling that the diameter of Ω is at most 1. We must estimate $|u(y)-u(z)|$ for $y, z \in \bar\Omega$. Initially, let us assume for simplicity that $z \in \partial\Omega$, so that $u(y) \leq u(z) = 0$. Let $z' \in \partial\Omega$ be a point such that $|y-z'| = \operatorname{dist}(y, \partial\Omega) =: d$. In particular $u(z') = u(z) = 0$ and $|y - z'| \leq |y - z|$. We may assume after a translation that $z' = 0$ and $y = (d, 0, \ldots, 0)$. Then, since Ω is convex and has diameter less than 1, and by the choice of z',

$$\Omega \subset \{x : 0 < x_1 < 1, |x_j| < 1 \text{ for } j = 2, \ldots, n\} =: R.$$

The point is that, since z' minimizes the distance from $\partial\Omega$ to y, the line joining y to z' must be normal to a supporting hyperplane at z'.

We write $a := u(y)$, and we define the linear functions

$$l_i^{\pm}(x) = a(1 \pm x_i) \quad i = 2, \ldots, n$$

and

$$l_1^+ = a\frac{x_1}{d}, \quad l_1^- = \frac{a(1-x_1)}{1-d}.$$

Then $l_i^+(y) = l_i^-(y) = a$ for all i. Also, since $a \le 0$, $l_i^+(x) \le 0$ and $l_i^-(x) \le 0$ for all x in the rectangle R, and hence in Ω, for all i.

Thus if l is any convex combination of the above linear functions, then $l(y) = a = u(y)$, and $l(x) \le 0 = u(x)$ on $\partial\Omega'$. It follows that $l - u$ attains its maximum at some point $x_0 \in \Omega$. In other words, $l(x) - u(x) \ge l(x_0) - u(x_0)$ for all $x \in \Omega$, which says exactly that $\mathrm{D}l \in \partial u(x)$.

If we let $p_i^\pm$ denote the (constant) gradient of $l_i^\pm$, it follows that every convex combination of $\{p_i^\pm\}$ belongs to $\partial u(x)$ for some $x \in \Omega$. It is easy to check that the convex hull $\mathrm{co}\{p_i^\pm\}$ has measure greater than $c|a|^n/d = c|u(y)|^n/d$. Thus

$$\operatorname{Det} \mathrm{D}^2 u(\Omega) \ge \mathcal{L}^n(\mathrm{co}\{p_i^\pm\}) = c\frac{|u(y)|^n}{|y-z'|} \ge c\frac{|u(y)-u(z)|^n}{|y-z|}$$

since $u(z) = 0$ and $|y - z| \ge |y - z'|$.

We finally claim that the same estimate holds if $z \notin \partial\Omega$. Indeed, if $u(y) \le u(z) < 0$, we define $\Omega' := \{x \in \Omega : u(x) < u(z)\}$, and for $x \in \bar{\Omega}'$ we define $u'(x) = u(x) - u(z)$. Then the above arguments apply to u' on Ω' and yield the estimate $\operatorname{Det} \mathrm{D}^2 u(\Omega) \ge \operatorname{Det} \mathrm{D}^2 u(\Omega') \ge c|u(y) - u(z)|^n/|y - z|$. Thus we can take the supremum over all $y, z \in \Omega$ to complete the proof of the (3.3).

(2) We now construct a convex function that is not $C^{0,\gamma}$ for any $\gamma > 1/n$, but such that $\operatorname{Det} \mathrm{D}^2 u$ is a finite measure. Let $f\colon [0,1] \to \mathbb{R}$ denote a strictly convex function such that $f(0) = f(1) = 0$, and to be further specified later. We will write points in $\mathbb{R}^n$ in the form $x = (x', x_n)$ with $x' \in \mathbb{R}^{n-1}, x_n \in \mathbb{R}$. Let $\Omega := \{(x', x_n) : 0 < x_n \text{ and } x_n + |x'| < 1\}$, and define

$$u(x', x_n) = (1 - |x'|) f\left(\frac{x_n}{1-|x'|}\right) \quad \text{for } (x', x_n) \in \Omega.$$

We first prove that u is convex. To do this, for $p \in \mathbb{R}^{n-1}$ and $s \in [0,1]$ define

$$u_{p,s}(x', x_n) = f(s) + f'(s)(x_n - s) - x' \cdot p[f(s) - sf'(s)].$$

We claim that

$$u(x) := \max_{|p| \le 1, s \in [0,1]} u_{p,s}(x) \quad \text{for } x \in \Omega. \tag{3.5}$$

This will imply in particular that u is convex. Note also that, since f is convex, $0 = f(0) \ge f(s) - sf'(s)$ for all $s \in [0,1]$, which implies that

$$\begin{aligned}
\max_{|p| \le 1, s \in [0,1]} u_{p,s}(x) &= \max_{s \in [0,1]} \left[f(s) + f'(s)(x_n - s) + \max_{|p| \le 1} (x' \cdot p)\big(sf'(s) - f(s)\big) \right] \\
&= \max_{s \in [0,1]} [f(s)(1 - |x'|) + f'(s)(x_n - s(1 - |x'|)] \\
&= (1 - |x'|) \max_{s \in [0,1]} \left[f(s) + f'(s)\left(\frac{x_n}{1 - |x'|} - s\right)\right].
\end{aligned}$$

Since f is convex, (3.5) follows.

Next we claim that

$$\{\partial u(x) : x \in \Omega\} = \{(q', q_n) : q_n = f'(s),\ |q'| \le g(s) \text{ for some } s \in (0,1)\}. \tag{3.6}$$

where we write $g(s) = sf'(s) - f(s)$. The right-hand side of (3.6) is exactly $\{\mathrm{D}u_{p,s} : |p| \le 1, s \in (0,1)\}$. For every $|p| \le 1$ and every s, $u_{p,s}(0, s) = f(s) = u(0, s)$.

Then (3.5) implies that $u_{p,s}$ is a supporting hyperplane at $(0, s)$, which implies the inclusion $\supset$ in (3.6).

To prove the other inclusion, suppose that $q \in \partial u(x)$ for some $x = (x', x_n)$. If $x' \neq 0$, then u is smooth and $\partial u(x) = \{\mathrm{D}u(x)\} = \{\mathrm{D}u_{p,s}\}$ for $p = x'/|x'|$ and $s = x_n/(1-|x'|)$, by a short calculation. And if $q = (q', q_n)$ belongs to $\partial u(x)$ for $x = (0, x_n)$, then

$$(3.7) \qquad (1-|y'|)f\left(\frac{y_n}{1-|y'|}\right) \geq f(x_n) + q' \cdot y' + q_n(y_n - x_n)$$

for all $y \in \Omega$. Taking y of the form $y = (0, y_n)$, we find that $q_n = f'(x_n)$. And considering y of the form $(y', (1-|y'|)x_n)$, we find after some calculations that $q' \cdot (y'/|y'|) \leq g(x_n)$. This completes the proof of the inclusion $\subset$ in (3.6).

We will eventually take f to be strictly convex and C^2 on $(0, 1]$, so that f' is invertible and $f'(1)$ makes sense. Then we deduce from (3.6) that

$$\{\partial u(x) : x \in \Omega\} = \big\{(q', q_n) : f'(0) < q_n < f'(1), |q'| \leq g\big(f'^{-1}(q_n)\big)\big\}.$$

(By $f'(0)$ we mean $\lim_{s \searrow 0} f'(s)$, which will turn out to be $-\infty$ for us.) Let us write h for f'^{-1}. From the above we deduce that

$$\operatorname{Det} \mathrm{D}^2 u(\Omega) = c_n \int_{f'(0)}^{f'(1)} \big(g \circ h(t)\big)^{n-1} \, \mathrm{d}t = c_n \int_0^1 g(s)^{n-1} f''(s) \, \mathrm{d}s$$

where we have made the change of variables $f'(s) = t$, so that $h(t) = s$. Observe also that $f''(s) = (1/s)g'(s)$, so

$$(3.8) \qquad \operatorname{Det} \mathrm{D}^2 u(\Omega) = c_n \int_0^1 g(s)^{n-1} g'(s) \frac{\mathrm{d}s}{s} = \frac{c_n}{n} \int_0^1 \frac{\mathrm{d}}{\mathrm{d}s}[g(s)^n] \frac{\mathrm{d}s}{s}.$$

Finally, we take f so be a strictly convex function, smooth in $(0, 1]$, such that

$$f(s) = -\frac{s^{1/n}}{|\log s|} \quad \text{for } 0 < s \leq \tfrac{1}{2}$$

and such that $f(1) = 0$ and f', f'' are bounded on $[\frac{1}{2}, 1]$. (One can verify by differentiating that $s^{1/n}(\log s)^{-1}$ is convex for $s \in (0, 1)$.)

Since $u(0, x_n) = f(x_n)$ it is clear that (3.4) holds. To finish the proof we must therefore check that $\operatorname{Det} \mathrm{D}^2 u(\Omega)$ is finite. Because f is smooth away from $s = 0$, we see from (3.8) that it suffices to check that

$$\int_0^{1/2} g(s)^{n-1} g'(s) \frac{1}{s} \, \mathrm{d}s < \infty.$$

To do this, we check by explicit computations that for $0 < s \leq \frac{1}{2}$, there exists C such that

$$0 < g(s) \leq -Cf(s), \quad 0 < g'(s) \leq -C\frac{f(s)}{s}.$$

Thus

$$\int_0^{1/2} g(s)^{n-1} g'(s) \frac{1}{s} \, \mathrm{d}s \leq C \int_0^{1/2} f(s)^n \frac{1}{s^2} \, \mathrm{d}s = C \int_0^{1/2} \frac{1}{s|\log s|^n} \, \mathrm{d}s < \infty$$

since $n \geq 2$. $\square$

Remark 1. One can construct a similar example that is smooth in Ω by taking u of the form

$$u(x', x_n) = \rho(x') f\left(\frac{x_n}{\rho(x')}\right) \quad \text{for } \rho(x') = 1 - |x'|^2 \text{ and } f \text{ as above.}$$

It is also not hard to modify u to construct an example in a smooth domain, since the corners in $\partial\Omega$ occur far from the point $(0,0)$ at which various Hölder seminorms blow up.

Remark 2. It is much easier to prove that no estimate of the form

$$[v]_{C^{0,\gamma}} \leq |\operatorname{Det} D^2 v|(\Omega) \quad v \in C_0(\bar{\Omega}) \cap C^\infty(\Omega), u \text{ convex in } \bar{\Omega} \tag{3.9}$$

can hold for any $\gamma > 1/n$, where the subscript in C_0 indicates functions that vanish on $\partial\Omega$. Indeed, this follows from an easy scaling considerations. To see this, let Ω be a set of diameter at most 1 and let $v \in C_0(\bar{\Omega}) \cap C^\infty(\Omega)$ be a convex function. Suppose for concreteness that $0 \in \Omega$ with $v(0) < 0$, and that $(a, 0, \ldots, 0) \in \partial\Omega$, so that $v(a, 0, \ldots, 0) = 0$. Necessarily $a < 1$.

Fix a parameter $q > 0$ and let $\Omega_\varepsilon = \{(\varepsilon x_1, x_2, \ldots, x_n\}) : x \in \Omega\}$. Define $v_{\varepsilon,q}(x_1, \ldots, x_n) = \varepsilon^q v(x_1/\varepsilon, x_2, \ldots, x_n)$. Then it is easy to check that

$$\det D^2 v_{\varepsilon,q}(x_1, \ldots, x_n) = \varepsilon^{nq-2} \det D^2 v(x_1/\varepsilon, x_1, \ldots, x_n),$$

and hence that

$$\int |\det D^2 v_{\varepsilon,p}| = \varepsilon^{nq-1} \int |\det D^2 v|.$$

Note also that

$$[v_{\varepsilon,q}]_{C^{0,\gamma}} \geq \frac{|v_{\varepsilon,q}(0) - v_{\varepsilon,q}(a\varepsilon, 0, \ldots, 0)|}{(a\varepsilon)^\gamma} = \varepsilon^{q-\gamma} \frac{|v(0)|}{a^\gamma}.$$

This easily implies that (3.9) fails when $\gamma > 1/n$.

4. Local representations of $\operatorname{Det} D^2 u$

In this section, we write $\operatorname{Det} D^2 u$ for $u \in MA_{\mathrm{loc}}(\Omega)$ to denote the measure in $\Omega \times \mathbb{R}^n$ as defined in (1.14).

It follows from general geometric measure theory considerations that, given a Monge–Ampère function u, there exists a function $d \colon \Omega \times \mathbb{R}^n \to \mathbb{Z}$, defined for *every* $x \in \Omega$ and $\mathcal{L}^n$ a.e. ξ, such that

$$\int \phi(x,\xi) \operatorname{Det} D^2 u(\mathrm{d}x, \mathrm{d}\xi) = \int_{\mathbb{R}^n} \left(\int_\Omega \phi(x,\xi) d(x,\xi) \, \mathrm{d}\mathcal{H}^0(x) \right) \mathcal{L}^n(\mathrm{d}\xi). \tag{4.1}$$

In general, $d(x,\xi)$ should reflect the local behavior of the level sets of

$$y \mapsto u(y) - \xi \cdot y =: u_\xi(y)$$

near $y = x$. Theorems of this sort have been established in the different but related context of normal cycles associated with certain subsets of $\mathbb{R}^n$, see for example [1, 6, 7, 19, 20]; this list is far from exhaustive. In these theorems, roughly speaking, $\operatorname{Det} D^2 u$ in (4.1) is replaced by a measure in $\mathbb{R}^n \times S^{n-1}$ associated with the Gaussian curvature and coming from an integral current in $\mathbb{R}^n \times S^{n-1}$ that roughly speaking corresponds to the (generalized) graph of the Gauss map of a possibly rather rough subset of $\mathbb{R}^n$. These results strongly suggest that for quite general Monge–Ampère functions, $d(x,\xi)$ in (4.1) should admit the characterization

$$(4.2)\quad d(x,\xi) = \lim_{r\searrow 0}\lim_{\varepsilon\searrow 0}[\chi(\{y : |y-x|<r, u_{x,\xi}(y)<\varepsilon\}) - \chi(\{y : |y-x|<r, u_{x,\xi}(y)<-\varepsilon\})]$$

where χ denotes the Euler characteristic and $u_{x,\xi}(y) = u(y) - [u(x)+\xi\cdot(y-x)]$. Note that when (4.2) holds, $d(x,\xi)$ is nonzero exactly when the local topology of level sets of $y\mapsto u(y)-\xi\cdot y$ changes at $y=x$.

The proofs of results about normal cycles cited above can almost surely be transposed to yield corresponding results about classes of Monge–Ampère functions. For example, [6, 19, 20] establish analogs of (4.1), (4.2) when the set underlying the normal cycle is a union of sets of positive reach. A set of positive reach is roughly analogous to a semiconvex function, and a union of such sets is by the same token analogous to a function of the form

$$(4.3)\qquad u(x) = \inf\{u_i(x)\}_{i=1}^M \quad \text{for } u_1,\dots,u_M \text{ semiconvex functions.}$$

(Fu [4] proves that functions u of this form are locally Monge–Ampère.) Presumably (4.2) can be established for such functions following arguments[2] developed in [6, 19] or other references.

In any case, our modest goal in this section is simply to illustrate (4.1), (4.2) by giving a reasonably elementary proof in a simple but nontrivial situation, that of homogeneous functions on $\mathbb{R}^2$. We will prove

Proposition 4.1. *Suppose that $u\in \mathrm{MA}_{\mathrm{loc}}(\mathbb{R}^2)$ is homogeneous of degree 1 Then (4.1) holds with*

$$(4.4)\quad d(x,\xi) = \begin{cases} 1-\frac{1}{2}\,\#[\textit{sign changes of } (u(x)-\xi\cdot x)|_{S^1}]\ \mathcal{L}^2 \textit{ a.e. } \xi & \textit{if } x=0,\\ 0 & \textit{if not.}\end{cases}$$

Note that (4.4) is consistent with (4.2).

We will write 2-forms in $\mathbb{R}^2\times\mathbb{R}^2$ as

$$(4.5)\qquad \psi = \psi_0\,\mathrm{d}x^1\wedge\mathrm{d}x^2 + \psi_1^{ij}\,\mathrm{d}x^i\wedge\mathrm{d}\xi^j + \psi_2\,\mathrm{d}\xi^1\wedge\mathrm{d}\xi^2$$

The main part of the proof of Proposition 4.1 is contained in the following

Lemma 4.1. *Suppose that u is homogeneous of degree 1 and smooth away from the origin. Then u is Monge–Ampère. Moreover, for ψ of the form (4.5),*

$$(4.6)\quad [\mathrm{d}u](\psi) = \int_{\mathbb{R}^2}[\psi_0(x,\mathrm{D}u(x))\,\mathrm{d}x^1\wedge\mathrm{d}x^2 + \psi_1^{ij}(x,\mathrm{D}u)\,\mathrm{d}x^i\wedge\mathrm{d}(u_{x_j})] + \int_{\mathbb{R}^2}\psi_2(0,\xi)\tilde{d}_0(\xi)\,\mathcal{L}^2(\mathrm{d}\xi)$$

where

$$(4.7)\qquad \tilde{d}_0(\xi) := \frac{1}{2\pi}\int_0^{2\pi}\frac{\gamma-\xi}{|\gamma-\xi|}\wedge\frac{\mathrm{d}}{\mathrm{d}\theta}\left(\frac{\gamma-\xi}{|\gamma-\xi|}\right)\mathrm{d}\theta,\quad \gamma(\theta) := \mathrm{D}u(r e^{\mathrm{i}\theta})$$

[2] The basic idea in many of these proofs is to establish the result first for $C^{1,1}$ sets; then to use an approximation argument to deduce that it still holds for some rougher class of sets, such as sets of positive reach; and finally to use additivity properties of the Euler characteristic to extend the result to a larger class of sets such as those generated by suitable unions of sets of positive reach. Some aspects of this line of argument are easier in the case of Monge–Ampère functions than in the geometric setting.

for $\xi \in \mathbb{R}^2 \setminus \mathrm{Image}(\gamma)$. *Finally, if* $K \subset \mathbb{R}^2$ *is any compact set then*

$$\big\|[\mathrm{d}u]\big\|(K \times \mathbb{R}^2) \le C \quad \text{for } C \text{ depending only on } K \text{ and } \|\gamma\|_{W^{1,1}(\mathbb{T})}. \tag{4.8}$$

Remark 3. Note that $\tilde{d}_0(\xi)$ is just the winding number of the curve γ about the point ξ.

Remark 4. Observe also that

$$\int_{\mathbb{R}^2} \tilde{d}_0(\xi)\, \mathcal{L}^2(\mathrm{d}\xi) = \frac{1}{2}\int_0^{2\pi} \gamma(\theta) \wedge \frac{\mathrm{d}\gamma}{\mathrm{d}\theta}\,\mathrm{d}\theta =: A. \tag{4.9}$$

This number is just the area (with sign and multiplicity) enclosed by the curve γ. This is proved by a well-known calculation that we recall below. It is a classical fact (see [9, 10] for example) that for a homogeneous function as above, the distributional determinant of $\mathrm{D}^2 u$ (which is a distribution on $\mathbb{R}^2$ rather than a measure in the product space) is exactly $A\delta_0$, which by (4) is just the marginal on the horizontal $\mathbb{R}^2$ of the measure $\mathrm{Det}\,\mathrm{D}^2 u$ in the product space. To prove (4), fix R such that the image of γ is contained in B_R, so that $d_0(\xi) = 0$ for $|\xi| > R$. Then

$$\begin{aligned}\int_{\mathbb{R}^2} \tilde{d}_0(\xi)\, \mathcal{L}^2(\mathrm{d}\xi) &= \frac{1}{2\pi}\int_{B_R}\int_0^{2\pi} \frac{\gamma-\xi}{|\gamma-\xi|} \wedge \frac{\mathrm{d}}{\mathrm{d}\theta}\left(\frac{\gamma-\xi}{|\gamma-\xi|}\right)\mathrm{d}\theta\, \mathcal{L}^2(\mathrm{d}\xi) \\ &= \frac{1}{2\pi}\int_0^{2\pi}\left(\int_{B_R} \frac{\gamma-\xi}{|\gamma-\xi|^2}\,\mathcal{L}^2(\mathrm{d}\xi)\right) \wedge \frac{\mathrm{d}}{\mathrm{d}\theta}\Big(\gamma-\xi\Big)\,\mathrm{d}\theta\end{aligned}$$

However, it is a classical fact, due to Newton, that $\int_{B_R} (\gamma-\xi)/|\gamma-\xi|^2\,\mathrm{d}\xi = \gamma$ when $R > |\gamma|$, so (4) follows.

We first assume Lemma 4.1 and use it to give the

PROOF OF PROPOSITION 4.1. Since u is homogeneous, we can write $u(re^{i\theta}) = f(\theta)$ for some $f\colon \mathbb{T} \to \mathbb{R}$, where $\mathbb{T} = \mathbb{R}/(2\pi\mathbb{Z})$.

We first prove the theorem under the assumption that f is smooth. Then it follows directly from Lemma 4.1 that for $\phi \in C_c^\infty(\Omega \times \mathbb{R}^2)$,

$$\int \phi(x,\xi)\,\mathrm{Det}\,\mathrm{D}^2 u(\mathrm{d}x, \mathrm{d}\xi) = [\mathrm{d}u](\phi\, \mathrm{d}\xi^1 \wedge \mathrm{d}\xi^2) = \int_{\mathbb{R}^2} \phi(0,\xi)\tilde{d}_0(\xi)\,\mathrm{d}\xi. \tag{4.10}$$

So to prove the proposition for smooth f, it suffices to show that

$$\tilde{d}_0(\xi) = 1 - \tfrac{1}{2}\,\#[\text{sign changes of } (u(x) - \xi\cdot x)|_{S^1}] \tag{4.11}$$

for $\xi \notin \mathrm{Image}(\gamma)$. The smoothness assumption will be dropped at the last step of the proof. We will write $d_0(\xi)$ to denote the right-hand side of (4.11). We will occasionally write $\tilde{d}_0(\xi; u)$ or $d_0(\xi; u)$ when we need to indicate the dependence of $\tilde{d}_0$ or d_0 on u.

(1) Initially for simplicity we assume that $0 \notin \mathrm{Image}(\gamma)$, and we claim that $\tilde{d}_0(0) = d_0(0)$. Let us write $\nu(\theta) = (\cos\theta, \sin\theta)$ and $\tau(\theta) = \nu'(\theta)$. Then easy computations show that

$$\gamma(\theta) = \mathrm{D}u(re^{\mathrm{i}\theta}) = \nu f(\theta) + \tau f'(\theta).$$

Also, $\gamma'(\theta) = \tau(f + f'') + \nu f' + \tau' f' = \tau(f + f'')$, since $\tau' = -\nu$. Thus

$$2\pi\tilde{d}_0(0) = \int_{\mathbb{T}} \frac{\gamma \times \gamma'}{|\gamma|^2} = \int_{\mathbb{T}} \frac{f^2 + ff''}{f^2 + f'^2} = \int_{\mathbb{T}}\left(1 + \frac{ff'' - f'^2}{f^2 + f'^2}\right).$$

The assumption that $0 \notin \mathrm{Image}(\gamma)$ implies that that $f^2 + f'^2$ never vanishes, so that $f' \neq 0$ whenever $f = 0$. Then f vanishes at a finite collection of points $\theta_1 < \cdots < \theta_{2k} < \theta_{2k+1} = \theta_1 + 2\pi$. Let I_j denote the open interval (θ_j, θ_{j+1}). Note that f must change sign at every θ_j. (This is why the number of intervals must be even.) Then we have

$$\int_{\mathbb{T}} \frac{ff'' - f'^2}{f^2 + f'^2} = \sum_{j=1}^{2k} \int_{\theta_j}^{\theta_{j+1}} \frac{f^2}{f^2 + f'^2} \frac{ff'' - f'^2}{f^2} \, \mathrm{d}\theta.$$

On I_j, we make the change of variables $u = f'/f$, so that $\mathrm{d}u = (f''f - f'^2)/f^2 \, \mathrm{d}\theta$. Note that $\lim_{\theta \nearrow \theta_{j+1}} f'(\theta)/f(\theta) = -\infty$ and $\lim_{\theta \searrow \theta_j} f'(\theta)/f(\theta) = +\infty$ for every j, so that

$$\int_{\theta_j}^{\theta_{j+1}} \frac{f^2}{f^2 + f'^2} \frac{ff'' - f'^2}{f^2} \, \mathrm{d}\theta = \int_{\infty}^{-\infty} \frac{1}{1 + u^2} \, \mathrm{d}u = \arctan(u)|_{\infty}^{-\infty} = -\pi$$

for every j. Combining these, we find that

$$\tilde{d}_0(0) = 1 - \tfrac{1}{2} \#(\text{sign changes of } f) = 1 - \tfrac{1}{2} \#(\text{sign changes of } u|_{S^1}) = d_0(0).$$

(2) For general $\xi \in \mathbb{R}^2$ such that $\xi \notin \mathrm{Image}(\gamma)$, define $v(x) = u(x) - \xi \cdot x$, and note that $Dv(re^{\mathrm{i}\theta}) = \gamma(\theta) - \xi$. Then

$$d_0(\xi; u) = d_0(0; v) = \tilde{d}_0(0; v) = \tilde{d}_0(\xi; u).$$

In view of (4.10), we have proved the proposition if f is smooth.

(3) Finally, assume only that u is homogeneous and a Monge–Ampère function. As remarked elsewhere, the fact that u is a Monge–Ampère function implies that $\mathrm{D}u$ has bounded variation, and hence that f' has bounded variation. Let f_k be a sequence of smooth functions $\mathbb{T} \to \mathbb{R}$ such that $\|f_k\|_{W^{2,1}} \leq C$ and such that $f_k \to f$ uniformly. It then follows from (4.8) that for every compact $K \subset \mathbb{R}^2$, there exists a constant C such that $\|[\mathrm{d}u_k]\|(K \times \mathbb{R}^2) \leq C$. Hence Corollary 1.1 implies that $\mathrm{Det}\,\mathrm{D}^2 u_k \rightharpoonup \mathrm{Det}\,\mathrm{D}^2 u$ weakly as measures. The uniform convergence $f_k \to f$ implies that $d_0(\xi; u_k) \to d_0(\xi, u)$ for a.e. ξ. Also, it follows from (4.16), established in the proof of Lemma 4.1 below, that $\|Dd_0(\cdot; u_k)\|_{L^1} = \|D\tilde{d}_0(\cdot; u_k)\|_{L^1} \leq C\|f_k\|_{W^{2,1}}$, so that in fact $d_0(\cdot; u_k) \to d_0(\cdot; u)$ in L^1. Thus

$$\begin{aligned} \int \phi(x,\xi) \,\mathrm{Det}\,\mathrm{D}^2 u(\mathrm{d}x, \mathrm{d}\xi) &= \lim_k \int \phi(x,\xi) \,\mathrm{Det}\,\mathrm{D}^2 u_k(\mathrm{d}x, \mathrm{d}\xi) \\ &= \lim_k \int \phi(0,\xi) d_0(\xi; u_k) \,\mathcal{L}^2(\mathrm{d}\xi) \\ &= \int \phi(0,\xi) d_0(\xi; u) \,\mathcal{L}^2(\mathrm{d}\xi). \end{aligned}$$

□

Next we present the

PROOF OF LEMMA 4.1. It is convenient to write

$$G_{\mathrm{D}u}(\psi) = \int_{\mathbb{R}^2} \left[\psi_0\big(x, \mathrm{D}u(x)\big) \,\mathrm{d}x^1 \wedge \mathrm{d}x^2 + \psi_1^{ij}(x, \mathrm{D}u) \,\mathrm{d}x^i \wedge \mathrm{d}(u_{x_j})\right],$$

$$T(\psi) = \int_{\mathbb{R}^2} \psi_2(0,\xi) \tilde{d}_0(\xi) \,\mathcal{L}^2(\mathrm{d}\xi)$$

for ψ of the form (4.5). Then we must show that

$$[\mathrm{d}u] = G_{\mathrm{D}u} + T \tag{4.12}$$

$G_{\mathrm{D}u}$ is just the current associated with integration over the graph of $\mathrm{D}u$. In general one also expects a term of the form $\int \psi_2(x, \mathrm{D}u)\,\mathrm{d}(u_{x_1}) \wedge \mathrm{d}(u_{x_2})$, but in this case it vanishes, since $\mathrm{d}(u_{x_1}) \wedge \mathrm{d}(u_{x_2}) = \det \mathrm{D}^2 u\,\mathrm{d}x^1 \wedge \mathrm{d}x^2 = 0$, due to the homogeneity of u.

To verify (4.12), we must check that $G_{\mathrm{D}u} + T$ is an integral current and satisfies the defining properties (1.3)–(1.6). The last condition (1.6) follows directly from the definitions. Condition (1.5) states that $[\mathrm{d}u]$ has finite mass in $K \times \mathbb{R}^2$ for any compact subset K of $\mathbb{R}^2$. The mass of $G_{\mathrm{D}u}$ in $K \times \mathbb{R}^2$ is bounded by

$$C \int_K (1 + |\mathrm{D}^2 u| + |\det \mathrm{D}^2 u|) = C \int_K (1 + |\mathrm{D}^2 u|) \tag{4.13}$$

and hence is finite if K is compact. And the mass of T is just the L^1 norm of $\tilde{d}_0$. From the definition one sees rather easily that $\tilde{d}_0(\xi) = 0$ if $|\xi| \geq \|\gamma\|_{L^\infty}$. Writing $R = \|\gamma\|_{L^\infty}$, it follows that

$$\|\tilde{d}_0\|_{L^1} \leq \frac{1}{2\pi} \int_{|\xi| \leq R} \int_0^{2\pi} \frac{|\gamma'(\theta)|}{|\gamma(\theta) - \xi|}\,\mathrm{d}\theta\,\mathcal{L}^2(\mathrm{d}\xi) \leq CR \int_0^{2\pi} |\gamma'|\,\mathrm{d}\theta.$$

Since $R = \|\gamma\|_{L^\infty} \leq C\|\gamma\|_{W^{1,1}}$, we conclude that $\|\tilde{d}_0\|_{L^1} \leq C\|\gamma\|^2_{W^{1,1}}$. Thus (1.5) holds, and (4.8) also follows, since the right-hand side of (4.13) can clearly be estimated in term of $\|\gamma\|_{W^{1,1}}$ and K.

For the Lagrangian condition (1.4) we must check that if ϕ is any smooth compactly supported function, then $(G_{\mathrm{D}u} + T)(\phi\omega) = 0$, where $\omega = \mathrm{d}x^1 \wedge \mathrm{d}\xi^1 + \mathrm{d}x^2 \wedge \mathrm{d}\xi^2$. The definition easily implies that $G_{\mathrm{D}u}(\phi\omega) = 0$—this is just the fact that $\mathrm{D}^2 u$ is symmetric. And it is clear from the definition of T that $T(\phi\omega) = 0$.

It is rather clear that both $G_{\mathrm{D}u}$ and T are integer multiplicity rectifiable, and hence that $G_{\mathrm{D}u} + T$ has the same property.

Finally, note that (1.3) is equivalent to

$$\partial T = -\partial G_{\mathrm{D}u}. \tag{4.14}$$

To check this, recall first that if we write $\eta = \eta_0^i(x,\xi)\,\mathrm{d}x^i + \eta_1^j(x,\xi)\,\mathrm{d}\xi^j = \eta_0 + \eta_1$, then

$$\partial G_{\mathrm{D}u}(\eta) = \int_0^{2\pi} \eta_1^j(0, \gamma(\theta))\frac{\mathrm{d}}{\mathrm{d}\theta}\gamma^j(\theta)\,\mathrm{d}\mathcal{H}^1(\theta).$$

This is well-known; for example, it is a special case of results in [10, Section 3.2.2]. For the sake of completeness, however, we give a proof in Lemma 4.2 below. Note also that, in view of the definition of T, for η as above we have

$$\partial T(\eta) = T(\mathrm{d}\eta) = T(\eta^j_{1,\xi_k}\,\mathrm{d}\xi^k \wedge \mathrm{d}\xi^j) = \int_{\{0\}\times\mathbb{R}^2} (\eta^2_{1,\xi_1} - \eta^1_{1,\xi_2})\tilde{d}_0(\xi)\,\mathrm{d}\mathcal{H}^2. \tag{4.15}$$

We write $\eta^j(\xi) = \eta_1^j(0,\xi)$, and we identify η with the vector field (η^1, η^2). We must now check that

$$\int_0^{2\pi} \eta(\gamma(\theta)) \cdot \gamma'(\theta)\,\mathrm{d}\theta = \int_{\mathbb{R}^2} (\eta^2_{\xi_1} - \eta^1_{\xi_2})\tilde{d}_0(\xi)\,\mathrm{d}\xi, \quad \eta \in C_c^\infty(\mathbb{R}^2; \mathbb{R}^2). \tag{4.16}$$

Fix $\eta \in C_c^\infty$ and let $f(\xi) = -\mathrm{D}^\perp \cdot \eta$, where $\mathrm{D}^\perp = (-\partial_{\xi_2}, \partial_{\xi_1})$. We use the notation $\Delta^{-1} f = (1/2\pi)\ln|\cdot| * f$, and we write

$$\zeta(\xi) = \mathrm{D}^\perp \Delta^{-1} f(\xi) = \int_{\mathbb{R}^2} \frac{1}{2\pi}\frac{(\xi - a)^\perp}{|\xi - a|^2} f(a)\,\mathrm{d}a.$$

Then $D^\perp \cdot \zeta = D^\perp \cdot D^\perp \Delta^{-1} f = -f = D^\perp \cdot \eta$. Thus η and ζ differ by a gradient, which implies that

$$\int_0^{2\pi} \eta\big(\gamma(\theta)\big) \cdot \gamma'(\theta)\, \mathrm{d}\theta = \int_0^{2\pi} \zeta\big(\gamma(\theta)\big) \cdot \gamma'(\theta)\, \mathrm{d}\theta.$$

On the other hand, expanding ζ yields

$$\begin{aligned}\int_0^{2\pi} \zeta\big(\gamma(\theta)\big) \cdot \gamma'(\theta)\, \mathrm{d}\theta &= \int_0^{2\pi} \int_{\mathbb{R}^2} f(a) \frac{1}{2\pi} \frac{(\gamma - a)^\perp}{|\gamma - a|^2} \cdot \gamma'(\theta)\, \mathrm{d}a\, \mathrm{d}\theta \\ &= \int_{\mathbb{R}^2} f(a) \tilde{d}_0(a)\, \mathrm{d}a\end{aligned}$$

proving (4.16). □

We conclude this discussion by giving a proof of the standard lemma used above.

Lemma 4.2. *Suppose that $v\colon \mathbb{R}^2 \to \mathbb{R}^2$ is homogeneous of degree 0, and smooth away from the origin. Let us write $v(re^{i\theta}) = \gamma(\theta)$, for $\gamma\colon \mathbb{R} \to \mathbb{R}^2$ smooth and 2π periodic. Let $V(x) = \big(x, v(x)\big)$, and let $G_v = V_\#([\mathbb{R}^n \setminus \{0\}])$, so that if ϕ has the form (1.1) (with $n = 2$) then*

$$G_v(\phi) = \int_{\mathbb{R}^2\setminus\{0\}} V^\# \phi = \int_{\mathbb{R}^2\setminus\{0\}} \sum_{|\alpha|+|\beta|=2} \phi^{\alpha\beta}\big(x, v(x)\big)\sigma(\alpha, \bar\alpha) M_{\bar\alpha}^{\beta}(Dv).$$

Then for $\eta = \eta_0^i(x,\xi)\, \mathrm{d}x^i + \eta_1^j(x,\xi)\, \mathrm{d}\xi^j = \eta_0 + \eta_1$,

$$\partial G_v(\eta) = \int_0^{2\pi} \eta_1^j\big(0, \gamma(\theta)\big) \frac{\mathrm{d}}{\mathrm{d}\theta} \gamma^j(\theta)\, \mathrm{d}\mathcal{H}^1(\theta).$$

Proof. Let $\chi\colon [0,\infty) \to [0,1]$ be a smooth nondecreasing function such that $\chi(r) = 0$ for $r \le 1$ and $\chi(r) = 1$ for $r \ge 2$. Let $\chi_s(r) = \chi(r/s)$, and let $\zeta_s(x) = \chi_s(|x|)$.

Then

$$\partial G_v(\eta) = G_v(\mathrm{d}\eta) = \lim_{r\to 0} \big[G_v(\zeta_s d\eta) + G_v\big((1 - \zeta_s)\, \mathrm{d}\eta\big)\big].$$

We first claim that $\lim_{s\to 0} G_v\big((1 - \zeta_s)\, \mathrm{d}\eta\big) = 0$. Indeed, $G_v\big((1 - \zeta_s)\, \mathrm{d}\eta\big)$ has the form

$$G_v\big((1 - \zeta_s)\, \mathrm{d}\eta\big) = \int_{|x|\le 2s} (1 - \zeta_s)[\phi_0(x, v) + \phi_1^{ij}(x, v) v_{x_j}^i + \phi_2(x, v) \det Dv]\, \mathrm{d}x.$$

for certain functions made out of derivatives of the components of η. The homogeneity of v implies that $0 = x_i v_{x_i}^j$, and this implies that $\det \mathrm{D}v = 0$. It further implies that $|v_{x_j}^i(x)| \le C|x|^{-1}$ whence

$$\big|G_v\big((1 - \zeta_s)\, \mathrm{d}\eta\big)\big| \le C \int_{|x|\le 2s} [1 + |x|^{-1}]\, \mathrm{d}x \to 0 \quad \text{as } s \to 0.$$

Next,

$$G_v(\zeta_s\, \mathrm{d}\eta) = G_v\big(\mathrm{d}(\zeta_s\eta)\big) - G_v(\mathrm{d}\zeta_s \wedge \eta) = \partial G_v(\zeta_s\eta) - G_v(\mathrm{d}\zeta_s \wedge \eta).$$

Note that

$$\partial G_v(\zeta_s\eta) = \partial V_\#([\mathbb{R}^2 \setminus \{0\}])(\zeta_s\eta) = V_\#\partial([\mathbb{R}^2 \setminus \{0\}])(\zeta_s\eta) = 0.$$

The exchange of ∂ and $V_\#$ is justified because $\zeta_s\eta$ is supported away from the origin (the only point where V fails to be smooth), and the last equality is geometrically clear. Alternately, the above identity can be checked via an integration by parts. Either way we conclude that

$$\partial G_v(\eta) = \lim_{r\to 0} G_v(\mathrm{d}\zeta_s \wedge \eta).$$

The expression on the right-hand side can be expanded as

$$G_v(\mathrm{d}\zeta_s \wedge \eta) = \int_{s\le|x|\le 2s} \chi_s'(|x|)\,\mathrm{d}|x| \wedge [\eta_0^i(x,v)\mathrm{d}x^i + \eta_1^j(x,v)\,\mathrm{d}v^j].$$

As above, simple scaling considerations show that the terms involving η_0 vanish as $s \to 0$. And the homogeneity implies that $\mathrm{d}v^j = \mathrm{d}\gamma^j(\theta)/\mathrm{d}\theta\,\mathrm{d}\theta$, so that rewriting the remaining terms in polar coordinates (and writing $\mathrm{d}r \wedge \mathrm{d}\theta$ as $\mathrm{d}r\,\mathrm{d}\theta$, as is standard) yields

$$\partial G_v(\eta) = \lim_{s\to 0} G_v(\mathrm{d}\zeta_s \wedge \eta_1) = \lim_{s\to 0}\int_{s\le|x|\le 2s} \chi_s'(r)\eta_1^j\big(re^{\mathrm{i}\theta},\gamma(\theta)\big)\frac{\mathrm{d}}{\mathrm{d}\theta}\gamma^j(\theta)\,\mathrm{d}r\,\mathrm{d}\theta.$$

The lemma follows from this identity by integrating first in the radial variable and using the elementary fact that

$$\lim_{s\to 0}\int_0^\infty \chi_s'(r)f(r)\,\mathrm{d}r = \lim_{s\to 0}\int_s^{2s} \chi_s'(r)f(r)\,\mathrm{d}r = f(0). \qquad \square$$

5. Open problems

5.1. Regularity. Investigate the continuity properties of Monge–Ampère functions. One might hope, based on the examples presented above, that a Monge–Ampère function in $\Omega \subset \mathbb{R}^n$ is $C^{0,2/(n+1)}$. Note that this is true when $n = 1$.

Lemma 3.2 implies that no proof relying only on control of $\operatorname{Det} \mathrm{D}^2 u$ can control the $C^{0,\alpha}$ seminorm for $\alpha > 1/n$, so that if the exponent $\alpha = 2/(n+1)$ is correct, it will require a more subtle proof.

We remark that, as far as we know, it has not even been shown that Monge–Ampère functions are continuous.

5.2. Local representations of $\operatorname{Det} \mathrm{D}^2 u$. As far as I know, every result along the lines of (4.1), (4.2), giving a local representation of $\operatorname{Det} \mathrm{D}^2 u$, is proved under some a priori regularity assumptions, such as those in (4.3) (or analogous assumptions in the context of geometric sets and normal cycles, the setting in which most results I know of are stated and proved).

Can one prove that (4.1), (4.2) hold for arbitrary Monge–Ampère functions, without any additional regularity assumptions? One difficulty is that, as emphasized above, regularity properties of Monge–Ampère functions are not well understood, so that it is not clear whether one can for example expect the right-hand side of (4.2) to be well-defined $\mathcal{H}^0(\mathrm{d}x) \times \mathcal{L}^n(\mathrm{d}\xi)$ almost everywhere, for general $u \in \mathrm{MA}_{\mathrm{loc}}(\Omega)$.

5.3. Density. There are a number of natural density questions which are more or less completely open. For example

Are smooth functions weakly dense in the space of Monge–Ampère functions? For example, given $u \in \mathrm{MA}_{\mathrm{loc}}(\Omega)$, does there exist a sequence $\{u_k\} \subset C^\infty(\Omega)$ such that $u_k \to u$ in L^1 and

$$\sup_k \int_K \sum M^{\alpha\beta}(\mathrm{D}^2 u_k)\,\mathrm{d}x \le C(K) < \infty \tag{5.1}$$

for every compact $K \subset \Omega$? This question was raised by Fu [4].

A related question deals with *strong* density of smooth functions. A function $u \in \mathrm{MA}(\Omega)$ is said to be strongly approximable by smooth functions if there exists a sequence $\{u_k\}$ of smooth functions on Ω such that

$$u_k \to u \text{ in } L^1 \text{ and } \mathbf{M}([\mathrm{d}u_k]) \to \mathbf{M}([\mathrm{d}u]) \quad \text{as } k \to \infty.$$

If $u \in \mathrm{MA}(\Omega)$ is weakly approximable by smooth functions, is it also *strongly* approximable?

Very similar questions have been studied in depth for Cartesian currents. In particular, Giaquinta, Modica, and Souček present counterexamples to approximability of Cartesian currents in [8].

We expect that Proposition 4.1 can be used to show that any Monge–Ampère function on the plane that is homogeneous of degree 1 is strongly approximable by smooth functions. This would still be very far from any general density result, but these considerations are at least mildly encouraging, since in particular it suggests that the counterexamples from [8] have no obvious analog in the setting of Monge–Ampère functions.

5.4. A variational problem arising in statistics. Suppose that Ω is a bounded, open subset of $\mathbb{R}^n$, and that we are given points $x_1, \ldots, x_M \in \Omega$ and numbers $y_1, \ldots, y_M \in \mathbb{R}$. Does the functional

$$I[u] := \sum_{i=1}^{M} |u(x_i) - y_i| + \int |\det \mathrm{D}^2 u(x)|\,\mathrm{d}x \tag{5.2}$$

admit a minimizer in the space of functions that vanish on $\partial\Omega$?

When $n = 1$, this functional and variants were studied by Koenker, Ng, and Portnoy in [16], in the context of nonparametric regression. These authors prove that the infimum of this functional is attained in $\{u \in W^{1,1}(\mathbb{R}) : u' \text{ has bounded variation}\}$, where $\int |u''|$ is interpreted as the total variation of u'. There is in general no uniqueness, but it is shown that one can find minimizers that are piecewise linear, with any discontinuities in the slope only occuring at the some subset of the points $\{x_i\}$. As a result, they show, one can compute minimizers very efficiently via linear programming.

The functional (5.2) was discussed [18] by Koenker, Mizera, and Portnoy as one of a number of possible higher-dimensional extensions of the one-dimensional version. They rapidly concluded that this functional has a number of pathological features that make it of questionable value for applications in statistics. These same features, however, make it an interesting model problem in the calculus of variations.

The space of Monge–Ampère functions, or something rather like it, might provide a suitable functional setting for studying this problem. This was one of the author's motivations for studying spaces of functions without any sort of convexity conditions in which $\mathrm{Det}\,\mathrm{D}^2 u$ makes sense as a measure.

5.5. Degenerate Monge–Ampère equations. One of the mains results of [13] is that, if $u\colon \Omega \subset \mathbb{R}^2 \to \mathbb{R}$ is a Monge–Ampère function such that $\operatorname{Det} \mathrm{D}^2 u = 0$ in the sense that

$$[\mathrm{d}u](\phi \,\mathrm{d}\xi^1 \wedge \mathrm{d}\xi^2) = 0 \quad \text{for all } \phi \in C_c^\infty(\Omega \times \mathbb{R}^2)$$

then for every $x \in \Omega$, at least one of the following must hold:

(1) u is affine in an open neighborhood of x; or

(2) There exists a line segment l_x, passing through x and meeting $\partial\Omega$ at both endpoints, along which $\mathrm{D}u$ is constant in the sense that
- every point along l_x is a Lebesgue point of $\mathrm{D}u$, with the same Lebesgue value; or
- every point on l_x belongs to the jump set $J_{\mathrm{D}u}$ of $\mathrm{D}u$, with same approximate limits on both sides of l_x.

In particular, *every* point in Ω is either a Lebesgue point of $\mathrm{D}u$ or belongs to the jump set of $\mathrm{D}u$.

Similar results have been proved for smoother functions that satisfy the equation $\det \mathrm{D}^2 u \equiv 0$, including for example by Hartman and Nirenberg [11] for $u \in C^2$, Kirchheim [15] for $u \in W^{2,\infty}$ and Pakzad [17] for $u \in W^{2,2}$.

What analogous result can be established in dimensions $n \geq 3$?

5.6. Miscellaneous other problems.

(1) The main result of Fu [5] is that, if u is a *locally Lipschitz* Monge–Ampère function, then

$$[\mathrm{d}u] \llcorner \Big(\sum \xi_i \,\mathrm{d}x^i\Big) - [\mathrm{d}u] \llcorner (p^\# \,\mathrm{d}u) = 0 \quad \text{where } p(x,\xi) := x. \tag{5.3}$$

This rather opaque identity is then used to show for example that if w is $C^{1,1}$ and u is Monge–Ampère and locally Lipschitz, then $w \circ u$ is Monge–Ampère, and a suitable chain rule holds. The proof of (5.3) in [5] relies on some delicate approximation arguments that do not automatically extend to the case of Monge Ampère functions that are not locally Lipschitz. Thus, one might ask whether this or its corollaries are valid for general Monge–Ampère functions.

We remark that (5.3) is trivially true if u is smooth, and follows by approximation for any Monge–Ampère function that can be weakly approximated by smooth functions. So these questions could be answered by proving suitable density results.

References

1. L. Bröcker and M. Kuppe, *Integral geometry of tame sets*, Geom. Dedicata **82** (2000), no. 1-3, 285–323.
2. H. Federer, *Geometric measure theory*, Grundlehren Math. Wiss., vol. 153, Springer, New York, 1969.
3. I. Fonseca and J. Malý, *From Jacobian to Hessian: distributional form and relaxation*, Riv. Mat. Univ. Parma (7) **4*** (2005), 45–74.
4. J. H. G. Fu, *Monge–Ampère functions.* I, Indiana Univ. Math. J. **38** (1989), no. 3, 745–771.
5. ———, *Monge–Ampère functions.* II, Indiana Univ. Math. J. **38** (1989), no. 3, 773–789.
6. ———, *Curvature measures and generalized Morse theory*, J. Differential Geom. **30** (1989), no. 3, 619–642.
7. ———, *Curvature measures of subanalytic sets*, Amer. J. Math. **116** (1994), no. 4, 819–880.
8. M. Giaquinta, G. Modica, and J. Souček, *Graphs of finite mass which cannot be approximated in area by smooth graphs*, Manuscripta Math. **78** (1993), no. 3, 259–271.

9. ———, *Cartesian currents in the calculus of variations. I: Cartesian currents*, Ergeb. Math. Grenzgeb. (3), vol. 37, Springer, Berlin, 1998.
10. ———, *Cartesian currents in the calculus of variations. II: Variational integrals*, Ergeb. Math. Grenzgeb. (3), vol. 38, Springer, Berlin, 1998.
11. P. Hartman and L. Nirenberg, *On spherical image maps whose Jacobians do not change sign*, Amer. J. Math. **81** (1959), 901–920.
12. T. Iwaniec, *On the concept of the weak Jacobian and Hessian*, Papers on Analysis, Rep. Univ. Jyväskylä Dep. Math. Stat., vol. 83, Univ. Jyväskylä, Jyväskylä, 2001, pp. 181–205.
13. R. L. Jerrard, *Some rigidity results related to Monge–Ampère functions*, preprint.
14. R. L. Jerrard and N. Jung, *Strict convergence and minimal liftings in BV*, Proc. Roy. Soc. Edinburgh Sect. A **134** (2004), no. 6, 1163–1176.
15. B. Kirchheim, *Geometry and rigidity of microstructures*, Habilitation thesis, Leipzig, 2001.
16. R. Koenker, P. Ng, and S. Portnoy, *Quantile smoothing splines*, Biometrika **81** (1994), no. 4, 673–680.
17. M. R. Pakzad, *On the Sobolev space of isometric immersions*, J. Differential Geom. **66** (2004), no. 1, 47–69.
18. S. Portnoy, personal communication.
19. J. Rataj and M. Zähle, *Curvatures and currents for unions of sets with positive reach. II*, Ann. Global Anal. Geom. **20** (2001), no. 1, 1–21.
20. M. Zähle, *Curvatures and currents for unions of sets with positive reach*, Geom. Dedicata **23** (1987), no. 2, 155–171.

Department of Mathematics, University of Toronto, Toronto, ON M5S 3G3, Canada

E-mail address: rjerrard@math.toronto.edu

Centre de Recherches Mathématiques
CRM Proceedings and Lecture Notes
Volume **44**, 2008

Variational Versus PDE-Based Approaches in Mathematical Image Processing

B. Kawohl

ABSTRACT. In mathematical image processing we are often presented with amazing examples of image enhancement algorithms. Yet, when applied to different noisy images, they can produce unwanted effects. The analysis of such algorithms lags behind their intuitive development. Two essentially different models have found wide recognition: a variational approach according to Mumford and Shah and an approach via nonlinear diffusion equations. One of these equations is nonparabolic and was suggested by Perona and Malik. In this short survey I point out a connection between these two seemingly unrelated approaches and explain some connections with total variation flow.

1. Introduction

Suppose $u_0\colon \Omega \to [0,1]$ represents the gray-scale distribution of a noisy image. *Mumford and Shah* suggested studying

$$J(v) = \alpha \int_\Omega |v-u_0|^2 \,\mathrm{d}x + \beta \int_{\Omega\setminus S_v} |\nabla v|^2 \,\mathrm{d}x + \mathcal{H}^{n-1}(S_v). \tag{1.1}$$

Here α and β are suitable positive constants which weigh the contributions of the three terms in the functional. A minimizer of this problem, if it exists, should have few sets of discontinuity, a relatively smooth appearance elsewhere and still resemble the original picture. One can replace the quadratic terms in the functional by strictly convex terms such as $|v-u_0|^p$ or $\sqrt{1+|\nabla v|^2}$ without changing essential features of the problem. For reasons of exposition, I will adhere to the simple functional above. Nowadays it is minimized on $\mathrm{SBV}(\Omega)$, a space of special functions of bounded variation. One of the analytical problems is a lack of semicontinuity of $\mathcal{H}^{n-1}(S)$ with respect to the Hausdorff topology.

To demonstrate the lack of semicontinuity consider the following sequence S_i of subsets of the unit interval $S_0 = (0,1)$. The first element of the sequence is $S_1 := (0,0.5)$, the second $S_2 := (0,0.25)\cup(0.5,0.75)$, $S_3 := (0,0.125)\cup(0.25,0.375)\cup(0.5,0.6125)\cup(0.75,0.875)$ etc.. Each element in this sequence has 1-dimensional

2000 *Mathematics Subject Classification.* Primary 68U10; Secondary 35J20, 35K55, 49J45, 94A08.

Key words and phrases. Perona – Malik equation, Mumford – Shah functional, Blake – Zisserman functional, total variation flow, staircasing.

This is the final form of the paper.

Hausdorff measure 0.5, but S_i has a Hausdorff-limit S_0, and $\mathcal{H}^1(S_0) = 1$. Incidentally, this "divide and conquer" algorithm is frequently used by politicians to prove that one can cut costs of public services in half by creating sufficiently small cost units, each of which is just a little short of what it really needs to function. Another application was given in [26], where a circle of radius 1 was apparently surrounded by a fence of total length $1 + \pi$.

Perona and Malik suggested taking the noisy image u_0 as initial datum for a diffusion equation such as

$$u_t - \operatorname{div}\left(\frac{\nabla u}{(1 + |\nabla u|^2)^2}\right) = 0, \tag{1.2}$$

or, more generally

$$u_t - \operatorname{div}(a(|\nabla u|^2)\nabla u) = 0, \tag{1.3}$$

with $a(s)$ positive and decreasing to zero as $s \to \infty$, and under no-flux boundary conditions.

Small diffusion near discontinuities in u_0 was supposed to lead to edge preservation, while large diffusion elsewhere would somehow mollify the brightness function and take out noise.

If $a(s) = s^{-1/2}$, then formally equation (1.3) becomes

$$u_t - \operatorname{div}\left(\frac{\nabla u}{|\nabla u|}\right) = 0. \tag{1.4}$$

This is commonly called total variation flow, *TV-flow* for short.

Let me explain why the Perona–Malik equations are called anisotropic. Upon differentiation, (1.3) becomes

$$u_t - a(|\nabla u|^2)\Delta u - 2a'(|\nabla u|^2)\nabla u D^2 u \nabla u = 0, \tag{1.5}$$

with D^2u denoting the Hessian matrix of second derivatives. We want to rewrite these second derivatives in intrinsic coordinates and set $\nu = -\nabla u/|\nabla u|$. Then

$$\Delta u = u_{\nu\nu} + (n-1)Hu_\nu = u_{\nu\nu} + \Delta_{n-1}u,$$

where H is the mean curvature of a level surface of u, or where Δ_{n-1} is the (n-1)-dimensional Laplace operator tangent to this level surface. Therefore (1.5) can be rewritten as

$$u_t - a(|\nabla u|^2)\Delta u - 2a'(|\nabla u|^2)|\nabla u|^2 u_{\nu\nu} = 0,$$

or even as

$$u_t - b(|\nabla u|^2)u_{\nu\nu} - a(|\nabla u|^2)\Delta_{n-1}u = 0.$$

Thus, for $a \neq b$ the diffusion in direction ν differs from the diffusion in the other directions and this is an anisotropic behavior. Moreover, in many cases $b(s)$ can even become negative for large values of s. Then there is a backward diffusion effect which leads to a steepening of profiles in direction of the gradient of u, but at the same time there is forward diffusion along level surfaces of u. Little is known about such diffusion equations.

This lecture is organized as follows. I present the Ambrosio–Tortorelli approximation of the Perona–Malik functional in Section 2. This approximation is amenable to numerical minimization algorithms, which in turn are related to the Perona–Malik equation. While such a formal analogy was also observed in [47],

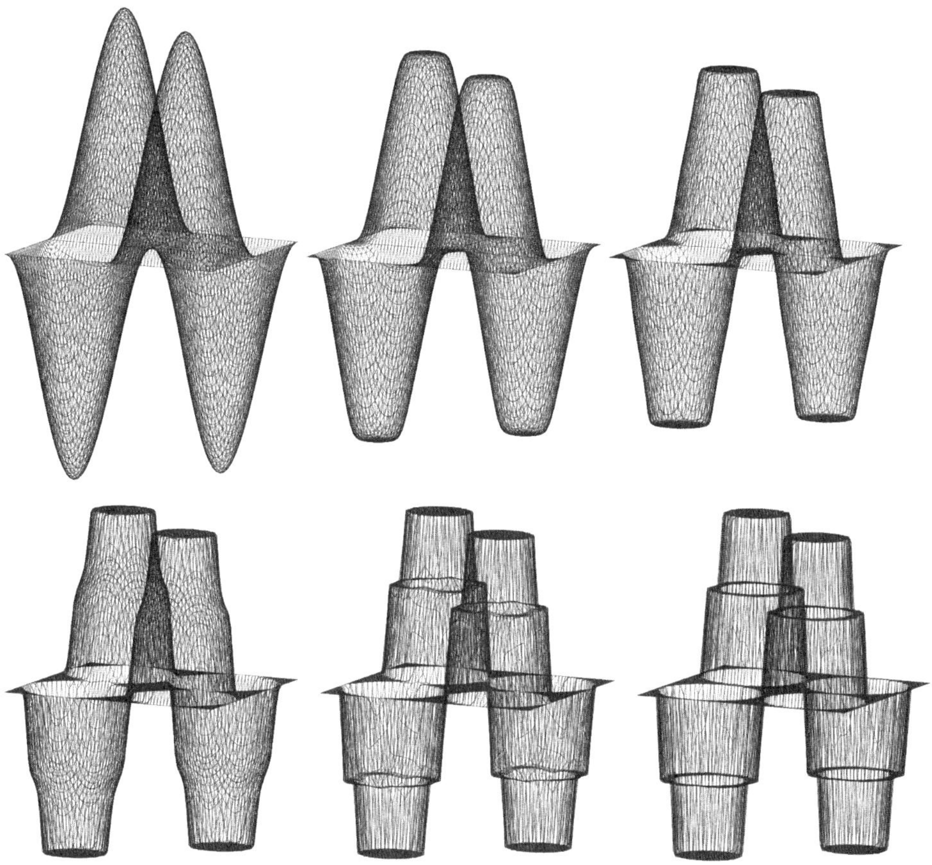

FIGURE 1. This figure shows the evolution of an academically chosen gray scale $u_0(x, y) = \sin x \sin y$ under Perona – Malik flow. One can see a clear formation of edges, the so-called staircasing effect.

Richardson and Mitterer could not reconcile the significant difference between the Perona – Malik equation and their equation (8) which reads

$$v_t - \beta \operatorname{div}(\eta^{\,2} \nabla v) = -\alpha(v - u_0)$$

in my notation, see also (2.3) below. I wish to point out that this problem was only overcome in [27]. In Section 3 I briefly survey some older but not as well-known results from [28] on forward-backward diffusion equations like the Perona – Malik equation. In Section 4 I comment on related but somehow more benign degenerate equations.

2. Mumford – Shah

While the Mumford – Shah functional (1.1) is well-defined for piecewise smooth functions u whose sets S_u of discontinuities are again piecewise smooth, it is difficult to evaluate it numerically. Any conceivable space of nonconforming finite elements would need to keep track of these discontinuities in its very definition. And to measure $\mathcal{H}^{n-1}(S_u)$ one would need to put a grid directly on the set S_u.

How can one approximate $\mathcal{H}^{n-1}(S_u)$? Modica & Mortola showed in [39] and [38] that $H_0(u) = \mathcal{H}^{n-1}(u=0)$ is the Γ-limit of the following family of functionals

$$H_\varepsilon(u) = \int_\Omega \varepsilon|\nabla u|^2 + \frac{1}{4\varepsilon}(u^2-1)^2\,\mathrm{d}x.$$

This suggests that the Hausdorff measure of a suitable level set can be approximated (in the sense of Γ-convergence) by a sequence of domain integrals.

And in fact, Ambrosio & Tortorelli showed in [3] that the Mumford–Shah functional

$$J_0(v) := \int_\Omega \alpha|v-u_0|^2\,\mathrm{d}x + \beta\int_{\Omega\setminus S_v}|\nabla v|^2\,\mathrm{d}x + \mathcal{H}^{n-1}(S_v) \tag{2.1}$$

is the Γ-limit of

$$J_\varepsilon(v,\eta) := \int_\Omega \alpha|v-u_0|^2 + \beta\eta^2|\nabla v|^2 + \varepsilon|\nabla\eta|^2 + \frac{1}{4\varepsilon}(\eta-1)^2\,\mathrm{d}x, \tag{2.2}$$

where $\{\eta=0\}$ stands for S_v. Therefore the set where η is close to zero can be used an edge detector. It indicates where $|\nabla v|$ is large. Notice that in contrast to the Mumford–Shah functional (2.1) the integration of the gradient term extends now over all of Ω and not over $\Omega\setminus S_v$.

For the reader's convenience I recall the definition of Γ-convergence. Let X be a metric space and $F_\varepsilon\colon X\mapsto[0,\infty]$ a family of mappings. Then F *is the* Γ*-limit of* F_ε as $\varepsilon\to 0$, iff (a) and (b) hold.

(a) For every $u\in X$ and every sequence $u_\varepsilon\to u$ in X

$$\liminf_{\varepsilon\to 0} F_\varepsilon(u_\varepsilon)\ge F(u).$$

(b) For every $u\in X$ there exists a sequence u_ε such that $u_\varepsilon\to u$ in X and

$$\lim_{\varepsilon\to 0} F_\varepsilon(u_\varepsilon) = F(u).$$

A convenient result in the theory of Γ-convergence states that minimizers of a Γ-convergent sequence of functionals converge to a minimizer of the limit functional.

Theorem 2.1. *If F is the Γ-limit of F_ε as $\varepsilon\to 0$ and u_ε minimizer of F_ε, then every cluster point u of the family $\{u_\varepsilon\}_{\varepsilon>0}$ minimizes F.*

This theorem provides a certain amount of confidence that for small ε a minimizer of the Ambrosio–Tortorelli functional (2.2) will be at least close to a minimizer of the original Mumford–Shah functional (2.1).

In fact one can show the following: If $(v_\varepsilon,\eta_\varepsilon)$ minimize J_ε, then (after possibly passing to a subsequence) $v_\varepsilon\to v$ and $\eta_\varepsilon\to 1$ in $L^2(\Omega)$ as $\varepsilon\to 0$, but $|\eta_\epsilon|\ll 1$ near S_{v_ε}. The associated Euler equations for minimizers of the Ambrosio-Tortorelli functional are obtained by taking first variations with respect to v and η. They are given by

$$\alpha(v_\varepsilon-u_0)-\beta\,\mathrm{div}(\eta_\varepsilon^2\nabla v_\varepsilon)=0 \tag{2.3}$$

and

$$\beta\eta_\varepsilon|\nabla v_\varepsilon|^2-\varepsilon\Delta\eta_\varepsilon+\frac{1}{4\varepsilon}(\eta_\varepsilon-1)=0,$$

or, upon rearranging the last equation, by

$$-\Delta\eta_\varepsilon+\frac{1+4\beta\varepsilon|\nabla v_\varepsilon|^2}{4\varepsilon^2}\left(\eta_\varepsilon-\frac{1}{1+4\beta\varepsilon|\nabla v_\varepsilon|^2}\right)=0. \tag{2.4}$$

"Now a miracle occurs..." Let us pretend for the moment, that the bracket in (2.4) is nonzero as $\varepsilon \to 0$. Then the second term in (2.4) is of order ε^{-2}. This is consistent with standard estimates in singular perturbation theory. However, Vese and Chan [50] set the round bracket equal to zero, i.e., $\eta = 1/(1 + 4\beta\varepsilon|\nabla v|^2)$ and neglect the term $\varepsilon|\nabla\eta|^2$ in J_ε. In this way they arrive at a modified Ambrosio–Tortorelli functional

$$\tilde{J}_\varepsilon(v) := \int_\Omega \alpha|v - u_0|^2 + \tilde{A}_\varepsilon(|\nabla v|^2)\,\mathrm{d}x \tag{2.5}$$

with $\tilde{A}_\varepsilon(|\nabla v|^2)$ given below in (2.6). On a continuous level this is an analytical crime, because one knows for instance that $\int_\Omega \varepsilon|\nabla v_\varepsilon|^2\,\mathrm{d}x$ is of order one, $O(1)$ and not $o(1)$. However, since the gray scale function v_ε can only take values in $[0, 1]$, upon discretization with grid-size h, its gradient cannot exceed $O(h^-1)$. Therefore, on a fixed grid one may indeed send ε to zero and justify the approximation of J_ε by $\tilde{J}_\varepsilon$ for $\varepsilon \ll h^2$.

Now I turn my attention to the modified functional (2.5) with

$$\tilde{A}_\varepsilon(s^2) := \frac{s^2}{1 + 4\beta\varepsilon s^2}. \tag{2.6}$$

The Euler equation for this modified functional reads

$$-\operatorname{div}\left(\frac{\nabla v}{1 + 4\beta\varepsilon|\nabla v|^2}\right) + \frac{\alpha}{4\beta\varepsilon}(v - u_0) = 0, \tag{2.7}$$

and this looks formally like the first step in an explicit Euler scheme for the Perona–Malik equation

$$v_t - \operatorname{div}\left(\frac{\nabla v}{1 + 4\beta\varepsilon|\nabla v|^2}\right) = 0 \tag{2.8}$$

with initial datum u_0. Alternatively it can be interpreted as a stationary solution of the ROF-model [21, 48] which was apparently first suggested by Nordström [44]:

$$v_t - \operatorname{div}\left(\frac{\nabla v}{1 + 4\beta\varepsilon|\nabla v|^2}\right) = -\lambda(v - u_0). \tag{2.9}$$

Remark 2.2. It is in this sense that the variational approach and the Perona–Malik approach are deeply interconnected. The diffusion approach is simply a Lyapunov flow for the nonconvex functional (2.5) and a solution of the Perona–Malik equation will lower the associated functional.

But there is more to learn from the functional (2.5). For nonnegative s the map $s \mapsto \tilde{A}_\varepsilon(s^2)$ is a monotone increasing nonconvex function of s which begins convex at zero and approaches a constant $(4\beta\varepsilon)^{-1}$ as $s \to \infty$. Calculus of variations tells us that an infimum of $\tilde{J}_\varepsilon$ must also minimize the convex hull of $\tilde{J}_\varepsilon$, but the convex hull of $\tilde{J}_\varepsilon$ would be just the first term in (2.5). Again it helps to realize that due to discretization we actually try to infimize $\tilde{J}_\varepsilon$ on those functions in $W^{1,\infty}(\Omega)$ the gradient of which does not exceed $1/h$ in modulus. The convex relaxation B_ε of $\tilde{A}_\varepsilon$ on $[0,\ 1/h]$ is then nontrivial, strictly convex near zero and affine near $1/h$. Consequently a function that realizes the infimum of (2.5) must have the modulus of its gradient either inside the interval $[0, s_1]$ where $\tilde{A}_\varepsilon$ coincides with the B_ε or at $1/h$. Therefore the gradient oscillates on a small scale between a number from $[0, s_1]$ and $1/h$; this effect is known as staircasing, see for example [34].

Remark 2.3. Staircasing is an effect that is caused by a nonconvex underlying functional which the solution tries to minimize.

Finally I wish to point out that there are many other functionals which are being used in image processing. Many of them lead to a qualitatively similar behavior. It is clear that the quadratic term $\alpha|v-u_0|^2$ can be replaced by a general power $p>1$ as, for instance in [21], or that the gradient term involving $\tilde{A}_\varepsilon$ term can be replaced $\phi(s^2)=\min\{s^2,\beta\}$. Then we speak of the Blake–Zisserman functional [9]. Its convex relaxation on $[0,1/h]$ is quadratic for small values of s and affine for s close to $1/h$. Even total variation falls into this class provided we are willing to accept that this functional is strictly convex only on the interval consisting of the single point zero.

Remark 2.4. This observation explains why the behavior of image processing via variational methods is fairly stable under a change of functionals and why many functionals work reasonably well as soon as one of them does. A word of caution though: tuning of constants like α,β and ε is required to obtain optimal results for a given set of images.

3. Perona – Malik

Consider the initial-boundary value problem

$$
\begin{aligned}
u_t-\operatorname{div}(a(|\nabla u|^2)\nabla u)&=0 &&\text{in } \Omega\times(0,T),\\
a(|\nabla u|^2)\frac{\partial u}{\partial\nu}&=0 &&\text{on } \partial\Omega\times(0,T),\\
u(x,0)&=u_0(x) &&\text{on } \Omega,
\end{aligned}
\tag{3.1}
$$

where $\Omega\subset\mathbb{R}^n$ is a bounded domain with piecewise smooth boundary $\partial\Omega$, and where u_0 represents a blurred image. Our structural assumptions on the coefficient a are that $a\in C^1([0,\infty))$, and $a(s)>0$, and that the ellipticity function $b(s):=a(s)+2sa'(s)$ is positive for s near 0 and changes sign exactly once at $s_0^2>0$. This includes the model cases: $a(s)=e^{-s}$ or $a(s)=(1+s)^{-1}$ that were studied by Perona and Malik.

As explained in the introduction, one speaks of the corresponding differential equation as anisotropic diffusion. In particular, if $\Omega\subset\mathbb{R}^2$ and if u is a classical solution and $|\nabla u|\neq 0$, one can rewrite the differential equation from (3.1) in intrinsic coordinates $\nu=-\nabla u/|\nabla u|$ and τ as

$$
u_t-b(|\nabla u|^2)u_{\nu\nu}-a(|\nabla u|^2)u_{\tau\tau}=0. \tag{3.2}
$$

Recall that the diffusion coefficient in direction τ is different from direction ν and that b changes sign while a remains positive.

What type of differential equation is this? For small values of $|\nabla u|$ it is a regular parabolic diffusion equation with a classical solution, see [33]. The stationary version resembles the classical Tricomi problem, a differential equation that switches type from elliptic to hyperbolic. If the last term is missing, it looks like a forward-backward diffusion equation. The last case occurs if $n=1$ and several results in [28] deal with this case.

The following questions can be raised:

- Can one describe initial or final data? In general, there is a negative answer [34].

• Can one prove existence? The answer to this question depends on what we understand by a solution and is in general negative. There are not even weak solutions [34]; but yes, there are some generalized solutions [53]; and if $|u_{0_x}| < s_0$ they are even regular [33].

• Can one prove uniqueness? This depends on the regularity of solutions. The answer is negative for $C^{0,1}$-solutions [53] and sometimes positive for C^1 solutions [28] or for C^2-solutions [35].

• Is there a maximum or comparison principle? This was dismissed in [2] because there is none for the backward heat equation. I shall now address these questions for C^1-solutions because for those we can evaluate the (sign of) the diffusion coefficients in every point (x,t) of the space-time cylinder.

Theorem 3.1 (Maximum principle in $\mathbb{R}^n$). *Suppose that $u \in C^{0,1}$ is a weak solution to the anisotropic diffusion problem. Then*

$$\max_{\overline{\Omega}\times[0,T]} |u(x,t)| = \max_{\overline{\Omega}} |u_0(x)|.$$

PROOF. For any $p \in [2,\infty)$ multiply the differential equation (3.1) with $|u|^{p-2}u$ and integrate over Ω. Integration by part gives

$$\begin{aligned} \frac{\mathrm{d}}{\mathrm{d}t}\int_\Omega |u(x,t)|^p \,\mathrm{d}x &= p\int_\Omega |u|^{p-2}u\; u_t \,\mathrm{d}x \\ &= p\int_\Omega |u|^{p-2}u \operatorname{div}(a(|\nabla u|^2)\nabla u)\,\mathrm{d}x \\ &= -p(p-1)\int_\Omega |u|^{p-2}a(|\nabla u|^2)|\nabla u|^2 \le 0. \end{aligned} \tag{3.3}$$

Therefore all $L^p(\Omega)$ norms are decreasing in t and after sending $p\to\infty$, so does the $L^\infty(\Omega)$ norm. To conclude the proof I should mention that Theorem 3.1 was also found by Weickert [51] independently from us. □

From now on I assume that $n = 1$. Assuming that a C^1 solution exists, we can partition $\Omega| \times \mathbb{R}^+ = (-1,1)\times\mathbb{R}^+$ into forward- and backward regimes, depending on the sign of $u_x^2 - s_0^2$.

Theorem 3.2 (Shrinking of backward regime, preservation of shape). *If u is a weak C^1 solution of (3.1) and if u_0 switches only finitely often from forward to backward intervals or vice versa, then the backward regime of u shrinks in time and does not migrate.*

While a proof of this result follows from a delicate analysis of the level lines $\{u_x = \pm s_0\}$ in [28, p. 119], Theorem 3.2 can be interpreted as follows: In image enhancement edges are characterized as backward regimes. Therefore edges do not move and become enhanced through anisotropic diffusion. This is the desired effect in image enhancement. Moreover, no new edges can appear as time evolves. To give you an idea of how this result is shown, take the differential equation

$$u_t - b(u_x^2)u_{xx} = 0, \tag{3.4}$$

Differentiate with respect to x

$$u_{xt} - b(u_x^2)u_{xxx} - 2b'(u_x^2)u_x u_{xx}^2 = 0 \tag{3.5}$$

and substitute $v =: u_x$

$$v_t - b(v^2)v_{xx} - c(v)v_x^2 = 0. \tag{3.6}$$

Now v satisfies a forward diffusion equation in the forward regime and a backward diffusion equation in the backward regime. Should a new backward regime be born in the evolution process, it would, without loss of generality, be a set where $u_x > s_0$ and it would be bounded by a set where $u_x = s_0$. Thus at some initial time in this set v would attain its minimum, a contradiction to the minimum principle for backward diffusion equations, which states that solutions attain their minimum at the final time.

In [28] N. Kutev and I gave an explicit counterexample to a general comparison principle in which $u_0(x) \leq v_0(x)$, and u_0 and v_0 touch each other in some point with slope $> s_0$. For any small but positive t the corresponding solutions u and v have crossed each other at and near this point, so that $u(t,x) \leq v(t,x)$ fails to hold in Ω. Nevertheless, one can sometimes compare different solutions, and a discrete version of the following theorem from [28] appeared only recently in [20].

Theorem 3.3 (Restricted comparison principle). *Suppose that u and v are C^1 solutions of* (3.1) *on $Q_T := (-1,1) \times (0,T)$ with C^2 initial data $u_0(x) \leq v_0(x)$ satisfying* (a) *or* (b):

(a) *the backward regimes of u_0 and v_0 have empty intersection, or*
(b) *there exists a datum w_0 between u_0 and v_0 such that $|w_{0_x}| < s_0$, i.e., the backward regime of w_0 is empty.*

Then $u(x,t) \leq v(x,t)$ in Q_T.

Again the proof is based on a simple idea. If $v - u$ has negative min in (y,t), then $v_x(y,t) = u_x(y,t)$ and by Theorem 3.2 y will be in the forward regime of both v and u. But then one is in a forward parabolic situation, where comparison results hold.

Theorem 3.4 (Uniqueness). *Suppose that u and v are weak C^1 solutions of* (3.1) *in $Q_T := (-1,1) \times (0,T)$ with identical and analytical initial data u_0. Moreover, assume that $(u_0)_x^2 - s_0^2$ has only simple zeroes and that the diffusion coefficient a is analytic. Then $u(x,t) \equiv v(x,t)$ in Q_T.*

The proof follows essentially from showing that the forward and backward regimes of u and v must coincide. There are many technicalities, e.g., on what happens in the interface where $u_x = \pm s_0$. In particular we use (and derive) the fact that the number of "edges" stays invariant.

Spoiling the beauty of these theoretical results on the Perona Malik equation, I should also mention that C^1 solutions must cease to exist as C^1 solutions after finite time. This is expressed in the following result from [28]

Theorem 3.5 (Nonexistence of global C^1-solutions). *Suppose that $u_0(x)$ satisfies certain technical but typical assumptions. Then there is no global weak solution in $C^1\big((-1,1) \times \mathbb{R}^+\big)$.*

The proof of this result uses a representation of u along level lines $\Gamma(t)$ of u_x and by showing that $I(t) = \int_0^{\Gamma(t)} u(y,t)\,dy$ blows up as $t \to \infty$.

4. TV flow and related equations

Since there is such a wide range of nonlinear diffusion equations that are used in mathematical image processing, there is not a single "right" equation. A popular

FIGURE 2. In this figure we see a noisy image of the Cologne cathedral under Perona–Malik flow taken from [30]. The image undergoes first a sharpening then a segmentation in which information gets lost. This illustrates the importance of choosing the right time to stop the flow. The stopping time is chosen interactively, like in the focusing of an overhead projector or beamer.

FIGURE 3. Our last sequence from [30] shows what the edge detector finds under Perona–Malik flow. When there are too many edges the image looks noisy, and when there are too few edges essential features are lost.

equation is total variation flow as in (1.4)

$$w_t - \operatorname{div}\left(\frac{\nabla w}{|\nabla w|}\right) = 0. \tag{4.1}$$

If the initial datum u_0 is chosen as the characteristic function of a disk, then its height decreases with constant speed proportional to the curvature of the disk [17]; and if the initial datum is the characteristic function of a convex set with corners, then the solution decays faster near the corners and its shape is rounded off, as displayed in [1]. How can one explain this behavior heuristically? The Ansatz $w(t,x) = T(t)u(x)$ with nonnegative T leads to

$$T'(t)u(x) - \operatorname{div}\left(\frac{\nabla u}{|\nabla u|}\right) = 0,$$

so that $T = -\lambda$ must be constant and $T(t) = \{T(0) - \lambda t\}_+$ decays linearly to zero in finite time. The constant λ is then supposed to solve the "eigenvalue" problem

$$\operatorname{div}\left(\frac{\nabla u}{|\nabla u|}\right) + \lambda u = 0. \tag{4.2}$$

In a radially decreasing setting this means $-1/r+\lambda u = 0$ or $\kappa = -\lambda u$. Here κ is the curvature of a level set. Thus on the line $\partial\{u > t\}$ the decay rate λ is proportional to curvature. This is why corners are rounded off as time evolves.

Some explicit solutions for (4.2) were calculated in [7], but notice that equation (4.2) is not homogeneous of degree one, so that a multiple cu of an eigenfunction u is no longer an eigenfunction for the eigenvalue λ.

For this aesthetic reason I suggest considering a general equation

$$w_t - Aw = 0$$

with A to be determined later, for which the Ansatz $w(t,x) = T(t)u(x)$ leads to

$$T'(t)u(x) - A[T(t)u(x)] = 0,$$

and if A is homogeneous of degree d, to

$$T'(t)u(x) - T^d(t)Au(x) = 0.$$

Now separation of variables gives

$$T^dT'(t) = -\lambda \quad \text{and} \quad Au + \lambda u = 0.$$

The "eigenfunction" for A should only be assigned that name if A is homogeneous of degree 1. For general $d \geq 0$ one easily calculates that

$$T(t) = \begin{cases} [(T^{1-d}(0) - (1-d)\lambda t)_+]^{1/(1-d)}, & \text{if } d \in [0,1), \\ T(0)\mathrm{e}^{-\lambda t}, & \text{if } d = 1, \\ [T^{1-d}(0) + (d-1)\lambda t]^{-1/(d-1)}, & \text{if } d \in (1,\infty), \end{cases} \tag{4.3}$$

so that for $d \in [0,1)$ there is finite lifetime (as in the TV case with $d = 0$), while for $d \geq 1$ there is polynomial decay, and for $d = 1$ even exponential decay of a solution.

Clearly there is more than one way to define an operator A that is homogeneous of decay 1. If for $p \in [1,\infty)$ the operator A is given by $Au = |u|^{2-p}\Delta_p u := |u|^{2-p}\operatorname{div}(|\nabla u|^{p-2}\nabla u)$, then the eigenvalue problem reads

$$\Delta_p u + \lambda |u|^{p-2}u = 0, \tag{4.4}$$

and at least for the first eigenvalue this is a well-understood problem, even when $p \to \infty$ and $p \to 1$, see for instance [8, 29, 31].

For reasons that will become clear below I now define

$$A_p u := \frac{1}{p}|\nabla u|^{2-p}\Delta_p u$$

and recall that $\Delta u = u_{\nu\nu} + (n-1)Hu_\nu$ and $\Delta_p u = |\nabla u|^{p-2}\{(p-1)u_{\nu\nu} + (n-1)Hu_\nu\}$ so that $A_p u = (1/p)|\nabla u|^{2-p}\Delta_p u = \big((p-1)/p\big)u_{\nu\nu} + \big((n-1)/p\big)Hu_\nu$. Thus

$$A_p u = \frac{1}{p}A_1 u + \frac{p-1}{p}A_\infty u$$

is a convex combination of A_1 and A_∞.

Special cases of $u_t - A_p u$ are then

(a) the case in which $p = \infty$. In this case $u_t - A_\infty u = u_t - u_{\nu\nu}$, an equation that is used in image processing and analyzed in [25],
(b) the case that $p = 2$, where $u_t - A_2 u = u_t - \frac{1}{2}\Delta u$, that is the linear heat equation, and
(c) the case $p = 1$, when $u_t - A_1 u = u_t - |\nabla u|\operatorname{div}(\nabla u/|\nabla u|) = 0$ represents the level set formulation of mean curvature flow, see [22].

Notice that all these equations are all *scaling invariant*, i.e., that brightness of initial data does not matter when it comes to determining the evolution of shapes. I take this and the fact that important special cases are included as an indication of the importance of this operator. For $p \in (1, \infty)$ the operator

$$A_p u := \frac{1}{p}|\nabla u|^{2-p}\Delta_p u = \sum_{i,j} a_{ij}(x) u_{x_i x_j}$$

is almost linear and has coefficients

$$a_{ij} = \frac{1}{p}\left(\delta_{ij} + (p-2)\frac{u_{x_i} u_{x_j}}{|\nabla u|^2}\right)$$

i.e., a bounded coefficient matrix

$$a = \frac{1}{p} I + \frac{p-2}{p}\frac{Du \otimes Du}{|Du|^2}$$

that is positive definite for $p \in (1, \infty)$. In fact

$$\min\left\{\frac{1}{p}, \frac{p-1}{p}\right\} I \leq a \leq \max\left\{\frac{1}{p}, \frac{p-1}{p}\right\} I.$$

In that respect the equation $u_t - A_p u = 0$ is benign, but quite degenerate. Nevertheless, my student K.Does has been able to show in [18] that viscosity solutions to this equation exist under initial datum u_0 and various boundary conditions. These solutions enjoy also comparison principles and a decay behavior that one expects from nondegenerate equations.

Acknowledgments. I thank S. Luckhaus for bringing [47] to my attention and M. Mester for his permission to use the figures from [30] in this article.

References

1. F. Alter, V. Caselles, and A. Chambolle, *Evolution of convex sets in the plane by minimizing total variation flow*, Interfaces Free Bound. **7** (2005), no. 1, 29–53.
2. L. Alvarez, F. Guichard, P.-L. Lions, and J.-M. Morel, *Axioms and fundamental equations of image processing*, Arch. Rational Mech. Anal. **123** (1993), no. 3, 199–257.
3. L. Ambrosio and V. M. Tortorelli, *Approximation of functionals depending on jumps by elliptic functionals via Γ-convergence*, Comm. Pure Appl. Math. **43** (1990), no. 8, 999–1036.
4. F. Andreu, C. Ballester, V. Caselles, and J. M. Mazón, *The Dirichlet problem for total variation flow*, J. Funct. Anal. **180** (2001), no. 2, 347–403.
5. F. Andreu, V. Caselles, J. I. Diaz, and J. M. Mazón, *Some qualitative properties for the total variation flow*, J. Funct. Anal. **188** (2002), no. 3, 516–547.
6. G. Aubert and P. Kornprobst, *Mathematical problems in image processing*, Appl. Math. Sci., vol. 147, Springer, New York, 2002.
7. G. Bellettini, V. Caselles, and M. Novaga, *Explicit solutions of the eigenvalue problem* $-\operatorname{div}(Du/|Du|) = u$ *in* $\mathbb{R}^2$, SIAM J. Numer. Anal. **36** (2005), no. 4, 1095–1129.
8. M. Belloni, B. Kawohl, and P. Juutinen, *The p-Laplace eigenvalue problem as* $p \to \infty$ *in a Finsler metric*, J. Eur. Math. Soc. (JEMS) **8** (2006), no. 1, 123–138.
9. A. Blake and A. Zisserman, *Visual reconstruction*, MIT Press Ser. Artificial Intelligence, MIT Press, Cambridge, MA, 1987.
10. A. Chambolle, *Image segmentation by variational methods: Mumford and Shah functionals and the discrete approximations*, SIAM J. Appl. Math. **55** (1995), no. 3, 827–863.
11. A. Chambolle and P.-L. Lions, *Image recovery via total variation minimization and related problems*, Numer. Math. **76** (1997), no. 2, 167–188.
12. T. F. Chan and S. Esedoḡlu, *Aspects of total variation regularized* L^1 *function approximation*, SIAM J. Appl. Math. **65** (2005), no. 5, 1817–1837.

13. T. F. Chan, J. Shen, and L. Vese, *Variational PDE models in image processing*, Notices Amer. Math. Soc. **50** (2003), no. 1, 14–26.
14. T. F. Chan and L. A. Vese, *Active contours without edges*, IEEE Trans. Image Process. **10** (2001), no. 2, 266–277.
15. M. Chipot, R. March, M. Rosati, and G. Vergara Caffarelli, *Analysis of a nonconvex problem related to signal selective smoothing*, Math. Methods Appl. Sci. **7** (1997), no. 3, 313–328.
16. G. Dal Maso, *An introduction to Γ-convergence*, Progr. Nonlinear Differential Equations Appl., vol. 8, Birkhäuser, Boston, 1993.
17. F. Dibos and G. Koepfler, *Global total variation minimization*, SIAM J. Numer. Anal. **37** (2000), no. 2, 646–664.
18. K. Does, *On a class of degenerate parabolic equations*, Ph.D. thesis, in preparation.
19. S. Esedoḡlu, *An analysis of the Perona–Malik scheme*, Comm. Pure Appl. Math. **54** (2001), no. 12, 1442–1487.
20. ———, *Stability properties of Perona–Malik scheme*, SIAM J. Numer. Anal. **44** (2006), no. 3, 1297–1313.
21. S. Esedoḡlu and S. Osher, *Decomposition of images by the anisotropic Rudin–Osher–Fatemi model*, Comm. Pure Appl. Math. **57** (2004), no. 12, 1609–1626.
22. L. C. Evans and J. Spruck, *Motion of level sets by mean curvature*. I, J. Differential Geom. **33** (1991), no. 3, 635–681.
23. A. Handlovičová and K. Mikula, *Analysis of a semidiscrete scheme for solving image smoothing equations of mean curvature flow type*, Acta Math. Univ. Comenian. (N.S.) **71** (2002), no. 1, 93–111.
24. K. Höllig, *Existence of infinitely many solutions for a forward backward heat equation*, Trans. Amer. Math. Soc. **278** (1983), no. 1, 299–316.
25. P. Juutinen and B. Kawohl, *On the evolution governed by the infinity Laplacian*, Math. Ann. **335** (2006), no. 4, 819–851.
26. B. Kawohl, *The opaque square and the opaque circle*, General Inequalities. 7 (Oberwolfach, 1995) (C. Bandle, W. N. Everitt, L. Losonczi, and W. Walter, eds.), Internat. Ser. Numer. Math., vol. 123, Birkhäuser, Basel, 1997, pp. 339–346.
27. ———, *From Mumford–Shah to Perona–Malik in image processing*, Math. Methods Appl. Sci. **27** (2004), no. 15, 1803–1814.
28. B. Kawohl and N. Kutev, *Maximum principle and comparison theorems for anisotropic diffusion*, Math. Ann. **311** (1998), no. 1, 107–123.
29. B. Kawohl and P. Lindqvist, *Positive eigenfunctions for the p-Laplace operator revisited*, Analysis (Munich) **26** (2006), no. 4, 545–550.
30. B. Kawohl and M. Mester, *Mathematische Bildverarbeitung*, Kooperative Informationsverarbeitung an der Universität zu Köln, Universität zu Köln, Cologne, 2003, pp. 86–91.
31. B. Kawohl and M. Novaga, *The p-Laplace eigenvalue problem as p approaches* 1 *and Cheeger sets in a Finsler metric*, J. Convex Anal. **15** (2008), no. 3, to appear.
32. B. Kawohl and F. Schuricht, *Dirichlet problems for the* 1*-Laplace operator, including the eigenvalue problem*, Commun. Contemp. Math. **9** (2007), no. 4, 515–543.
33. B. Kawohl and J. Stará, *Observations on a nonlinear evolution equation*, Appl. Anal. **85** (2006), no. 1-3, 143–152.
34. S. Kichenassamy, *The Perona–Malik paradox*, SIAM J. Appl. Math. **57** (1997), no. 5, 1328–1342.
35. A. V. Lair, *Uniqueness for a forward backward diffusion equation*, Trans. Amer. Math. Soc. **291** (1985), no. 1, 311–317.
36. R. March, *Visual reconstruction with discontinuities using variational methods*, Image and Vision Computing **10** (1992), no. 1, 30–38.
37. M. Mester, *Mathematische Methoden zur Bildbearbeitung*, Ph.D. thesis, Universität zu Köln, 2004.
38. L. Modica, *Γ-convergence to minimal surfaces problem and global solutions of* $\Delta u = 2(u^3 - u)$, Proceedings of the International Meeting on Recent Methods in Nonlinear Analysis (Rome, 1978) (E. De Giorgi, E. Mageners, and U. Mosco, eds.), Pitagora, Bologna, 1979, pp. 223–244.
39. L. Modica and S. Mortola, *Il limite nella Γ-convergenza di una famiglia di funzionali ellittici*, Boll. Un. Mat. Ital. A (5) **14** (1977), no. 3, 526–529.
40. J.-M. Morel and S. Solimini, *Variational methods in image segmentation*, Progr. Nonlinear Differential Equations Appl., vol. 14, Birkhäuser, Boston, 1995.

41. M. Morini and M. Negri, *Mumford – Shah functional as Γ-limit of discrete Perona – Malik energies*, Math. Models Methods Appl. Sci. **13** (2003), no. 6, 785 – 805.
42. D. Mumford and J. Shah, *Optimal approximations by piecewise smooth functions and associated variational problems*, Comm. Pure Appl. Math. **42** (1989), no. 5, 577 – 685.
43. M. Negri and M. Paolini, *Numerical minimization of the Mumford – Shah functional*, Calcolo **38** (2001), no. 2, 67 – 84.
44. K. N. Nordström, *Biased anisotropic diffusion: a unified regularization and diffusion approach to edge detection*, Image and Vision Computing **8** (1990), no. 4, 318 – 327.
45. S. Osher and M. Fedkiw, *Level set methods and dynamic implicit surfaces*, Appl. Math. Sci., vol. 153, Springer, New York, 2003.
46. P. Perona and J. Malik, *Scale-space and edge detection using anisotropic diffusion*, IEEE Transactions on Pattern Analysis and Machine Intelligence **12** (1990), no. 7, 629 – 639.
47. T. J. Richardson and S. K. Mitter, *A variational formulation-based edge focussing algorithm*, Sādhanā **22** (1997), no. 4, 553 – 574.
48. L. I. Rudin, S. Osher, and E. Fatemi, *Nonlinear total variation based noise removal algorithms*, Phys. D **60** (1992), no. 1-4, 259 – 268.
49. G. Sapiro, *Geometric partial differential equations and image analysis*, Cambridge Univ. Press, Cambridge, 2001.
50. L. Vese and T. F. Chan, *Reduced non-convex functional approximations for image restoration and segmentation*, Technical Report Report 97-56, UCLA Computational and Applied Mathematics, 1997.
51. J. Weickert, *Anisotropic diffusion in image processing*, European Consort. Math. Indust., Teubner, Stuttgart, 1998.
52. ———, *Applications of nonlinear diffusion in image processing and computer vision*, Acta Math. Univ. Comenian. (N.S.) **70** (2000), no. 1, 33 – 50.
53. K. Zhang, *Existence of infinitely many solutions for the one-dimensional Perona – Malik model*, Calc. Var. Partial Differential Equations **26** (2006), no. 2, 171 – 199.

Mathematisches Institut, Universität zu Köln, 50923 Köln, Germany
E-mail address: kawohl@math.uni-koeln.de

Centre de Recherches Mathématiques
CRM Proceedings and Lecture Notes
Volume **44**, 2008

On the Energy of a Chern – Simons – Higgs Vortex Lattice

Matthias Kurzke and Daniel Spirn

Abstract. We study the interaction of Chern – Simons – Higgs vortices with an external magnetic field. Away from self-duality, we give a detailed picture and compute critical fields using a Γ-convergence approach involving a Jacobian compactness result. Closer to the self-dual scaling, where the Γ-convergence approach fails, we analyze energy minimizers with supercritical magnetic fields and show that their energy scales asymptotically with the energy of a vortex lattice for most parameter ranges.

1. Introduction

We study the abelian Chern – Simons – Higgs energy

$$G_{\mathrm{csh}}(u, A; h_{\mathrm{ex}}) = \frac{1}{2}\int_U |\nabla_A u|^2 + \frac{\mu^2}{4}\frac{|\operatorname{curl} A - h_{\mathrm{ex}}|^2}{|u|^2} + \frac{1}{\varepsilon^2}|u|^2(1-|u|^2)^2\,\mathrm{d}x \tag{1.1}$$

for a complex order parameter u and a magnetic potential A. We assume that U is a two-dimensional, simply connected domain and h_{ex} a given external magnetic field. The parameter ε will be assumed to be small, while several possible scalings of μ will be considered.

This functional arises by considering time-independent ($\partial_t u \equiv 0$) extremals of the Chern – Simons – Higgs Lagrangian

$$\mathcal{L}_{\mathrm{CSH}} = \int D_\alpha \overline{D^\alpha u} + \frac{\mu}{4}\epsilon^{\alpha\beta\gamma} A_\alpha F_{\beta\gamma} - \frac{1}{\varepsilon^2}|u|^2(1-|u|^2)^2,$$

where we have written $D_\alpha = (-\partial_\Phi, \nabla_A)$, $\partial_\Phi = \partial_t - \mathrm{i}\Phi$ and $\nabla_A = \nabla - \mathrm{i}A$ for the potential $A_\alpha = (\Phi, A)$. Furthermore we use $\epsilon^{\alpha\beta\gamma}$ to denote the antisymmetric tensor, F_α for the Maxwell curvature tensor. The term $\epsilon^{\alpha\beta\gamma}A_\alpha F_{\beta\gamma}$ is known as the Chern – Simons term. This classical field theory, defined on $(2+1)$-dimensional Minkowski space, has been studied as an anyon model [3, 7, 8]. Anyon models allow for fractional values of charge and have applications to high temperature superconductivity and quantum Hall effects and have therefore caused much interest in the physics community, see the book of Yang [20] for an excellent overview of Chern – Simons – Higgs and related gauge theories. If we understand CSH as a

2000 *Mathematics Subject Classification.* Primary 35J50; Secondary 81T13.

D. Spirn was supported in part by NSF grant DMS-0510121.

This is the final form of the paper.

model for high temperature superconductors, we are naturally led to the inclusion of interaction with an applied magnetic field and so have included $h_{\rm ex}$ in (1.1).

We briefly describe some features of CSH topological vortices. First of all, magnetic fields concentrate in *annular regions* about each topological vortex. This is in contrast to Ginzburg–Landau vortices, where the magnetic field concentrates at the *site* of the vortex. This different concentration behavior is due to the $1/|u|^2$ part of the second term of (1.1). Second, in the $\varepsilon \to 0$ limit $|u_\varepsilon|$ relaxes to $\mathbb{S}^1 \cup \{0\}$, as opposed to $\mathbb{S}^1$ in the Ginzburg–Landau case.

Most attention has focused on the *self-dual* case where $\mu = \varepsilon$ and $h_{\rm ex} = 0$. In this case the CSH equations reduce, following Hong–Kim–Pac[7] and Jackiw–Weinberg [8], to a system of first order PDEs. Solutions can be recovered by solving (after a substitution) a Liouville-type elliptic equation, similar to the Jaffe–Taubes approach to solving self-dual solutions in Ginzburg–Landau theory [9]. We direct the reader to [3–5, 7, 8, 16, 19, 20] and the references therein for a partial accounting of the self-dual theory for Chern–Simons–Higgs equations.

The first rigorous results for non self-dual Chern–Simons–Higgs equations with ε and $\mu = O(1)$ are those of Han–Kim [6], who showed the existence of critical points of (1.1) and studied sequential minimizers of (1.1) with $A_\varepsilon \equiv 0$, i.e.,

$$E_{\rm csh}(u) = \frac{1}{2}\int_U |\nabla u|^2 + \frac{1}{\varepsilon^2}|u|^2(1-|u|^2)^2 \tag{1.2}$$

and Dirichlet boundary condition $u_\varepsilon = g$ on ∂U with $|g| = 1$. Their proofs are similar in spirit to Béthuel–Brezis–Hélein [1] for the simplified Ginzburg–Landau energy

$$E_{\rm gl}(u) = \frac{1}{2}\int_U |\nabla u|^2 + \frac{1}{2\varepsilon^2}(1-|u|^2)^2. \tag{1.3}$$

The limit of minimizers $u_\varepsilon \rightharpoonup u_\star$ of (1.2) is a harmonic map into $\mathbb{S}^1$ with singularities of unit degree located at a minimizing configuration of the BBH renormalized energy $W(\{a_j\})$, just as for minimizers of (1.3) in [1]. In [6] the proof makes use of the maximum principle to establish $|u|_{L^\infty} \le 1$, which is central to their argument; however, the maximum principle fails when A is nontrivial.

In [13, 14] the authors initiated study of (1.1) with $A_\varepsilon \not\equiv 0$ in the Γ-convergence framework under various energy scalings for $\varepsilon \to 0$ and $\mu_\varepsilon \to \mu \in (0, +\infty]$. One consequence of our Γ-limit theorem, see Theorem 2 below, is that it allows for a calculation of the first critical field (the external magnetic field where minimizers first start to have vortices). We prove that for $\mu \in (0, +\infty]$

$$H_{c_1} = C(U, \mu)|\log \varepsilon|, \tag{1.4}$$

where $C(U, +\infty) > 0$ is a finite, nonzero constant, see (2.4) for the precise form of H_{c_1}. These results are reviewed in Section 2 below.

On the other hand if $\mu_\varepsilon \to 0$, parts of the compactness result fail, as was shown in [13]. This failure seems to follow from the lessened effect of the energy reduction via vortex nucleation. Hence, we cannot employ the same variational argument to compute the critical magnetic field when $\mu_\varepsilon \to 0$. On the other hand formal considerations point to a critical nucleation field of

$$H_{c_1} = \frac{C}{\mu_\varepsilon^2}|\log \varepsilon| \tag{1.5}$$

as $\mu_\varepsilon \to 0$, see Remark 5 in Section 2. Note this implies that the critical field should be asymptotically larger than $|\log \varepsilon|$, the Ginzburg–Landau critical field strength. Magnetic field strengths below (1.5) should yield minimizers that satisfy the Meissner solution $u \equiv 1$ and $A \equiv 0$; whereas, magnetic fields asymptotically above (1.5) should yield something akin to a vortex lattice, as was shown by Sandier–Serfaty [18] for Ginzburg–Landau minimizers.

In this paper we aim to show evidence of the first critical field satisfying (1.5) by showing the transition from the Meissner state $u \equiv 1$ and $A \equiv 0$ with energy $(|U|/8)\mu_\varepsilon^2 h_{\text{ex}}^2$ to the lattice state with energy $(|U|/2)h_{\text{ex}} \log\big(1/(\varepsilon\sqrt{h_{\text{ex}}})\big)$ occurs through the transition field strength (1.5). Hence we should expect a vortex lattice for $h_{\text{ex}} \gg |\log \varepsilon|/\mu_\varepsilon^2$ when $\mu_\varepsilon \to 0$, where throughout the article we denote $f_\varepsilon \ll g_\varepsilon$ if $f_\varepsilon/g_\varepsilon \to 0$ as $\varepsilon \to 0$.

The purpose of this article is to establish that global minimizers with field strengths asymptotically above (1.5) attain the *same* energy as that of a vortex lattice. We allow for a large regime of rates of $\mu_\varepsilon \to 0$. In particular we have

Theorem 1. *Assume that $\{u_\varepsilon, A_\varepsilon\}$ is a sequence of minimizers of G_{csh} under the condition $|u_\varepsilon| = 1$ on ∂U. If there exists a small $\gamma > 0$ such that*

$$\max\left\{|\log \varepsilon|, \frac{|\log \varepsilon|}{\mu_\varepsilon^2}\right\} \ll h_{\text{ex}} \ll \varepsilon^{-2/3+\gamma} \tag{1.6}$$

and μ_ε satisfies

$$\mu_\varepsilon^2 \gg \varepsilon^{1/3-\gamma} \quad \text{and} \quad \mu_\varepsilon^{-4} \le C h_{\text{ex}} \tag{1.7}$$

for some constant C, then as $\varepsilon \to 0$, the minimal energy satisfies

$$G_{\text{csh}}(u_\varepsilon, A_\varepsilon) = \frac{1}{2}|U| h_{\text{ex}} \log \frac{1}{\varepsilon\sqrt{h_{\text{ex}}}}\big(1 \pm o_\varepsilon(1)\big), \tag{1.8}$$

where $o_\varepsilon(1)$ signifies a quantity that vanishes as $\varepsilon \to 0$.

The energy scaling (1.8) is the same energy scaling as for supercritical magnetic fields in the Ginzburg–Landau equations, see Sandier–Serfaty [18], and we employ a similar approach to establishing Theorem 1.

Remark 1. The lower bound on the supercritical field strength in (1.6) follows from combining (1.4) and (1.5). The rest of the range conditions for h_{ex} and μ_ε follow from the proof of the theorem, see Remark 7.

Remark 2. When $\mu_\varepsilon \to +\infty$, the energy scales in the same way as when $\mu_\varepsilon \equiv C$, a finite, nonzero constant. Hence, there seems to be little benefit to energy reduction via vortex nucleation from having an asymptotically large μ_ε.

Remark 3. When $\mu_\varepsilon \to 0$ and $h_{\text{ex}} \ll \mu_\varepsilon^{-4}$, we have slightly weaker results concerning only an upper or a lower bound for the energy, see Propositions 4 and 5, respectively. We note there is an energy mismatch:

$$\frac{1}{2}|U| h_{\text{ex}} \log \frac{\mu_\varepsilon^2}{\varepsilon} \le G_{\text{csh}}(u, A) \le \frac{1}{2}|U| h_{\text{ex}} \log \frac{1}{\varepsilon\sqrt{h_{\text{ex}}}},$$

and we are unsure of whether the lower bound or the upper bound is optimal.

Remark 4. We can adapt the techniques presented here to understanding the Ginzburg–Landau energy with asymptotically large or small domains, [15].

The paper is organized as follows. In Section 2 we review the Γ-convergence results of [13, 14] and provide a sketch of their proofs. Section 3 establishes the energy upper bound of the vortex lattice in the supercritical regime. Finally, in Section 4 we prove the corresponding energy lower bound in the supercritical regime. The results found in Sections 3 and 4 are new.

2. The Γ-convergence result and corollaries

In this section we recall results from our previous articles [13, 14] about the asymptotic behavior of the CSH energy in the $\varepsilon \to 0$ regime. The results are stated in the spirit of Γ-convergence, that is, separated into a compactness result combined with a lower bound for the energy and a construction that shows the essential optimality of the lower bound.

The techniques used are related to the approach of Jerrard–Soner [11, 12] combined with a version of the vortex ball construction method of Jerrard [10] and Sandier [17]. Similar to their approach to studying (1.3), we first study the simplified functional

$$E_{\mathrm{csh}}(u) = \frac{1}{2}\int_U |\nabla u|^2 + \frac{1}{\varepsilon^2}|u|^2(1-|u|^2)^2, \tag{2.1}$$

which helps us express the energetic limit of $G_{\mathrm{csh}}(u, A)$. Similar to the approach in [12], it is possible to consider various rescalings of the terms in the energy. Choosing scaling factors r_ε, s_ε, t_ε and considering $v_\varepsilon = j(u_\varepsilon)/r_\varepsilon$, $a_\varepsilon = A_\varepsilon/s_\varepsilon$ and $h_{\mathrm{ex}} = H/t_\varepsilon$ then the energy approximately decomposes as

$$G_{\mathrm{csh}}(u_\varepsilon, A_\varepsilon; h_{\mathrm{ex}}) \approx E_{\mathrm{csh}}(|u_\varepsilon|) + \frac{r_\varepsilon^2}{2}\left[\left\| v_\varepsilon - \frac{s_\varepsilon}{r_\varepsilon} a_\varepsilon \right\|^2_{L^2(U\setminus \bigcup B_{r_j})} + \frac{|\log\varepsilon|}{r_\varepsilon}\|\operatorname{curl} v_\varepsilon\|_{\mathcal{M}(\bigcup B_{r_j})}\right.$$
$$\left. + \frac{\mu_\varepsilon^2}{4}\left\| \frac{(s_\varepsilon/r_\varepsilon)\operatorname{curl} a_\varepsilon - (t_\varepsilon/r_\varepsilon)H}{|u_\varepsilon|}\right\|^2_{L^2(U)}\right].$$

As in [12], we discovered in [14] that the critical scaling where the order parameter interacts at the same order as the induced magnetic field is given by choosing $r_\varepsilon = s_\varepsilon = t_\varepsilon = |\log\varepsilon|$ and $G_{\mathrm{csh}} \approx |\log\varepsilon|^2$.

In [13], the authors extended these results to scaling limits where $\mu_\varepsilon \to \infty$ and gave an example for a noncompact sequence for any $\mu_\varepsilon \to 0$. Unlike the case of $\mu_\varepsilon \equiv \mu$, these cases can't be reduced to $\mu = 1$ by a simple rescaling of the domain.

Our general assumptions are that $U \subset \mathbb{R}^2$ is a bounded, simply connected domain with smooth boundary. For simplicity, we state the result in the Coulomb gauge, which amounts to considering only pairs (u, A) with $\nabla \cdot A = 0$ in U and $A \cdot \nu = 0$ on ∂U. These conditions can always be satisfied by an appropriate gauge transformation replacing (u, A) by $(ue^{\mathrm{i}\chi}, A + \nabla\chi)$ without changing the energy.

Theorem 2 ([13, 14]). *Assume that the external field satisfies $h_{\mathrm{ex}} = H|\log\varepsilon|$ for some $H > 0$, and consider a sequence $\{u_\varepsilon, A_\varepsilon\}$ that satisfies the Coulomb gauge condition and*

$$G_{\mathrm{csh}}(u_\varepsilon, A_\varepsilon; h_{ex}) \leq K|\log\varepsilon|^2.$$

Set $a_\varepsilon = (1/|\log\varepsilon|)A_\varepsilon$, then $\{a_\varepsilon\}$ is weakly precompact in $W^{1,p}$ for all $p < 2$, and for a subsequence such that $a_\varepsilon \rightharpoonup a$ there holds $(\operatorname{curl} a_\varepsilon - H)/|u_\varepsilon| \rightharpoonup \operatorname{curl} a - H$ in L^2.

Additionally, the compactness assertions of the last theorem hold: $v_\varepsilon = (1/|\log\varepsilon|)j(u_\varepsilon)$ *converges to* v *weakly in all* L^p *with* $p<2$, $v_\varepsilon/|u_\varepsilon| \rightharpoonup v$ *in* L^2, *and* $w_\varepsilon = J(u_\varepsilon)/|\log\varepsilon| \rightharpoonup w = \frac{1}{2}\operatorname{curl} v$. *Taking a subsequence, the modulus* $\rho_\varepsilon = |u_\varepsilon|$ *satisfies* $\rho_\varepsilon \to \rho$ *strongly in* L^p *for* $p<+\infty$ *where* ρ *is either identically* 0 *or identically* 1. *If* $\rho = 0$, *then* $\operatorname{curl} a = H$ *and* $v=0$. *Furthermore, the energy satisfies*

$$\liminf_{\varepsilon\to 0}\frac{1}{|\log\varepsilon|^2}G_{\mathrm{csh}}(u_\varepsilon, A_\varepsilon; h_{\mathrm{ex}}) \geq G^\mu_\rho(v,a;H), \tag{2.2}$$

with

$$G^\mu_1(v,a;H) = \frac{1}{2}\left(\int_U |v-a|^2 + \frac{\mu^2}{4}|\operatorname{curl} a - H|^2 + \|\operatorname{curl} v\|_{\mathcal{M}}\right)$$
$$G^\mu_0(v,a;H) = 0.$$

when $\mu<\infty$. *Finally,*

$$G^\infty_1(v,a;H) = \begin{cases} \frac{1}{2}\left(\int_U |v-a|^2 + \|\operatorname{curl} v\|_{\mathcal{M}}\right) & \text{if } \operatorname{curl} a = H \\ +\infty & \text{else}\end{cases} \tag{2.3}$$
$$G^\infty_0(v,a;H) = 0.$$

Conversely, for any $a\in H^1(U;\mathbb{R}^2)$ *and* $v\in L^2(U;\mathbb{R}^2)$ *such that* $w = \frac{1}{2}\operatorname{curl} v$ *is a Radon measure, there exists a sequence* $\{u_\varepsilon\}$ *in* $H^1(U;\mathbb{C})$ *with* $|u_\varepsilon| = 1$ *on* ∂U *and a sequence* $\{A_\varepsilon\}\in H^1(U;\mathbb{C})$ *satisfying the Coulomb gauge conditions such that* $v_\varepsilon = (1/|\log\varepsilon|)j(u_\varepsilon) \rightharpoonup v$ *in* L^2, $w_\varepsilon = (1/|\log\varepsilon|)J(u_\varepsilon) \rightharpoonup w$ *in* $(C^{0,\beta})^*$, $a_\varepsilon = (1/|\log\varepsilon|)A_\varepsilon \rightharpoonup a$ *in* H^1, *and such that* (2.2) *holds with equality for* $\rho = 1$. *For* $\rho = 0$, *there exists a sequence* $(u_\varepsilon, A_\varepsilon)$ *with* $u_\varepsilon \to 0$ *and* $(1/|\log\varepsilon|)\operatorname{curl} A_\varepsilon \to H$ *such that* $G_{\mathrm{csh}}(u_\varepsilon, A_\varepsilon; h_{\mathrm{ex}}) \to 0$.

As an application of Theorem 2, we calculate the critical field H_{c_1} for which vortices appear in nonzero minimizers of $G_{\mathrm{csh}}(u_\varepsilon, A_\varepsilon; h_{\mathrm{ex}})$.

Corollary 1. *As* $\varepsilon\to 0$, *the critical field* H_{c_1} *for nontrivial local minimizers is given asymptotically by* $H_{c_1}|\log\varepsilon|$, *where*

$$H_{c_1}(\mu) = \frac{2}{\mu^2\max_{\overline{U}}|z_\mu|} \tag{2.4}$$

and z_μ *is the solution of*

$$-\frac{\mu^2}{4}\Delta z_\mu + z_\mu + 1 = 0 \text{ in } U \quad z_\mu = 0 \text{ on } \partial U.$$

Concerning the dependence on μ, *we have that*

$$\mu^2 H_{c_1}(\mu) \to 2 \tag{2.5}$$

as $\mu\to 0$. *Furthermore,* $H_{c_1}(\mu)$ *is decreasing in* μ *and converges to a limit* $\overline{H}(U) > 0$ *as* $\mu\to\infty$, *which is given by* $\overline{H}(U) = 2/\max_{\overline{U}}|z_\infty|$, *where* z_∞ *is the solution of* $-\Delta z_\infty + 1 = 0$ *with Dirichlet boundary conditions, which is the same as the critical field for letting* $\mu\to\infty$ *in* (2.3).

The corollary follows from Theorem 2 using some analysis of the limit functional, see [13, 14].

We obtain bounds for the energy of minimizers in these specific scalings by a simple application of the "construction" part of the Γ-convergence result of Theorem 2:

Proposition 1. *The limiting energy functional satisfies the following bounds:*

• **Vortex state.** *With* $v = a$ *and* $\operatorname{curl} v = \operatorname{curl} a = H$ *we obtain (even for* $\mu = \infty$*)*

$$G^\mu(v, a; H) = \tfrac{1}{2} H|U|$$

so for an appropriate approximating sequence

$$G_{\text{csh}}(u_\varepsilon, A_\varepsilon; h_{\text{ex}}) \leq \frac{1}{2} H|U||\log \varepsilon|^2 \big(1 + o_\varepsilon(1)\big) = \frac{|U|}{2} h_{\text{ex}} |\log \varepsilon| \big(1 + o_\varepsilon(1)\big).$$

• **Meissner state.** *Choosing* $v = a = 0$*, we obtain*

$$G^\mu(v, a; H) = \frac{\mu^2}{8} H^2 |U|,$$

so there exists an approximating sequence satisfying

$$G_{\text{csh}}(u_\varepsilon, A_\varepsilon; h_{\text{ex}}) \leq \frac{\mu^2 |U|}{8} H^2 |\log \varepsilon|^2 \big(1 + o_\varepsilon(1)\big) = \frac{\mu^2 |U|}{8} h_{\text{ex}}^2 \big(1 + o_\varepsilon(1)\big).$$

Remark 5. We can now see the evidence of the proposed critical field (1.5). First, if one compares the two upper bounds given in Proposition 1, one finds that they give the same value if $h_{\text{ex}} = H|\log \varepsilon| = 4|\log \varepsilon|/\mu^2$. Second, the evidence also comes from the asymptotic limit (2.5).

In the following section, we will extend these results outside of the critical scaling treated using the Γ-convergence result. In particular, we extend it to cases where $\mu_\varepsilon \to 0$ slowly enough.

Sketch of proof of Theorem 2. The main step in the proof of Theorem 2 is a compactness theorem for the Jacobians $J(u_\varepsilon)$ under a bound on the reduced energy $E_{\text{csh}}(u)$. The main ingredients in the compactness proof are estimates on the amount of level sets of an auxiliary function where the degree of u satisfies certain bounds.

Set $\phi \in C_c^{0,1}(U)$ a Lipschitz function vanishing on ∂U. We define $\Omega(t) = \{x \in U \text{ such that } \phi(x) > t\}$ then $\partial\Omega(t)$ is a level set ϕ. Let

$$\operatorname{Reg}(\phi) := \{t \in [0, \|\phi\|_{L^\infty}] \text{ such that } \partial\Omega(t) = \phi^{-1}(t),\ \partial\Omega(t) \text{ rectifiable}, \\ \text{and } \mathcal{H}^1\big(\partial\Omega(t)\big) < \infty\}.$$

By the co-area formula $|\operatorname{Reg}(\phi)| = \|\phi\|_{L^\infty}$ and $t \in \operatorname{Reg}(\phi)$ implies $\partial\Omega(t)$ is a union of finite Jordan curves, $\Gamma_i(t)$. We set, as in [11],

$$\Gamma(t) := \bigcup \Big\{ \text{components of } \partial\Omega(t) \text{ such that } \min_{x \in \Gamma_i(t)} |u| > \frac{1}{2} \Big\}$$

We set $d \in \mathbb{Z}^+$ and define

$$D_d := \big\{ t \in \operatorname{Reg}(\phi) : \Gamma(t) \text{ is nonempty and } \big|\deg\big(u; \Gamma(t)\big)\big| \geq d + 1 \big\}$$

Let us define

$$E_\phi(u) = \int_{\operatorname{spt}(\phi)} e_{\text{csh}}(u)\, \mathrm{d}x$$

for short. With these definitions we have

Proposition 2. *Suppose $U \subset \mathbb{R}^2$ and $u \in H^1(U;\mathbb{C})$ then*

$$\left|\int_U \phi J(u)\,\mathrm{d}x\right| \le (\pi d + \varepsilon^{1/2})\|\phi\|_{L^\infty} + \varepsilon^{1/3}\|\nabla\phi\|_{L^\infty}\left[2E_\phi^2(u) + 3E_\phi(u) + \frac{|\operatorname{spt}(\phi)|}{4}\right] + \frac{|D_d|}{4}E_\phi(u) \tag{2.6}$$

for any $\varepsilon \le 1$.

In order to use (2.6), we need to estimate $|D_d|$. This is controlled by the following result.

Proposition 3. *Suppose $|u| = 1$ on ∂U and let $\varepsilon \in (0, \mathrm{e}^{-2})$. Define*

$$\varepsilon^{\alpha_\varepsilon} = \varepsilon|\log\varepsilon|^2 \int_U e_{\mathrm{csh}}(u)\,\mathrm{d}x \quad \eta_\varepsilon = \frac{1}{1 - 2/|\log\varepsilon|} > 1,$$

then

$$|D_d| \le 8\varepsilon^{\alpha_\varepsilon}\|\nabla\phi\|_{L^\infty}\left(1 + \exp\left(\frac{\eta_\varepsilon}{\pi d_\star}\int_V e_{\mathrm{csh}}(u)\,\mathrm{d}x\right)\right)$$

where $d_\star = d + 1$. Here $V \equiv \dot{\bigcup}_j B_{r_j}$ is the union of disjoint balls B_{r_j} with $B_{r_j} \subseteq \operatorname{spt}(\phi) \subseteq U$ and $\sum_j r_j = |D_d|/(2\|\nabla\phi\|_{L^\infty})$.

Theorem 3. *Suppose*

$$K_2^U = \sup_{0<\varepsilon<\varepsilon_0} \int_{\operatorname{spt}(\phi)} \frac{e_{\mathrm{csh}}(u_\varepsilon)}{|\log\varepsilon|^2}\,\mathrm{d}x < \infty.$$

Set $v_\varepsilon = j(u_\varepsilon)/|\log\varepsilon|$ and $w_\varepsilon = J(u_\varepsilon)/|\log\varepsilon| = \frac{1}{2}\operatorname{curl} v_\varepsilon$, then w_ε is precompact in $(C^{0,\beta})^*$ for any $\beta > 0$. Furthermore, $\|v_\varepsilon\|_{L^p} \le C$ for $1 \le p < (\gamma+2)/(\gamma+1)$ and all $\varepsilon > 0$. Finally, $v_\varepsilon/|u_\varepsilon| \rightharpoonup v$ weakly in L^2.

The main step in the proof is the bound

$$\left|\int \phi w_\varepsilon\,\mathrm{d}x\right| \le \pi d_\lambda\|\phi\|_{L^\infty} + C\varepsilon^{(\lambda-1)/(2\lambda)}\|\phi\|_{C^{0,1}}, \tag{2.7}$$

for $d_\lambda = \lfloor(\lambda/\pi)\int_V e_{\mathrm{csh}}(u)/|\log\varepsilon|^2\rfloor$ and $\lambda > 1$. Inequality (2.7) is a consequence of Propositions 2 and 3. The $(C^{0,\beta})^*$ compactness then follows by interpolation, see [11]. The energy lower bound also follows from (2.7) and an appropriate choice of a test function. On the other hand the energy upper bound follows from an argument similar to the argument in Section 3 below. □

3. Upper bound construction

In this section, we generalize the results of Proposition 1 to more general energy and magnetic field scalings, by giving a more direct construction instead of relying on our previous Γ-convergence argument.

Depending on the scaling of h_{ex} and μ_ε, we can again consider two cases corresponding to "Meissner" and "vortex" state.

Proposition 4. *The following asymptotic bounds hold for sequences of minimizers $(u_\varepsilon, A_\varepsilon)$:*

- **Vortex state.** *If* $h_{\rm ex} \ll 1/\varepsilon$ *and* $\mu_\varepsilon \gg \varepsilon^{\alpha/2}$ *for some* $\alpha < 1$*, then*

$$G_{\rm csh}(u_\varepsilon, A_\varepsilon) \le \frac{1}{2}|U| h_{\rm ex} \log \frac{1}{\varepsilon\sqrt{h_{\rm ex}}} \tag{3.1}$$

- **Meissner state.** *For any value of* $h_{\rm ex}$*, there holds*

$$G_{\rm csh}(u_\varepsilon, A_\varepsilon) \le \frac{|U|}{8}\mu_\varepsilon^2 h_{\rm ex}^2 \tag{3.2}$$

Proof of Proposition 4. The second claim is trivially verified by choosing $(u_\varepsilon, A_\varepsilon) = (1, 0)$. For the first, we adopt a construction due to Sandier – Serfaty [18] and construct a periodic lattice of vortices that will have the claimed energy bound. The proof uses Lemmas 1 and 2 below. First we construct h, which will be close to $\operatorname{curl} A$. For $L = \sqrt{2\pi/h_{\rm ex}}$ and $K_L = (-L/2, L/2)^2$, we define h as the solution of

$$\begin{aligned} -\frac{\mu_\varepsilon^2}{4}\Delta h + h &= 2\pi\delta_\varepsilon && \text{in } K_L \\ \frac{\partial h}{\partial \nu} &= 0 && \text{on } \partial K_L, \end{aligned}$$

where δ_ε is an approximation of a δ-distribution given by

$$\delta_\varepsilon(x) = \begin{cases} 1/(\pi\varepsilon^2) & |x| < \varepsilon \\ 0 & \text{else.} \end{cases}$$

Due to the Neumann boundary condition, we can periodically extend h to $\mathbb{R}^2$, so in particular, h is defined in U. We let A_0 be the solution of the system

$$\begin{aligned} \operatorname{curl} A_0 &= h && \text{in } U \\ \operatorname{div} A_0 &= 0 && \text{in } U \\ A_0 \cdot \nu &= 0 && \text{on } \partial U. \end{aligned} \tag{3.3}$$

Next we define $u = \rho e^{i\varphi}$, where ρ is the periodic extension of a function in K_L such that

$$\rho(x) = \begin{cases} 0 & |x| < \varepsilon \\ |x|/\varepsilon - 1 & \varepsilon \le |x| \le 2\varepsilon \\ 1 & x > 2\varepsilon, \end{cases} \tag{3.4}$$

and φ is given by $\nabla\varphi = A_0 - (\mu_\varepsilon^2/4)\nabla^\perp h$. For any closed C^1 curve $\Gamma \subset U \cap \operatorname{spt}\rho$ bounding a domain I with $\partial I = \Gamma$ there holds

$$\int_\Gamma \nabla\varphi \cdot \tau = \int_\Gamma \left(-\frac{\mu_\varepsilon^2}{4}\frac{\partial h}{\partial \nu} + A_0 \cdot \tau \right) = \int_I \left(-\frac{\mu_\varepsilon^2}{4}\Delta h + h \right) = \int_I \delta_\varepsilon \in 2\pi\mathbb{Z},$$

so $e^{i\varphi}$ is continuous in $U \cap \operatorname{spt}\rho$, and $\rho = 0$ elsewhere. Therefore, u is a well-defined $W^{1,\infty}$ function everywhere.

We define A as the solution of

$$\begin{aligned} \operatorname{curl} A &= \rho(\operatorname{curl} A_0 - h_{\rm ex}) + h_{\rm ex} && \text{in } U \\ \operatorname{div} A &= 0 && \text{in } U \\ A \cdot \nu &= 0 && \text{on } \partial U. \end{aligned} \tag{3.5}$$

We calculate the energy of (u, A), ignoring the slight mismatch between the lattice and the domain, which can be dealt with by choosing an offset just as in [18].

$$G_{\mathrm{csh}}(u,A) = \frac{1}{2}\int_U |(\nabla - \mathrm{i}A)u|^2 + \frac{\mu_\varepsilon^2}{4}\frac{|\mathrm{curl}\, A - h_{\mathrm{ex}}|^2}{|u|^2} + \frac{1}{\varepsilon^2}|u|^2(1-|u|^2)^2$$
$$= \frac{1}{2}\int_U \rho^2|\nabla\varphi - A|^2 + |\nabla\rho|^2 + \frac{\mu_\varepsilon^2}{4}|h - h_{\mathrm{ex}}|^2 + \frac{1}{\varepsilon^2}\rho^2(1-\rho^2)^2. \tag{3.6}$$

On K_L and each of its periodic copies we have $\int_{K_L} |\nabla\rho|^2 + (1/\varepsilon^2)\rho^2(1-\rho^2)^2 \le C$, so on U these terms can be bounded by NC, where $N \approx |U|/|K_L| = |U|h_{\mathrm{ex}}/(2\pi)$ is the number of squares. We decompose the first term in (3.6) as follows:

$$\rho^2|\nabla\varphi - A|^2 = \rho^2|\nabla\varphi - A_0|^2 + \rho^2(|\nabla\varphi - A|^2 - |\nabla\varphi - A_0|^2)$$
$$= \frac{\rho^2\mu_\varepsilon^4}{16}|\nabla h|^2 + \rho^2\big(|A|^2 - |A_0|^2 + 2\nabla\varphi\cdot(A_0 - A)\big) \tag{3.7}$$

The last term in (3.7) can be further rewritten as

$$\rho^2(|A|^2 - |A_0|^2 + 2(A_0 - \frac{\mu_\varepsilon^2}{4}\nabla^\perp h)\cdot(A_0 - A)$$
$$= \rho^2\Big(|A - A_0|^2 + \frac{\mu_\varepsilon^2}{2}(A - A_0)\cdot\nabla^\perp h\Big). \tag{3.8}$$

Using Lemma 1 and (3.6), we can estimate

$$\frac{1}{2}\int_U \rho^2|\nabla\varphi - A_0|^2 + \frac{\mu_\varepsilon^2}{4\rho^2}|\mathrm{curl}\, A - h_{\mathrm{ex}}|^2 + \frac{1}{\varepsilon^2}\rho^2(1-\rho^2)^2$$
$$\le N\pi\Big(\log\frac{L}{\varepsilon} + C\Big) \le \frac{1}{2}h_{\mathrm{ex}}|U|\log\frac{1}{\varepsilon\sqrt{h_{\mathrm{ex}}}}\big(1 + o_\varepsilon(1)\big). \tag{3.9}$$

To estimate the missing terms in (3.7), we use elliptic regularity. Since $\mathrm{curl}\, A - \mathrm{curl}\, A_0 = (h - h_{\mathrm{ex}})(\rho - 1)$, it follows using Lemma 2 and $\rho = 1$ outside $B_\varepsilon(a_i)$ that

$$\|A - A_0\|^2_{L^2(U)} \le C\int_{\bigcup B_{2\varepsilon}(a_i)} h_{\mathrm{ex}}^2 + \Big(\frac{C}{\mu_\varepsilon^2}\log\frac{1}{\varepsilon\sqrt{h_{\mathrm{ex}}}}\Big)^2 \tag{3.10}$$

and we can estimate this as

$$\|A - A_0\|^2_{L^2(U)} \le C(N\varepsilon^2)h_{\mathrm{ex}}^2 + \frac{C}{\mu_\varepsilon^4}\Big|\log\frac{1}{\varepsilon\sqrt{h_{\mathrm{ex}}}}\Big|^2(N\varepsilon^2). \tag{3.11}$$

The first of these terms is $\ll h_{\mathrm{ex}}$, since $N \approx h_{\mathrm{ex}}$ and $\varepsilon h_{\mathrm{ex}} \ll 1$. For the second term, we use that $\mu_\varepsilon^2 \gg \varepsilon^\alpha$, from which it follows that also this term is $\ll h_{\mathrm{ex}}\log\big(1/(\varepsilon\sqrt{h_{\mathrm{ex}}})\big)$. The final term in (3.8) can then be also bounded by

$$C\mu_\varepsilon^2\|\nabla h\|_{L^2(U)}\|A - A_0\|_{L^2(U)},$$

which is small by Lemma 1 and the smallness we established for (3.11). □

We finish this section with two technical lemmas used in the previous proof.

Lemma 1. *Suppose $L > 2\varepsilon$, then on K_L, we have the estimate*

$$\int_{K_L} \frac{\mu_\varepsilon^4}{32}|\nabla h|^2 + \frac{\mu_\varepsilon^2}{8}|h - h_{\mathrm{ex}}|^2 \le \Big(1 + \frac{C}{\mu_\varepsilon^2 h_{\mathrm{ex}}}\Big)\Big(\pi\log\frac{L}{\varepsilon} + C\Big). \tag{3.12}$$

PROOF. Following closely Sandier – Serfaty [18], we first consider the problem on a ball instead of a square. Let f be the solution of

$$\begin{aligned} -\frac{\mu_\varepsilon^2}{4}\Delta f + f &= 2\pi\delta_\varepsilon && B_L \\ \frac{\partial f}{\partial \nu} &= 0 && \partial B_L. \end{aligned}$$

Using the same techniques as Sandier – Serfaty, we obtain that f is radial, $f \in H^2$, $f \geq 0$, and the estimates $|\nabla f| \leq 4/(\mu_\varepsilon^2 |x|)$ for $|x| > \varepsilon$ and $|\nabla f| \leq 4r/(\mu_\varepsilon^2 \varepsilon^2)$ for $|x| \leq \varepsilon$, from which we deduce

$$\int_{B_L} |\nabla f|^2 \leq \frac{16}{\mu_\varepsilon^4}\left(2\pi \log\frac{L}{\varepsilon} + \pi\right)$$

and for any $q < 2$, $\|\nabla f\|_{L^q} \leq C_q$.

To obtain information about h from this, we introduce $g = h - f$, which solves

$$\begin{aligned} -\frac{\mu_\varepsilon^2}{4}\Delta g + g &= 0 && K_L \\ \frac{\partial g}{\partial \nu} &= -\frac{\partial f}{\partial \nu} && \partial K_L. \end{aligned}$$

It is easy to see that $0 \leq \bar{g} = (1/K_L)\int_{K_L} g \leq h_{\rm ex}$. By the equation for g, we have

$$\int_K \left(-\frac{\mu_\varepsilon^2}{4}\Delta g + g - \bar{g}\right)(g - \bar{g}) = -\int_K \bar{g}(g - \bar{g}).$$

Integrating by parts, this shows

$$-\frac{\mu_\varepsilon^2}{4}\int_{\partial K_L} \frac{\partial g}{\partial \nu}(g - \bar{g}) + \int_{K_L} \frac{\mu_\varepsilon^2}{4}|\nabla g|^2 + |g - \bar{g}|^2 = -\int_{K_L} \bar{g}(g - \bar{g}).$$

As in [18], there holds by Poincaré and Cauchy – Schwarz

$$\left|\bar{g}\int_{K_L} (g - \bar{g})\right| \leq C\|\nabla g\|_{L^2(K_L)}.$$

From the boundary trace theorem, we obtain

$$\int_{\partial K_L} |g - \bar{g}|^2 \leq \frac{C}{\sqrt{h_{\rm ex}}}\int_{K_L} |\nabla g|^2.$$

Furthermore, due to the estimates on ∇f and $L > 2\varepsilon$,

$$\int_{\partial K_L} \left|\frac{\partial f}{\partial \nu}\right|^2 \leq C\mu_\varepsilon^{-4}\sqrt{h_{\rm ex}},$$

so

$$\int_{\partial K_L} \left|\frac{\partial f}{\partial \nu}\right||g - \bar{g}| \leq C\mu_\varepsilon^{-2} h_{\rm ex}^{1/4} h_{\rm ex}^{-1/4}\|\nabla g\|_{L^2(K_L)} \leq C\mu_\varepsilon^{-2}\|\nabla g\|_{L^2(K_L)}.$$

We therefore have the estimate

$$\int_{K_L} \frac{\mu_\varepsilon^2}{4}|\nabla g|^2 + |g - \bar{g}|^2 \leq C\sqrt{\int_{K_L} |\nabla g|^2},$$

which shows $\int_{K_L} |\nabla g|^2 \leq C\mu_\varepsilon^{-4}$.

We obtain for h

$$\int_{K_L} |\nabla h|^2 = \int_{K_L} |\nabla f|^2 + |\nabla g|^2 + 2\nabla f \cdot \nabla g.$$

To estimate the $\nabla f \cdot \nabla g$ term, we integrate by parts:

$$\int_{K_L} \nabla f \cdot \nabla g = \int_{\partial K_L} \frac{\partial f}{\partial \nu}(g - \bar{g}) - \int_{K_L} \Delta f (g - \bar{g}).$$

Using the equation, we obtain

$$\left| \frac{\mu_\varepsilon^2}{4} \int_{K_L} \Delta f (g - \bar{g}) \right| = \left| \int_{K_L} (f - 2\pi\delta_\varepsilon)(g - \bar{g}) \right|$$

As in [18], we deduce using Hölder's inequality that $|\int \nabla f \cdot \nabla g| \leq C\mu_\varepsilon^{-4}$. Summing up, we obtain

$$\int_{K_L} |\nabla h|^2 \leq 32\mu_\varepsilon^{-4}\left(\pi \log \frac{L}{\varepsilon} + C\right)$$

so using Poincaré inequality

$$\int_{K_L} |h - h_{\mathrm{ex}}|^2 \leq \frac{C}{h_{\mathrm{ex}}} \int_{K_L} |\nabla h|^2.$$

Combining everything, we see that

$$\frac{1}{2} \int_{K_L} \frac{\mu_\varepsilon^4}{16} |\nabla h|^2 \leq \pi \log \frac{L}{\varepsilon} + C$$

and

$$\frac{\mu_\varepsilon^2}{8} \int_{K_L} |h - h_{\mathrm{ex}}|^2 \leq \frac{C}{\mu_\varepsilon^2 h_{\mathrm{ex}}} \left(\log \frac{L}{\varepsilon} + C\right)$$

so the claim follows. □

Lemma 2. *The function h as defined above satisfies*

$$h \leq C\left(h_{\mathrm{ex}} + \frac{C}{\mu_\varepsilon^2}\left(\log \frac{1}{\varepsilon\sqrt{h_{\mathrm{ex}}}} + 1\right)\right).$$

PROOF. From the pointwise estimates for ∇f and $\int f$, we deduce $|f| \leq Ch_{\mathrm{ex}} + (C/\mu_\varepsilon^2)\big(\log\big(1/(\varepsilon\sqrt{h_{\mathrm{ex}}})\big) + 1\big)$. Using the maximum principle for g, we obtain that $|g| \leq ch_{\mathrm{ex}}$. This proves the claim. □

4. Lower bound estimate

We complete the energy asymptotics with the following lower bound for minimizers.

Proposition 5. *Assume $|u_\varepsilon| \geq 1 - 1/(2|\log \varepsilon|)$ on ∂U, and assume further that*

$$\max\{1, \mu_\varepsilon^{-2}\}|\log \varepsilon| \ll h_{\mathrm{ex}} \ll \varepsilon^{-2/3}|\log \varepsilon|^{-5/3}, \tag{4.1}$$

then minimizing sequence $(u_\varepsilon, A_\varepsilon)$ satisfies:

- *when $\varepsilon^{1/3}|\log \varepsilon|^{5/6} \ll \mu_\varepsilon^2$ and $\mu_\varepsilon^{-4} \leq Ch_{\mathrm{ex}}$ then*

$$G_{\mathrm{csh}}(u_\varepsilon, A_\varepsilon) \geq \frac{1}{2}|U|h_{\mathrm{ex}} \log \frac{1}{\varepsilon\sqrt{h_{\mathrm{ex}}}}\big(1 - o_\varepsilon(1)\big) \tag{4.2}$$

- *when $\varepsilon^{2/3}|\log \varepsilon|^{8/3} \ll \mu_\varepsilon^2 \ll 1/|\log \varepsilon|$ and $h_{\mathrm{ex}} \leq C\mu_\varepsilon^{-4}$ then*

$$G_{\mathrm{csh}}(u_\varepsilon, A_\varepsilon) \geq \frac{1}{2}|U|h_{\mathrm{ex}} \log \frac{\mu_\varepsilon^2}{\varepsilon}\big(1 - o_\varepsilon(1)\big). \tag{4.3}$$

for some constant C.

Remark 6. Note that the first case allows for $\mu_\varepsilon \to 0$. The bifurcation between the two cases arise from the definition of $V_\varepsilon = \log \min\{\mu_\varepsilon^2/\varepsilon, 1/(\varepsilon\sqrt{h_{\rm ex}})\}$, see (4.1) and (4.5).

Remark 7. The exponent $\frac{2}{3}$ seems to be an artifact of the proof, see (4.20). Formal considerations point to an upper bound of the form $h_{\rm ex} \ll \varepsilon^{-1}$, a field strength associated to the formation of domain walls. This failure to reach ε^{-1} seems to do with our method, which is fundamentally different from [18], see Remark 8. Since $\mu_\varepsilon \gg \varepsilon^{1/3} \gg \varepsilon$, we fail to reach the self-dual regime.

We emulate the method of Sandier – Serfaty. Let $K \subset U$, define

$$G_K = \int_K g_{\rm csh}(u_\varepsilon, A_\varepsilon)\,\mathrm{d}x.$$

We choose squares K of sidelength $\delta = \delta(\varepsilon) \to 0$ so that the following scalings hold:

$$V_\varepsilon \ll h_{\rm ex}\delta^2 \ll \min\{h_{\rm ex}, V_\varepsilon^2\} \tag{4.4}$$

where

$$V_\varepsilon = \log\left(\min\left\{\frac{\mu_\varepsilon^2}{\varepsilon}, \frac{1}{\varepsilon\sqrt{h_{\rm ex}}}\right\}\right). \tag{4.5}$$

Finally, we assume we have a good energy bound on the square:

$$G_K \le h_{\rm ex}\delta^2 V_\varepsilon, \tag{4.6}$$

otherwise our lower bound would directly follow.

In order to proceed with the lower bound argument, we choose a small square $\widetilde{K} \subset K$ such that $\rho\big|_{\partial\widetilde{K}} \ge 1 - 1/|\log\varepsilon|$. In order to assure that $\widetilde{K}$ is not too much smaller than K, we have the following global bound on the size of the set of U in which $|u|$ dips below $1 - 1/|\log\varepsilon|$. We note that the Ginzburg – Landau potential has nice features that do not necessitate a shrinking of the square, see Remark 8.

Lemma 3. *Set $U_t = \{x \in U$ such that $\rho(x) < t\}$. Suppose $\rho \ge 1 - 1/(2|\log\varepsilon|)$ on ∂U then*

$$r\big(U_{1-1/|\log\varepsilon|}\big) \le 2\varepsilon|\log\varepsilon|^2 G_{\rm csh}(|u|, 0) \tag{4.7}$$

for $\varepsilon \le 1/34$, where (following [17])

$$r(S) = \inf\left\{\sum 2r_j \text{ such that } S \subseteq \bigcup B_{r_j}\right\} \tag{4.8}$$

for any $S \subset U$. Furthermore, there exists $\widetilde{K} \subseteq K$, a nested square of side length $\tilde{\delta}$ with

$$\delta - r(U_{1-1/|\log\varepsilon|}) \le \tilde{\delta} \le \delta, \tag{4.9}$$

such that $\rho\big|_{\partial\widetilde{K}} \ge 1 - 1/|\log\varepsilon|$.

Proof. (1) Repeating the Modica – Mortola argument from [14], which in turn was adapted from the Ginzburg – Landau case in [17]:

$$\int_U g_{\rm csh}(\rho, 0)\,\mathrm{d}x = \int_U \frac{|\nabla\rho|^2 2}{+} \frac{1}{2\varepsilon^2}\rho^2(1-\rho^2)^2\,\mathrm{d}x$$
$$= \int_0^\infty \int_{\gamma_t \cap U} \frac{|\nabla\rho|}{2} + \frac{1}{2\varepsilon^2}\frac{t^2(1-t^2)^2}{|\nabla\rho|}\,\mathrm{d}\mathcal{H}^1\,\mathrm{d}t$$

$$\geq \frac{1}{\varepsilon}\int_0^1 t(1-t^2)\mathcal{H}^1\big(\rho^{-1}(t)\big)\,\mathrm{d}t$$
$$\geq \frac{1}{\varepsilon}\int_{1-1/|\log\varepsilon|}^{1-1/(2|\log\varepsilon|)} t(1-t^2)\mathcal{H}^1\big(\rho^{-1}(t)\big)\,\mathrm{d}t.$$

Since $\rho \geq 1-1/(2|\log\varepsilon|)$ on ∂U, then $\rho^{-1}(t)\cap\partial U = \varnothing$ for t in the range of integration in the last line. Hence the level set curves are closed, and by the isoperimetric inequality $\mathcal{H}^1(\rho^{-1}(t)) \geq r(U_t)$ for this range of t. Finally, for $t \leq s$, $U_t \subseteq U_s$; and hence $r(U_t) \leq r(U_s)$ and so

$$\begin{aligned}
\int_U g_{\mathrm{csh}}(\rho,0)\,\mathrm{d}x &\geq \frac{1}{\varepsilon}\int_{1-1/|\log\varepsilon|}^{1-1/(2|\log\varepsilon|)} t(1-t^2)\mathcal{H}^1\big(\rho^{-1}(t)\big)\,\mathrm{d}t \\
&\geq \frac{1}{\varepsilon}\int_{1-1/|\log\varepsilon|}^{1-1/(2|\log\varepsilon|)} t(1-t^2)r(U_t)\,\mathrm{d}t \\
&\geq \frac{r(U_{1-1/|\log\varepsilon|})}{\varepsilon}\int_{1-1/|\log\varepsilon|}^{1-1/(2|\log\varepsilon|)} t-t^3\,\mathrm{d}t \\
&= \frac{r(U_{1-1/|\log\varepsilon|})}{\varepsilon}\left(\frac{t^2}{2}-\frac{t^4}{4}\right)\Bigg|_{1-1/|\log\varepsilon|}^{1-1/(2|\log\varepsilon|)}.
\end{aligned}$$

We evaluate the integral, and since $(t^2/2-t^4/4)|_{1-1/|\log\varepsilon|}^{1-1/(2|\log\varepsilon|)} \geq 1/(4|\log\varepsilon|^2)\times \big(3-7/(2|\log\varepsilon|)\big)$ then

$$r(U_{1-1/|\log\varepsilon|}) \leq 4\varepsilon\frac{|\log\varepsilon|^2}{3-7/(2|\log\varepsilon|)}\int_U g_{\mathrm{csh}}(|u|,0)\,\mathrm{d}x \leq 2\varepsilon|\log\varepsilon|^2 G_{\mathrm{csh}}(|u|,0)$$

so long as $|\log\varepsilon| \geq \frac{7}{2}$.

(2) Therefore, if $\{x\in K \text{ such that } \rho(x) < 1-1/|\log\varepsilon|\}\cap\partial K \neq \varnothing$ then we shrink the size of the square slightly to get a good boundary. In particular from (4.7) we see that $r(U_{1-1/|\log\varepsilon|}) < C\varepsilon|\log\varepsilon|^2 G_{\mathrm{csh}}(\rho,0)$. Therefore, there exists

$$\tilde\delta \in [\delta - r(U_{1-1/|\log\varepsilon|}),\delta]$$

and a nested square $\widetilde{K}\subset K$ of sidelength $\tilde\delta$ such that $\rho\big|_{\partial\widetilde{K}} \geq 1-1/|\log\varepsilon|$. □

We now work on the square $\widetilde{K}$ defined in Lemma 3. Set $\Omega_t = \{x\in\widetilde{K} : \rho(x) < t\}$ and let $\Theta(t) = \int_{\widetilde{K}\setminus\Omega_t}|\nabla\varphi - A|^2\,\mathrm{d}x$ where $u=\rho e^{\mathrm{i}\varphi}$. Following the spirit of [17]

$$\begin{aligned}
&\int_K g_{\mathrm{csh}}(u,A)\,\mathrm{d}x \\
&\quad\geq \int_{\widetilde{K}} g_{\mathrm{csh}}(u,A)\,\mathrm{d}x \\
&\quad= \frac{1}{2}\int_0^\infty\left[\int_{\partial\Omega_t\cap\widetilde{K}}|\nabla\rho| + \frac{t^2(1-t^2)^2}{\varepsilon^2|\nabla\rho|}\,\mathrm{d}l - t^2\Theta'(t)\right]\mathrm{d}t + \frac{\mu_\varepsilon^2}{8}\int_{\widetilde{K}}\frac{|h-h_{\mathrm{ex}}|^2}{\rho^2}\,\mathrm{d}x
\end{aligned}$$

by the co-area formula. By Lemma 10 in Section 4.1

$$\frac{\mu_\varepsilon^2}{8}\int_{\widetilde{K}}\frac{|h-h_{\mathrm{ex}}|^2}{\rho^2}\,\mathrm{d}x \geq \frac{\mu_\varepsilon^2}{8}\int_{\widetilde{K}}|h-h_{\mathrm{ex}}|^2\,\mathrm{d}x - o_\varepsilon(1).$$

Therefore, integration by parts on the third term yields

$$\int_K g_{\mathrm{csh}}(u,A)\,\mathrm{d}x \geq \int_0^1 \left[\int_{\partial\Omega_t\cap\widetilde{K}} \frac{|\nabla\rho|}{2} + \frac{t^2(1-t^2)^2}{2\varepsilon^2|\nabla\rho|}\,\mathrm{d}l \right.$$
$$\left. + t\left[\int_{\widetilde{K}\setminus\Omega_t} |\nabla\varphi - A|^2\,\mathrm{d}x + \frac{\mu_\varepsilon^2}{8}\int_{\widetilde{K}} |h-h_{\mathrm{ex}}|^2\,\mathrm{d}x\right]\right]\mathrm{d}t.$$

or

$$G_K = \int_K g_{\mathrm{csh}}(u,A)\,\mathrm{d}x \geq \int_0^1 a(t) + 2tb(t)\,\mathrm{d}t \tag{4.10}$$

with

$$a(t) = \int_{\partial\Omega_t\cap\widetilde{K}} \frac{|\nabla\rho|}{2} + \frac{t^2(1-t^2)^2}{2\varepsilon^2|\nabla\rho|}\,\mathrm{d}l \tag{4.11}$$

$$b(t) = \frac{1}{2}\int_{\widetilde{K}\setminus\overline{\Omega}_t} |\nabla\varphi - A|^2\,\mathrm{d}x + \frac{\mu_\varepsilon^2}{8}\int_{\widetilde{K}} |h-h_{\mathrm{ex}}|^2\,\mathrm{d}x. \tag{4.12}$$

We start with an estimate on $a(t)$ that uses the Modica–Mortola trick.

Lemma 4. *Let $\widetilde{K}$ be the square chosen in Lemma 3 with sidelength $\tilde{\delta}$ satisfying (4.9), then there exists a constant C such that*

$$\int_0^1 a(t)\,\mathrm{d}t \geq C\int_0^{1-1/|\log\varepsilon|} \frac{r(\Omega_t)t(1-t^2)}{\varepsilon}\,\mathrm{d}t \tag{4.13}$$

where $\Omega_t = \{x\in\widetilde{K}$ such that $\rho(x)<t\}$ and where $r(U_{1-1/|\log\varepsilon|})$ is bounded in Lemma 3.

Proof. This lemma follows from an argument similar to Sandier–Serfaty [18], and we sketch the argument here. The method is inspired from Modica–Mortola. Recall that $\omega_t = \{x\in K$ such that $\rho(x)<t\}$.

(1) If $\rho(x)\big|_{\partial K} \geq 1-1/|\log\varepsilon|$ then for all $0\leq t<1-1/|\log\varepsilon|$ we have

$$\partial\{x\in K \text{ such that } \rho(x)<t\}\cap\partial K = \varnothing.$$

Define $\Omega_t = \{x\in K$ such that $\rho(x)<t\}$ and $\partial\Omega_t = \gamma_t \equiv (\gamma_t\cap K)\cup(\gamma_t\cap\partial K)$, then we have

$$\begin{aligned}\int_0^1 a(t)\,\mathrm{d}t &= \int_0^1\left[\int_{\gamma_t\cap K}\frac{|\nabla\rho|}{2} + \frac{t^2(1-t^2)}{2\varepsilon^2|\nabla\rho|}\,\mathrm{d}l\right]\mathrm{d}t\\ &\geq \int_0^1\left[\int_{\gamma_t\cap K}\frac{t(1-t^2)}{\varepsilon}\,\mathrm{d}l\right]\mathrm{d}t\\ &= \int_0^1\left[\frac{t(1-t^2)}{\varepsilon}\mathcal{H}^1(\gamma_t\cap K)\right]\mathrm{d}t.\end{aligned}$$

The isoperimetric inequality implies $\mathcal{H}^1(\gamma_t\cap K)\geq r(\Omega_t)$ for $t\leq 1-1/|\log\varepsilon|$ since these level sets are closed. Therefore,

$$\int_0^1 a(t)\,\mathrm{d}t \geq C\int_0^{1-1/|\log\varepsilon|}\frac{r(\Omega_t)t(1-t^2)}{\varepsilon}\,\mathrm{d}t. \qquad \square$$

Remark 8. In the Ginzburg–Landau case [18] one can find an improved lower bound by making use of an interesting property. If $|\Omega_t|$ is the area of Ω_t then one finds

$$\int_K \frac{|\nabla\rho|^2}{2} + \frac{(1-\rho^2)^2}{4\varepsilon^2}\,\mathrm{d}x \geq C\int_{\varepsilon/\delta}^1 \frac{1-t^2}{\varepsilon}\left[\mathcal{H}^1(\gamma_t\cap K) + \frac{|\Omega_t|}{\delta}\right]\mathrm{d}t.$$

The bracketed sum bounds $r(\Omega_t)$ from above, regardless of whether the level set curves hit ∂K. Hence one gets to the upper bound without restriction on h_{ex}. If one tries to replicate the Sandier–Serfaty argument with the Chern–Simons–Higgs potential, one finds

$$\int_K \frac{|\nabla\rho|^2}{2} + \frac{\rho^2(1-\rho^2)^2}{2\varepsilon^2}\,\mathrm{d}x \geq C\int_0^1 \frac{t(1-t^2)}{\varepsilon}\left[\mathcal{H}^1(\gamma_t\cap K) + \frac{(3t^2-1)|\Omega_t|}{\varepsilon}\right]\mathrm{d}t,$$

and since the second term is negative for $0\leq t<3^{-1/2}$, it is not entire obvious how to use the bracketed terms to bound $r(\Omega_t)$ from above.

In order to bound $b(t)$ we use the following variation of a covering argument developed in [18]. In our case we have a modified regime of radii in which our function $f(r,R)$ is increasing.

Lemma 5. *Assume $V\subset\mathbb{R}^2$ is open and $\omega\subset\mathbb{R}^2$ is compact. Assume $v\colon V\setminus\omega\to\mathbb{S}^1$ and $A\colon V\to\mathbb{R}^2$. Then for any $\sigma\geq r(\omega)$ such that $\sigma\leq\min\{\mu_\varepsilon^2/3, 1/(3\sqrt{h_{\mathrm{ex}}}), 1\}$ there exists a family $\{B_i\}$ of disjoint disks of radii r_i such that*

(1) $\sum r_i = \sigma$
(2) $\omega\subset\cup\overline{B}_i\subset\Omega_1$
(3) *Letting $h=\operatorname{curl}A$ and $v=\mathrm{e}^{\mathrm{i}\varphi}$, then*

$$\frac{1}{2}\int_{B_i\setminus\omega}|\nabla\varphi-A|^2 + \frac{\mu_\varepsilon^2}{8}\int_{B_i}|h-h_{\mathrm{ex}}|^2 \geq \pi|d_i|\left(\log\frac{\min\{1,\mu_\varepsilon^2,1/\sqrt{h_{\mathrm{ex}}}\}}{r(B_i)} - C\right)^+$$

where d_i is the winding number of v restricted to ∂B_i, if $\overline{B}_i\Subset\Omega_1\cap V$, and zero otherwise.

The proof of Lemma 5 follows from the two lemmas below.

Lemma 6. *Let $v\colon B_R\setminus\overline{B}_r\to\mathbb{S}^1$ and $A\colon B_R\to\mathbb{R}^2$. Then for all $h_{\mathrm{ex}}>0$ and $0<r<R<\min\{\mu_\varepsilon^2/2,1\}$,*

$$\frac{1}{2}\int_{B_R\setminus\overline{B}_r}|\nabla\varphi-A|^2 + \frac{R-r}{2}\frac{\mu_\varepsilon^2}{4}\int_{B_R}|h-h_{\mathrm{ex}}|^2 \geq \pi|d|\left(\log\frac{R}{r} - \frac{2}{\mu_\varepsilon^2}(R-r) - h_{\mathrm{ex}}\frac{R^2-r^2}{2}\right)$$

where d is the winding number of $v=\mathrm{e}^{\mathrm{i}\varphi}$ restricted to any circle ∂B_s, $r<s<R$. Furthermore, the function

$$f(r,R) = \log\frac{R}{r} - \frac{2}{\mu_\varepsilon^2}(R-r) - h_{\mathrm{ex}}\frac{R^2-r^2}{2}$$

is increasing in R for $0<r<R\leq\min\{1,\mu_\varepsilon^2/3,1/(3\sqrt{h_{\mathrm{ex}}})\}$.

Proof. This proof is based on the argument in [17].

(1) Set

$$g(t) = \frac{1}{2}\int_{\partial B_t}\left|\frac{\partial\varphi}{\partial\tau} - A\cdot\tau\right|^2 + \frac{\mu_\varepsilon^2}{8}\int_{B_t}|h - h_{\rm ex}|^2$$

then

$$\begin{aligned}\int_{B_R\setminus\overline{B_r}} g_{\rm csh}(u,A)\,\mathrm{d}x &\geq \frac{1}{2}\int_{B_R\setminus\overline{B}_r}|\nabla\varphi - A|^2 + \frac{\mu_\varepsilon^2}{8}(R-r)\int_{B_r}|h-h_{\rm ex}|^2\\ &\geq \int_r^R g(t)\,\mathrm{d}t.\end{aligned}$$

Let $\alpha_t = \int_{B_t} h$ then

$$2\pi d = \alpha_t - \int_{\partial B_t}\frac{\partial\varphi}{\partial\tau} - A\cdot\tau$$

then by Cauchy–Schwarz implies

$$\int_{\partial B_t}\left|\frac{\partial\varphi}{\partial\tau} - A\cdot\tau\right|^2 \geq \frac{1}{2\pi t}(2\pi d - \alpha_t)^2$$

Second,

$$\int_{B_t}|h - h_{\rm ex}|^2 \geq \frac{1}{\pi t^2}(\alpha_t - \pi t^2 h_{\rm ex})^2$$

so

$$g(t) \geq \frac{1}{4\pi t}(2\pi d - \alpha)^2 + \frac{\mu_\varepsilon^2}{8\pi t^2}(\alpha - \pi t^2 h_{\rm ex})^2$$

which we minimize over α. Note that if $f(x) = (1/a)(c-x)^2 + (1/b)(d-x)^2$ then the minimum occurs at $x = (bc+ad)/(a+b)$ or $f(x) \geq (c-d)^2/(a+b)$. Applying to the above inequality yields

$$\begin{aligned}g(t) &\geq \frac{(2\pi d - \pi t^2 h_{\rm ex})^2}{4\pi t + 8\pi t^2/\mu_\varepsilon^2} = \frac{(2\pi d)^2 - 4\pi^2 dt^2 h_{\rm ex} + \pi^2 t^4 h_{\rm ex}^2}{4\pi t(1 + (2/\mu_\varepsilon^2)t)}\\ &\geq \left(\frac{\pi d^2}{t} - \pi t|d|h_{\rm ex}\right)\left(1 + \frac{2}{\mu_\varepsilon^2}t\right)^{-1}.\end{aligned}$$

We compute the lower bound. Suppose $0 < t < 1$ then $0 < (2/\mu_\varepsilon^2)t < 2/\mu_\varepsilon^2$, and if $0 < t \leq \min\{\mu_\varepsilon^2/2, 1\}$ then $0 < (2/\mu_\varepsilon^2)t \leq 1$. This implies

$$\left(1 + \frac{2}{\mu_\varepsilon^2}t\right)^{-1} \geq 1 - \frac{2}{\mu_\varepsilon^2}t > 0$$

for all $0 < t \leq \min\{\mu_\varepsilon^2/2, 1\}$. Therefore,

$$g(t) \geq \left(\frac{\pi d^2}{t} - \pi t|d|h_{\rm ex}\right)\left(1 - \frac{2}{\mu_\varepsilon^2}t\right) \geq \pi|d|\left(\frac{1}{t} - \frac{2}{\mu_\varepsilon^2} - th_{\rm ex}\right)$$

and so

$$\int_r^R g(t)\,\mathrm{d}t \geq \pi|d|\left(\log\frac{R}{r} - \frac{2}{\mu_\varepsilon^2}(R-r) - \frac{h_{\rm ex}}{2}(R^2 - r^2)\right).$$

(2) We now show that $f(r,R)$ is increasing on the claimed interval. Set $\gamma = 1/\mu_\varepsilon^2$, then $f'(r,\cdot) = 1/x - 2\gamma - h_{\rm ex}x \geq 0$ for all x up to

$$\frac{\sqrt{4\gamma^2/h_{\rm ex}^2 + 4/h_{\rm ex}} - 2\gamma/h_{\rm ex}}{2} = \frac{\sqrt{\gamma^2 + h_{\rm ex}} - \gamma}{h_{\rm ex}} = \frac{1}{\sqrt{\gamma^2 + h_{\rm ex}} + \gamma}.$$

If $\gamma \geq \sqrt{h_{\mathrm{ex}}}$, then $\gamma^2 > h_{\mathrm{ex}}$ and $f(x)$ is increasing up to $x = 1/(\sqrt{\gamma^2 + h_{\mathrm{ex}}} + \gamma) \geq 1/(\sqrt{2\gamma^2} + \gamma) = (1/\gamma)(1/(\sqrt{2}+1)) \geq 1/(3\gamma)$. On the other hand if $\gamma \leq \sqrt{h_{\mathrm{ex}}}$, then $\gamma^2 \leq h_{\mathrm{ex}}$ and $f(x)$ is increasing up to $x = 1/(\sqrt{\gamma^2 + h_{\mathrm{ex}}} + \gamma) \geq 1/(\sqrt{2h_{\mathrm{ex}}} + \sqrt{h_{\mathrm{ex}}}) = 1/\sqrt{h_{\mathrm{ex}}}(1/(\sqrt{2}+1)) \geq 1/(3\sqrt{h_{\mathrm{ex}}})$. Combining both estimates yields the lemma. □

The function $f(r, R)$ described in Lemma 6 has the following nice properties that will be used in the proof of Lemma 5.

Lemma 7. *Let $f(r, R)$ satisfy the following properties:*

(1) *$f(r, \cdot)$ is increasing on $[0, \frac{1}{3}\min\{\mu_\varepsilon^2, 1/\sqrt{h_{\mathrm{ex}}}, 3\}$.*
(2) *$f(r, r) = 0$, thus $f(r, R) \geq 0$ for all $r \leq R \leq \frac{1}{3}\min\{\mu_\varepsilon^2, 1/\sqrt{h_{\mathrm{ex}}}, 3\}]$.*
(3) *$f(r, s) + f(s, R) = f(r, R)$.*
(4) *If $(0 \leq r_i \leq R_i)$ are $2k$ positive real numbers and d_i are integers, and*

$$\frac{R_i}{r_i} = \frac{R_j}{r_j} = \alpha > 1$$

for all i, j. Then

$$\sum_{i=1}^{k} |d_i| f(r_i, R_i) \geq \left(\sum_{i=1}^{k} |d_i|\right) f\left(\sum_{i=1}^{k} r_i, \sum_{i=1}^{k} R_i\right).$$

PROOF. Given Lemma 6, only (4) is nontrivial, and the proof of a similar result can be found in [17]. □

Finally, we complete the proof of Lemma 5 by quoting a similar result.

PROOF OF LEMMA 5. The proof is almost identical to the proof of [18, Proposition 4.1] so we leave it out. The argument uses the vortex ball growing / merging process of [10, 17]. There are a few differences since the growth function $f(r, R)$ has two parameters that can blowup as $\varepsilon \to 0$. However, within the parameter range $0 < r < R < \min\{1, \mu_\varepsilon^2/3, 1/(3\sqrt{h_{\mathrm{ex}}})\}$, one finds that the lower bound does indeed hold. □

We can now complete the lower bound on $b(t)$. We use $|\log\varepsilon| \ll h_{\mathrm{ex}}$ implicitly at many steps.

Lemma 8. *Let $\widetilde{K}$ be the square chosen in Lemma 3 with sidelength $\tilde{\delta}$ and recall $\Omega_t = \{x \in \widetilde{K}$ such that $\rho(x) < t\}$. Then for all $0 < t < 1$*

$$b(t) \geq \pi\Bigg(\sum_{i:B_i \Subset \widetilde{K}} |d_{B_i}|\Bigg)\left(\log\frac{\min\{\mu_\varepsilon^2, 1/\sqrt{h_{\mathrm{ex}}}\}}{r(\Omega_t)} - C\right)^+ \tag{4.14}$$

and

$$2\pi d_t \geq h_{\mathrm{ex}}\tilde{\delta}^2\left[\left(1 - \frac{\alpha}{\tilde{\delta}}\right)^2 - \frac{\chi}{t}\sqrt{\frac{\tilde{\delta}}{\alpha}} - \frac{2\chi}{t}\sqrt{\frac{\tilde{\delta}^2}{\mu_\varepsilon^2}}\right] \tag{4.15}$$

where $\chi^2 = G_K/(h_{\mathrm{ex}}^2\tilde{\delta}^4)$ and $\min\{1/\sqrt{h_{\mathrm{ex}}}, \mu_\varepsilon^2\} \leq \alpha \leq \tilde{\delta}/4$.

PROOF. If we take $\omega = \Omega_t$, $v = u/|u|$, $V = \widetilde{K}$ and $\sigma = \min\{1, 1/\sqrt{h_{\mathrm{ex}}}, \mu_\varepsilon^2\}$ then Lemmas 5 and 7 imply

$$b(t) \geq \pi\Bigg(\sum_{i:B_i \Subset \widetilde{K}} |d_{B_i}|\Bigg)\left(\log\frac{\min\{1, \mu_\varepsilon^2, 1/\sqrt{h_{\mathrm{ex}}}\}}{r(\Omega_t)} - C\right)^+$$

and the difficulty comes in controlling $(\sum_{i:B_i\Subset\widetilde{K}}|d_{B_i}|)$. Here the interaction between magnetic field and degree is nontrivial. We now provide an estimate on the degree via energy bounds.

(1) Consider the following minimization problem:

$$\min_{\substack{A\in H^1(\widetilde{K},\mathbb{R}^2)\\ \operatorname{div} A=0}} \frac{1}{2}\int_{\widetilde{K}\setminus\overline{\Omega}_t} |\nabla\varphi - A|^2\,\mathrm{d}x + \frac{\mu_\varepsilon^2}{8}\int_{\widetilde{K}} |\operatorname{curl} A - h_{\mathrm{ex}}|^2\,\mathrm{d}x$$

where $\varphi = \arg u$. The minimizer is achieved for some $\bar{A}$ with $\bar{h} = \operatorname{curl}\bar{A}$, and it satisfies

$$-\frac{\mu_\varepsilon^2}{4}\nabla^\perp\bar{h} = \nabla\varphi - \bar{A} \quad \text{in } \widetilde{K}\setminus\overline{\Omega}_t$$

with $\bar{h} = h_{\mathrm{ex}}$ on $\partial\widetilde{K}$ and $\bar{h} = $ constant in $\overline{\Omega}_t$. On each component ω_i of Ω_t we compute the degree around $\partial\omega_i$,

$$\begin{aligned} 2\pi d_i &= \int_{\partial\omega_i} \frac{\partial\varphi}{\partial\tau}\,\mathrm{d}l = \int_{\partial\omega_i} \bar{A}\cdot\tau - \frac{\mu_\varepsilon^2}{4}\tau\cdot\nabla^\perp\bar{h}\,\mathrm{d}l \\ &= \int_{\omega_i} \bar{h}\,\mathrm{d}x - \frac{\mu_\varepsilon^2}{4}\int_{\partial\omega_i}\frac{\partial\bar{h}}{\partial n}\,\mathrm{d}l. \end{aligned}$$

Furthermore, in $\widetilde{K}\setminus\overline{\Omega}_t$ then

$$-\frac{\mu_\varepsilon^2}{4}\Delta\bar{h} + \bar{h} = 0$$

and so for some $\widetilde{K}'\subset\widetilde{K}$ we have

$$\begin{aligned} -\frac{\mu_\varepsilon^2}{4}\int_{\partial\widetilde{K}'}\frac{\partial\bar{h}}{\partial n}\,\mathrm{d}l + \int_{\widetilde{K}'\setminus\overline{\Omega}_t}\bar{h} &= -\sum_i \frac{\mu_\varepsilon^2}{4}\int_{\partial\omega_i}\frac{\partial\bar{h}}{\partial n}\,\mathrm{d}l \\ &= -\sum_i\int_{\omega_i}\bar{h}\,\mathrm{d}x + 2\pi\sum_{i:\omega_i\subset\widetilde{K}'} d_i \end{aligned}$$

or

$$2\pi\sum_{i:\omega_i\subset\widetilde{K}'} d_i = -\frac{\mu_\varepsilon^2}{4}\int_{\partial\widetilde{K}'}\frac{\partial\bar{h}}{\partial n}\,\mathrm{d}l + \int_{\widetilde{K}'}\bar{h}\,\mathrm{d}x, \tag{4.16}$$

which controls the total degree by the magnetic field.

(2) Consider now for $|u|\geq t$ then $\nabla\varphi - \bar{A} = -(\mu_\varepsilon^2/4)\nabla^\perp\bar{h}$ and

$$\begin{aligned} G_K &\geq \frac{t^2}{2}\left[\int_{\widetilde{K}\setminus\Omega_t}|\nabla\varphi - \bar{A}|^2 + \frac{\mu_\varepsilon^2}{4}\int_{\widetilde{K}}|\bar{h} - h_{\mathrm{ex}}|^2\,\mathrm{d}x\right] \\ &\geq \frac{\mu_\varepsilon^2 t^2}{8}\int_{\widetilde{K}}\frac{\mu_\varepsilon^2}{4}|\nabla\bar{h}|^2 + |\bar{h} - h_{\mathrm{ex}}|^2\,\mathrm{d}x \end{aligned}$$

then

$$\int_{\widetilde{K}}\frac{\mu_\varepsilon^2}{4}|\nabla\bar{h}|^2 + |\bar{h} - h_{\mathrm{ex}}|^2 \leq \frac{8}{\mu_\varepsilon^2 t^2}G_K$$

Set $\widetilde{K}_t = \{x\in\widetilde{K} : \operatorname{dist}(x,\partial\widetilde{K}) > t\}$ then for any $\min\{1/\sqrt{h_{\mathrm{ex}}},\mu_\varepsilon^2\}\leq\alpha\leq\tilde{\delta}/4$, the set $T = \{t\in[0,\alpha]$ such that $\partial\widetilde{K}_t\cap\bigcup_i B_i = \varnothing\}$ has measure at least $\alpha/2$ since

$\sum r_i = \min\{\mu_\varepsilon^2/3, 1/(3\sqrt{h_{\rm ex}})\}$. If we define the set

$$T' = \left\{ t \in T \text{ such that } \int_{\partial \widetilde{K}_t} \left| \frac{\partial \bar{h}}{\partial n} \right|^2 \mathrm{d}l \le \frac{8}{\alpha \mu_\varepsilon^2 t^2} G_K \right\}$$

then by the energy estimate

$$\frac{\mu_\varepsilon^2}{4} \int_0^\alpha \int_{\partial \widetilde{K}_t} |\nabla \bar{h}|^2 \,\mathrm{d}l \,\mathrm{d}t = \frac{\mu_\varepsilon^2}{4} \int_{\widetilde{K} \setminus \widetilde{K}_\alpha} |\nabla \bar{h}|^2 \le \frac{8}{\mu_\varepsilon^2 t^2} G_K$$

implies there exists $t_0 \in (0, \alpha)$ such that $0 \le t_0 \le \alpha < \tilde{\delta}/4$ with $\partial \widetilde{K}_{t_0} \cap \bigcup B_i = \varnothing$ and $(\mu_\varepsilon^2/4) \int_{\partial \widetilde{K}_{t_0}} |\partial \bar{h}/\partial n|^2 \,\mathrm{d}l < 8 G_K/(\alpha \mu_\varepsilon^2 t^2)$. Thus

$$\begin{aligned} 2\pi d_t \ge 2\pi \sum_i d_{B_i} &= -\frac{\mu_\varepsilon^2}{4} \int_{\partial \widetilde{K}_{t_0}} \frac{\partial \bar{h}}{\partial n} \,\mathrm{d}l + \int_{\widetilde{K}_{t_0}} \bar{h} \,\mathrm{d}x \\ &\ge \int_{\widetilde{K}_{t_0}} h_{\rm ex} \,\mathrm{d}x - \left| \int_{\widetilde{K}_{t_0}} \bar{h} - h_{\rm ex} \,\mathrm{d}x \right| - \left| \frac{\mu_\varepsilon^2}{4} \int_{\partial \widetilde{K}_{t_0}} \frac{\partial \bar{h}}{\partial n} \,\mathrm{d}l \right|. \end{aligned}$$

Naively estimating the last two terms by

$$\left| \int_{\widetilde{K}_{t_0}} \bar{h} - h_{\rm ex} \,\mathrm{d}x \right| \le \|\bar{h} - h_{\rm ex}\|_{L^2(\widetilde{K}_{t_0})} |\widetilde{K}_{t_0}|^{1/2} \le \sqrt{\frac{4 G_K \tilde{\delta}^2}{\mu_\varepsilon^2 t^2}}$$

and

$$\left| \frac{\mu_\varepsilon^2}{4} \int_{\partial \widetilde{K}_{t_0}} \frac{\partial \bar{h}}{\partial n} \,\mathrm{d}x \right| \le \left| \frac{\mu_\varepsilon^2}{4} \frac{\partial \bar{h}}{\partial n} \right|_{L^2(\widetilde{K}_{t_0})} |\partial \widetilde{K}_{t_0}|^{1/2} \le \sqrt{\frac{G_K \tilde{\delta}}{\alpha t^2}},$$

then (4.16) yields

$$2\pi d_t \ge h_{\rm ex} \tilde{\delta}^2 \left[\left(1 - \frac{\alpha}{\tilde{\delta}}\right)^2 - \frac{1}{t} \sqrt{\frac{G_K}{h_{\rm ex}^2 \tilde{\delta}^4} \frac{\tilde{\delta}}{\alpha}} - \frac{2}{t} \sqrt{\frac{G_K}{h_{\rm ex}^2 \tilde{\delta}^4} \frac{\tilde{\delta}^2}{\mu_\varepsilon^2}} \right]. \qquad \square$$

Remark 9. In order to use (4.15) we should have the following estimates hold

$$\frac{\alpha}{\tilde{\delta}} \ll 1 \quad \tilde{\delta} \ll 1 \quad \chi \ll 1 \quad \frac{\chi^2 \tilde{\delta}}{\alpha} \ll 1 \quad \frac{\chi \tilde{\delta}}{\mu_\varepsilon} \ll 1, \tag{4.17}$$

otherwise the right-hand side of (4.15) is zero, and the lower bound fails. Note that (4.17) implies the rate $\tilde{\delta}/\alpha \ll 1/\chi^2$, which we use in Lemma 9 to choose α correctly.

We are now able to proof the essential lower bound on the square.

Lemma 9. *Suppose* (4.4) *and* (4.6) *hold, then*

$$G_K \ge \tfrac{1}{2} h_{\rm ex} \delta^2 V_\varepsilon \big(1 - o_\varepsilon(1)\big). \tag{4.18}$$

PROOF. This lower bound is established via estimates in the spirit of [18].

(1) We need to make sure that $\tilde{\delta}$ is on the same order as δ. In particular $\delta - r(U_{1-1/|\log \varepsilon|}) \le \tilde{\delta} \le \delta$; therefore, we want $r(U_{1-1/|\log \varepsilon|}) \ll \delta$. From Lemma 3

$$\frac{r(U_{1-1/|\log \varepsilon|})}{\delta} \ll C \frac{\varepsilon |\log \varepsilon|^2}{\delta} \int_U g_{\rm csh}(\rho, 0) \,\mathrm{d}x.$$

Using the upper bound $G_{\mathrm{csh}}(\rho,0) \le G_{\mathrm{csh}}(u,A) \le \frac{1}{2}|U|h_{\mathrm{ex}}\log\big(1/(\varepsilon\sqrt{h_{\mathrm{ex}}})\big) \le Ch_{\mathrm{ex}} \times |\log\varepsilon|$, and from (4.4) we have

$$\frac{1}{\delta} \ll \sqrt{\frac{h_{\mathrm{ex}}}{V_\varepsilon}} \ll C\sqrt{\frac{h_{\mathrm{ex}}}{|\log\varepsilon|}},$$

since $V_\varepsilon = \log(\min\{\mu_\varepsilon^2/\varepsilon, 1/(\varepsilon\sqrt{h_{\mathrm{ex}}})\}) \ge C|\log\varepsilon|$. Therefore,

$$\frac{r(U_{1-1/|\log\varepsilon|})}{\delta} \ll C\varepsilon|\log\varepsilon|^2\sqrt{\frac{h_{\mathrm{ex}}}{|\log\varepsilon|}}h_{\mathrm{ex}}|\log\varepsilon| \le C\varepsilon h_{\mathrm{ex}}^{3/2}|\log\varepsilon|^{5/2} \ll 1 \tag{4.19}$$

by taking

$$h_{\mathrm{ex}} \ll \frac{1}{\varepsilon^{2/3}|\log\varepsilon|^{5/3}}, \tag{4.20}$$

which establishes the upper bound on h_{ex}.

(2) We examine $a(t)+2tb(t)$, using (4.13) and (4.14), and note the sum is of the form $C_1 r + C_2(\log(C_3/r) - C_4)^+$ for $r = r(\Omega_t)$. The function $p(r) = C_1 r + C_2(\log(C_3/r) - C_4)^+$ has critical points at $r = \{C_2/C_1, C_3 \mathrm{e}^{-C_4}\}$. Therefore, $p(r)$ satisfies

$$p(r) \ge \min\left\{\frac{C_1 C_3}{\mathrm{e}^{C_4}}, C_2 + C_2\left(\log\frac{C_1C_3}{C_2} - C_4\right)^+\right\},$$

and since $x \mapsto -\log cx$ is decreasing in x, we have

$$p(r) \ge C_2\left(\log\frac{C_1C_3}{C_2} - C_4\right)^+. \tag{4.21}$$

We apply to our lower bound with

$$C_1 = \frac{t(1-t^2)}{\varepsilon}, \qquad C_2 = h_{\mathrm{ex}}\tilde{\delta}^2\left(\left(1-\frac{\alpha}{\tilde{\delta}}\right)^2 t - \chi\min\left\{\sqrt{\frac{\tilde{\delta}}{\alpha}}, \frac{\tilde{\delta}}{\mu_\varepsilon}\right\}\right)^+,$$

$$C_3 = \min\left\{\mu_\varepsilon^2, \frac{1}{\sqrt{h_{\mathrm{ex}}}}\right\} \qquad C_4 = C.$$

From Remark 9 if (4.17) is satisfied then C_2 is nontrivial. From the lower bound $(x-a)^+(y-b)^+ \ge xy - xb - ya$ we have

$$\begin{aligned} a(t) + 2tb(t) \ge{}& -C_2\log C_2 \\ &+ h_{\mathrm{ex}}\tilde{\delta}^2\left(1-\frac{\alpha}{\tilde{\delta}}\right)^2\left[t\log|t(1-t^2)| + t\log\left(\min\left\{\frac{\mu_\varepsilon^2}{\varepsilon}, \frac{1}{\varepsilon\sqrt{h_{\mathrm{ex}}}}\right\}\right) - Ct\right] \\ &- \chi\min\left\{\sqrt{\frac{\tilde{\delta}}{\alpha}}, \frac{\tilde{\delta}}{\mu_\varepsilon}\right\}\left[\log|t(1-t^2)| + \log\left(\min\left\{\frac{\mu_\varepsilon^2}{\varepsilon}, \frac{1}{\varepsilon\sqrt{h_{\mathrm{ex}}}}\right\}\right)\right]. \end{aligned}$$

First note that $-C_2\log C_2 \ge -C_2\log(h_{\mathrm{ex}}\tilde{\delta}^2)$ since

$$0 \le \left(\left(1-\frac{\alpha}{\tilde{\delta}^2}\right)^2 t - \chi\min\left\{\sqrt{\frac{\tilde{\delta}}{\alpha}}, \frac{\tilde{\delta}}{\mu_\varepsilon}\right\}\right)^+ \le 1.$$

Furthermore, since $1 \ll h_{\mathrm{ex}}\tilde{\delta}^2$, hence $-C_2 \log C_2 \geq -h_{\mathrm{ex}}\tilde{\delta}^2 \log(h_{\mathrm{ex}}\delta^2)t$. We integrate the above lower bound. Using

$$\left|\int_0^1 t\log\big(t(1-t^2)\big)\,\mathrm{d}t\right| = \frac{3}{4} \quad \text{and} \quad \left|\int_0^1 \log|t(1-t^2)|\,\mathrm{d}t\right| = 2\log 2 - 1,$$

we find

$$\begin{aligned}(4.22)\quad &\int_0^{1-1/|\log\varepsilon|} a(t) + 2tb(t)\,\mathrm{d}t \\ &\geq \frac{1}{2}\left(1-\frac{1}{|\log\varepsilon|}\right)^2\left(1-\frac{\alpha}{\tilde{\delta}}\right)^2 h_{\mathrm{ex}}\tilde{\delta}^2\log\left(\min\left\{\frac{\mu_\varepsilon^2}{\varepsilon},\frac{1}{\varepsilon\sqrt{h_{\mathrm{ex}}}}\right\}\right) \\ &\quad - \chi\min\left\{\sqrt{\frac{\tilde{\delta}}{\alpha}},\frac{\tilde{\delta}}{\mu_\varepsilon}\right\}\log\left(\min\left\{\frac{\mu_\varepsilon^2}{\varepsilon},\frac{1}{\varepsilon\sqrt{h_{\mathrm{ex}}}}\right\}\right) \\ &\quad - h_{\mathrm{ex}}\tilde{\delta}^2\log(h_{\mathrm{ex}}\delta^2) - C\left[\chi\min\left\{\sqrt{\frac{\tilde{\delta}}{\alpha}},\frac{\tilde{\delta}}{\mu_\varepsilon}\right\} + h_{\mathrm{ex}}\tilde{\delta}^2\right].\end{aligned}$$

We now show that most of the terms in (4.22) are $o_\varepsilon(1)$ compared to the leading order term $\frac{1}{2}h_{\mathrm{ex}}\tilde{\delta}^2 V_\varepsilon$.

(3) We proceed with several comparison arguments that will be grouped together at the end of this step. From (4.4) we see $h_{\mathrm{ex}}\delta^2 \ll V_\varepsilon^2$, hence $\log(h_{\mathrm{ex}}\delta) \ll C\log V_\varepsilon \ll V_\varepsilon$ since $V_\varepsilon \gg 1$. Therefore,

$$(4.23)\qquad h_{\mathrm{ex}}\tilde{\delta}^2\log(h_{\mathrm{ex}}\delta^2) \ll h_{\mathrm{ex}}\tilde{\delta}^2 V_\varepsilon$$

Next from (4.1) we have

$$(4.24)\qquad \frac{V_\varepsilon}{h_{\mathrm{ex}}\mu_\varepsilon^2} \leq \frac{V_\varepsilon}{h_{\mathrm{ex}}}\max\{\mu_\varepsilon^{-2},1\} \ll 1.$$

Recall that $\chi^2 = G_K/(h_{\mathrm{ex}}^2\tilde{\delta}^4)$; therefore, from (4.6) and (4.24) we have

$$\begin{aligned}(4.25)\qquad \frac{\tilde{\delta}^2\chi^2}{\mu_\varepsilon^2} &= \frac{G_K}{\tilde{\delta}^2 h_{\mathrm{ex}}^2\mu_\varepsilon^2} \leq C\frac{\delta^2 h_{\mathrm{ex}}V_\varepsilon}{\tilde{\delta}^2 h_{\mathrm{ex}}^2\mu_\varepsilon^2} \\ &= C\frac{\delta^2}{\tilde{\delta}^2}\frac{V_\varepsilon}{h_\varepsilon\mu_\varepsilon^2} \ll \frac{C}{\big(1-r(U_{1-1/|\log\varepsilon|})\big)^2} \leq C,\end{aligned}$$

where the last line follows from the definition of $\tilde{\delta}$ and Lemma 3.

Note that

$$(4.26)\quad \chi^2 = \frac{G_K}{h_{\mathrm{ex}}^2\tilde{\delta}^4} \leq C\frac{h_{\mathrm{ex}}\delta^2}{h_{\mathrm{ex}}^2\tilde{\delta}^4} \leq \frac{\delta^4}{\tilde{\delta}^4}\frac{1}{h_{\mathrm{ex}}\delta^2} \ll \frac{C}{V_\varepsilon}\frac{\delta^4}{\tilde{\delta}^4} \ll \frac{1}{\big(1-r(U_{1-1/|\log\varepsilon|})\big)^4} \leq C$$

where we used the definition of $\tilde{\delta}$ and Lemma 3, hence $\chi \to 0$. From Remark 9 we choose $\alpha = \tilde{\delta}\chi^{2\gamma}$ for $0 < \gamma < \frac{1}{2}$, then (4.26) implies both

$$(4.27)\qquad \frac{\alpha}{\tilde{\delta}} = \chi^{2\gamma} \ll 1$$

and

$$(4.28)\qquad \frac{\chi^2\tilde{\delta}}{\alpha} = \chi^{2(1-\gamma)} \ll 1.$$

Since

$$1 \ll V_\varepsilon \ll h_{\text{ex}}\delta^2 \leq h_{\text{ex}}\tilde{\delta}^2\big(1+r(U_{1-1/|\log\varepsilon|})\big)^2 \leq 2h_{\text{ex}}\tilde{\delta}^2 \tag{4.29}$$

then using (4.22), (4.23), (4.25), (4.27), (4.28), (4.29), along with $1/|\log\varepsilon| \ll 1$, we establish the lower bound at leading order:

$$\int_0^{1-1/|\log\varepsilon|} a(t)+2tb(t)\,\mathrm{d}t \geq \frac{1}{2}h_{\text{ex}}\tilde{\delta}^2 V_\varepsilon\big(1-o_\varepsilon(1)\big). \tag{4.30}$$

From Step 1 we see that $\tilde{\delta} \geq \delta(1-o_\varepsilon)$; therefore, (4.30) implies

$$\int_0^1 a(t)+2tb(t)\,\mathrm{d}t \geq \int_0^{1-1/|\log\varepsilon|} a(t)+2tb(t)\,\mathrm{d}t \geq \frac{1}{2}h_{\text{ex}}\delta^2 V_\varepsilon\big(1-o_\varepsilon(1)\big),$$

and hence (4.18). □

4.1. Energy minimizer estimate. We conclude with the following estimate used in the proof of the lower bound.

Lemma 10. *Let $\widetilde{K} \subset U$ satisfy $\rho\big|_{\partial\widetilde{K}} \geq 1-1/|\log\varepsilon|$ and $|\widetilde{K}| \leq \tilde{\delta}^2 \leq \delta^2$, where δ satisfies conditions* (4.4)*, then a minimizer (u,A) satisfies*

$$\mu_\varepsilon^2 \int_{\widetilde{K}} \frac{|h-h_{\text{ex}}|^2}{\rho^2}|1-\rho^2| \ll 1 \tag{4.31}$$

for $\varepsilon^{-1/3} \ll \mu_\varepsilon$ and $|\log\varepsilon| \ll h_{\text{ex}} \ll \varepsilon^{-2/3}$.

In order to establish (4.31), we need the following regularity estimates.

Lemma 11. *Let $\widetilde{K}$ satisfy the same hypotheses as in Lemma* 10*, then*

$$\|1-\rho\|_{L^p} \leq C_{p,\gamma}\varepsilon^{1/p-\gamma}|\log\varepsilon|^3 \tag{4.32}$$

$$\left\|\frac{h-h_{\text{ex}}}{\rho^2}\right\|_{W^{1,q}(\widetilde{K})} \leq C_q\sqrt{\frac{|\log\varepsilon|^3}{\mu_\varepsilon^2}}\max\left\{\sqrt{\frac{|\log\varepsilon|^3}{\mu_\varepsilon^2}},1\right\} \tag{4.33}$$

for any $2<p<\infty$, γ small, and $1 \leq q < 2$.

Proof. The proof entails using the Euler–Lagrange equations for (1.1)

$$-\frac{\mu_\varepsilon^2}{4}\frac{|h-h_{\text{ex}}|^2}{\rho^4}u = \nabla_A^2 u + u(1-\rho^2)(3\rho^2-1) \tag{4.34}$$

$$\frac{\mu_\varepsilon^2}{4}\nabla^\perp\left(\frac{h-h_{\text{ex}}}{\rho^2}\right) = j_A(u) \tag{4.35}$$

in order to get improved regularity for the magnetic field. Note that

$$\int_{\widetilde{K}} g_{\text{csh}}(u,A)\,\mathrm{d}x \leq h_{\text{ex}}\log\frac{1}{\varepsilon\sqrt{h_{\text{ex}}}}\delta^2 \leq Ch_{\text{ex}}|\log\varepsilon|\delta^2 \leq C|\log\varepsilon|^3 \tag{4.36}$$

from (4.1), (4.6), and (4.4). Finally, we note that

$$\delta^2 \leq C. \tag{4.37}$$

(1) First, we claim that $\|1-\rho\|^2_{L^2(\widetilde{K})} \le C\varepsilon|\log\varepsilon|^3$. Since $\rho\big|_{\partial\widetilde{K}} \ge 1 - 1/|\log\varepsilon|$ then following the argument of Lemma 3 we find

$$\int_{\widetilde{K}} g_{\mathrm{csh}} \ge \frac{r(\{\widetilde{K}\cap\rho\le\frac12\})}{\varepsilon}\int_{1/2}^{1-1/|\log\varepsilon|} t - t^3\,\mathrm{d}t$$
$$\ge \frac{r(\{\widetilde{K}\cap\rho\le\frac12\})}{\varepsilon}\left(\frac{9}{64} - \frac{1}{|\log\varepsilon|^2} - \frac{1}{2|\log\varepsilon|^4}\right) \ge \frac{r(\{\widetilde{K}\cap\rho\le\frac12\})}{8\varepsilon}$$

for $\varepsilon \le \frac45$, so

$$\{x\in\widetilde{K} \text{ such that } \rho\le\tfrac12\} \subset \cup B_{r_k}$$

where

$$2\sum r_k \le C\varepsilon\int_{\widetilde{K}} g_{\mathrm{csh}}(u,A)\,\mathrm{d}x \le C\varepsilon|\log\varepsilon|^3,$$

having used (4.36). Since

$$|\widetilde{K}\cap\{\rho<\tfrac12\}| \le C\sum r_k^2 \le C\Big(\sum r_k\Big)^2 \le C\varepsilon^2|\log\varepsilon|^6,$$

Therefore,

$$\begin{aligned}\|1-\rho^2\|^2_{L^2(\widetilde{K})} &= 2\|1-\rho^2\|^2_{L^2(\widetilde{K}\cap\{\rho\le 1/2\})} + 2\|1-\rho^2\|^2_{L^2(\widetilde{K}\cap\{\rho>1/2\})}\\ &\le 2|\widetilde{K}\cap\{\rho<\tfrac12\}| + 8\int_{\widetilde{K}\cap\{\rho>1/2\}}\rho^2(1-\rho^2)^2\\ &\le C\varepsilon^2|\log\varepsilon|^6 + C\varepsilon\int_{\widetilde{K}} g_{\mathrm{csh}}(\rho,0)\,\mathrm{d}x\\ &\le C\varepsilon^2|\log\varepsilon|^6 + C\varepsilon|\log\varepsilon|^3 \le C\varepsilon|\log\varepsilon|^3,\end{aligned}$$

which implies the claim.

(2) Next $\rho^6+\rho^2 = \rho^6 - 2\rho^4 + \rho^2 + 2\rho^4 \le \rho^2(1-\rho^2)^2 + \rho^6/3 + \frac83$; hence, $\rho^6/3+\rho^2 \le (\varepsilon/2)g_{\mathrm{csh}} + \frac83$. So $(1-\rho)^2 \le \rho^2+1 \le (\varepsilon/2)g_{\mathrm{csh}} + 11/3$; therefore, using (4.37)

$$\begin{aligned}\int_{\widetilde{K}} |\nabla(1-\rho)|^2 + (1-\rho)^2 &\le (1+\varepsilon/2)\int_{\widetilde{K}} g_{\mathrm{csh}} + 4|\widetilde{K}|\\ &\le 2|\log\varepsilon|^3 + 4\delta^2 \le C|\log\varepsilon|^3.\end{aligned} \tag{4.38}$$

We now interpolate between the norms. Since $\|f\|_{L^p} \le \|f\|^{\alpha}_{L^q}\|f\|^{1-\alpha}_{L^r}$ for $1/p = \alpha/q + (1-\alpha)/r$,

$$\begin{aligned}\|1-\rho\|_{L^p(\widetilde{K})} &\le \|1-\rho\|^{\alpha}_{L^2(\widetilde{K})}\|1-\rho\|^{1-\alpha}_{L^r(\widetilde{K})}\\ &\le C_r\|1-\rho\|^{\alpha}_{L^2(\widetilde{K})}\|1-\rho\|^{1-\alpha}_{H^1(\widetilde{K})}\\ &\le C_r(\varepsilon|\log\varepsilon|^3)^{\alpha/2}(|\log\varepsilon|^3)^{1-\alpha/2} \le C_r\varepsilon^{\alpha/2}|\log\varepsilon|^3\end{aligned}$$

where $1/p = \alpha/2 + (1-\alpha)/r$. Hence, we can take $\alpha/2 = 1/p - \gamma$ for some small γ. This proves (4.32).

(3) We now establish the bound on the magnetic field. Note $\|\rho\|_{H^1(\widetilde{K})} \le C|\log\varepsilon|^{3/2}$ from the same argument as for (4.38). From (4.35) we have

$$\left\|\frac{1}{\rho}\nabla\left(\frac{h-h_{\mathrm{ex}}}{\rho^2}\right)\right\|^2_{L^2(\widetilde{K})} \le C\frac{|\log\varepsilon|^3}{\mu_\varepsilon^4},$$

and so

$$\left\| \nabla \left(\frac{h - h_{\mathrm{ex}}}{\rho^2} \right) \right\|_{L^q(\widetilde{K})} \le \left\| \frac{1}{\rho} \nabla \left(\frac{h - h_{\mathrm{ex}}}{\rho^2} \right) \right\|_{L^2(\widetilde{K})} \|\rho\|_{L^{2q/(2-q)}(\widetilde{K})}$$
$$\le C_q \left(\frac{|\log \varepsilon|^3}{\mu_\varepsilon^4} \right)^{1/2} |\log \varepsilon|^{3/2} \le C_q \frac{|\log \varepsilon|^3}{\mu_\varepsilon^2}.$$

Next, $(h - h_{\mathrm{ex}})/\rho^2 = (h - h_{\mathrm{ex}})/\rho + \big((h - h_{\mathrm{ex}})/\rho^2\big)(1 - \rho)$ thus

$$\left\| \frac{h - h_{\mathrm{ex}}}{\rho^2} \right\|_{L^q(\widetilde{K})} \le \left\| \frac{h - h_{\mathrm{ex}}}{\rho} \right\|_{L^q(\widetilde{K})} + \left\| \frac{h - h_{\mathrm{ex}}}{\rho^2}(1 - \rho) \right\|_{L^q(\widetilde{K})}.$$

The first term can be estimated via

$$\left\| \frac{h - h_{\mathrm{ex}}}{\rho} \right\|_{L^q(\widetilde{K})} \le \left\| \frac{h - h_{\mathrm{ex}}}{\rho} \right\|_{L^2(\widetilde{K})} |\widetilde{K}|^{1/q - 1/2} \le C \left(\frac{|\log \varepsilon|^3}{\mu_\varepsilon^2} \right)^{1/2} \delta^{2(1/q - 1/2)}$$

and the second term can be estimated

$$\left\| \frac{h - h_{\mathrm{ex}}}{\rho^2}(1 - \rho) \right\|_{L^q(\widetilde{K})} \le \left\| \frac{h - h_{\mathrm{ex}}}{\rho^2} \right\|_{L^r(\widetilde{K})} \|1 - \rho\|_{L^s(\widetilde{K})}$$
$$\le C_{r,s,\gamma} \left\| \frac{h - h_{\mathrm{ex}}}{\rho^2} \right\|_{W^{1,q}(\widetilde{K})} \varepsilon^{1/s - \gamma} |\log \varepsilon|^3$$

for $1/q = 1/r + 1/s$. Therefore,

$$\left\| \frac{h - h_{\mathrm{ex}}}{\rho^2} \right\|_{W^{1,q}(\widetilde{K})} \le C_p \frac{|\log \varepsilon|^3}{\mu_\varepsilon^2} + C \left(\frac{|\log \varepsilon|^3}{\mu_\varepsilon^2} \right)^{1/2} \delta^{2(1/q - 1/2)}$$
$$+ C_{r,s,\gamma} \left\| \frac{h - h_{\mathrm{ex}}}{\rho^2} \right\|_{W^{1,q}(\widetilde{K})} \varepsilon^{1/s - \gamma} |\log \varepsilon|^3$$

or

$$\left\| \frac{h - h_{\mathrm{ex}}}{\rho^2} \right\|_{W^{1,q}(\widetilde{K})} \le \frac{C_p}{(1 - C_{r,s,\gamma} \varepsilon^{1/s - \gamma} |\log \varepsilon|^3)} \sqrt{\frac{|\log \varepsilon|^3}{\mu_\varepsilon^2}} \max\left\{ \sqrt{\frac{|\log \varepsilon|^3}{\mu_\varepsilon^2}}, \delta^{2/q - 1} \right\}.$$

Using (4.37), we find (4.33) for all ε small enough. □

Finally, we have the

Proof of Lemma 10. For $1 = 1/a + 1/b + 1/c$ then

$$\int_{\widetilde{K}} \frac{|h - h_{\mathrm{ex}}|^2}{\rho^4} \rho^2 |1 - \rho^2| \le \left\| \frac{|h - h_{\mathrm{ex}}|^2}{\rho^4} \right\|_{L^a(\widetilde{K})} \|\rho^2\|_{L^b(\widetilde{K})} \|1 - \rho^2\|_{L^c(\widetilde{K})}$$
$$\le \left\| \frac{h - h_{\mathrm{ex}}}{\rho^2} \right\|^2_{L^{2a}(\widetilde{K})} \|\rho\|^2_{L^{2b}(\widetilde{K})} \|1 - \rho^2\|_{L^c(\widetilde{K})}$$
$$\le C \left\| \frac{h - h_{\mathrm{ex}}}{\rho^2} \right\|^2_{W^{1,q}(\widetilde{K})} \|\rho\|^2_{H^1(\widetilde{K})} \|1 - \rho^2\|_{L^c(\widetilde{K})}$$

Using Lemma 11, we estimate for $p < 2$ and γ small,

$$\begin{aligned}\mu_\varepsilon^2 \int_{\widetilde{K}} \frac{|h-h_{\mathrm{ex}}|^2}{\rho^4}\rho^2|1-\rho^2| \\ &\le C_{q,\gamma}\mu_\varepsilon^2\left(\frac{|\log\varepsilon|^3}{\mu_\varepsilon^2}\max\left\{\frac{|\log\varepsilon|^3}{\mu_\varepsilon^2},1\right\}\right)|\log\varepsilon|^3(\varepsilon^{1/p-\gamma}|\log\varepsilon|^3) \\ &\le C_{p,\gamma}\varepsilon^{1/p-\gamma}|\log\varepsilon|^9\max\{|\log\varepsilon|^3\varepsilon^{-1/3},1\},\end{aligned}$$

since $\varepsilon^{1/3} \ll \mu_\varepsilon^2$ and $\delta \ll 1$. Therefore,

$$\mu_\varepsilon^2 \int_{\widetilde{K}} \frac{|h-h_{\mathrm{ex}}|^2}{\rho^4}\rho^2|1-\rho^2| \le C_{p,\gamma}\varepsilon^{1/p-\gamma-1/3}|\log\varepsilon|^{12} \ll 1$$

since we can choose p and γ so that $1/p > \frac{1}{3}$. □

PROOF OF PROPOSITION 5. We now consider the energy G_{csh} in the entire domain. From Lemma 9 and (4.10)

$$\begin{aligned}G_{\mathrm{csh}}(u_\varepsilon, A_\varepsilon) &\ge \sum_{K_i\subset U}\left[\sum_{G_{K_i}\le h_{\mathrm{ex}}\delta^2V_\varepsilon} G_{K_i} + \sum_{G_{K_i}>h_{\mathrm{ex}}\delta^2V_\varepsilon} G_{K_i}\right] \\ &\ge \sum_{K_i\subset U}\left[\sum_{G_{K_i}\le h_{\mathrm{ex}}\delta^2V_\varepsilon} \frac{1}{2}h_{\mathrm{ex}}\delta^2V_\varepsilon\big(1-o_\varepsilon(1)\big) + \sum_{G_{K_i}>h_{\mathrm{ex}}\delta^2V_\varepsilon} \frac{1}{2}h_{\mathrm{ex}}\delta^2V_\varepsilon\right] \\ &\ge \sum_{K_i\subset U}\frac{1}{2}h_{\mathrm{ex}}\delta^2V_\varepsilon\big(1-o_\varepsilon(1)\big) \ge \frac{1}{2}|U|h_{\mathrm{ex}}V_\varepsilon(1-o_\varepsilon),\end{aligned}$$

since for ε small, $\sum_{K_i\subset U}\delta^2 \ge |U|(1-o_\varepsilon)$. □

PROOF OF THEOREM 1. The upper bound in (1.8) follows from Proposition 4, since $\varepsilon^{1/6} \gg \varepsilon^{1/2}$. The lower bound is given in Proposition 5. □

References

1. F. Béthuel, H. Brezis, and F. Hélein, *Ginzburg–Landau vortices*, Progr. Nonlinear Differential Equations Appl., vol. 13, Birkhäuser, Boston, MA, 1994.
2. F. Béthuel and T. Rivière, *Vortices for a variational problem related to superconductivity*, Ann. Inst. H. Poincaré Anal. Non Linéaire **12** (1995), no. 3, 243–303.
3. L. A. Caffarelli and Y. S. Yang, *Vortex condensation in the Chern–Simons–Higgs model: An existence theorem*, Comm. Math. Phys. **168** (1995), no. 2, 321–336.
4. D. Chae and N. Kim, *Topological multivortex solutions of the self-dual Maxwell–Chern–Simons–Higgs system*, J. Differential Equations **134** (1997), no. 1, 154–182.
5. J. Han and J. Jang, *Self-dual Chern–Simons vortices on bounded domains*, Lett. Math. Phys. **64** (2003), no. 1, 45–56.
6. J. Han and N. Kim, *Nonself-dual Chern–Simons and Maxwell–Chern–Simons vortices on bounded domains*, J. Funct. Anal. **221** (2005), no. 1, 167–204.
7. J. Hong, Y. Kim, and P. Y. Pac, *Multivortex solutions of the abelian Chern–Simons vortices*, Phys. Rev. Lett. **64** (1990), no. 19, 2230–2233.
8. R. Jackiw and E. J. Weinberg, *Self-dual Chern–Simons vortices*, Phys. Rev. Lett. **64** (1990), no. 19, 2234–2237.
9. A. Jaffe and C. Taubes, *Vortices and monopoles: Structure of static gauge theories*, Progr. Phys., vol. 2, Birkhäuser, Boston, MA, 1980.
10. R. L. Jerrard, *Lower bounds for generalized Ginzburg–Landau functionals*, SIAM J. Math. Anal. **30** (1999), no. 4, 721–746.
11. R. L. Jerrard and H. M. Soner, *The Jacobian and the Ginzburg–Landau energy*, Calc. Var. Partial Differential Equations **14** (2002), no. 2, 151–191.

12. ______, *Limiting behavior of the Ginzburg–Landau functional*, J. Funct. Anal. **192** (2002), no. 2, 524–561.
13. M. Kurzke and D. Spirn, *Scaling limits of the Chern–Simons–Higgs energy*, Commun. Contemp. Math., to appear.
14. ______, *Gamma limit of the nonself-dual Chern–Simons–Higgs energy*, J. Funct. Anal., submitted.
15. ______, *On the energy of superconductors in large and small domains*, preprint.
16. T. Ricciardi and G. Tarantello, *Vortices in the Maxwell–Chern–Simons theory*, Comm. Pure Appl. Math. **53** (2000), no. 7, 811–851.
17. É. Sandier, *Lower bounds for the energy of unit vector fields and applications*, J. Funct. Anal. **152** (1998), no. 2, 379–403.
18. É. Sandier and S. Serfaty, *On the energy of type-*II *superconductors in the mixed phase*, Rev. Math. Phys. **12** (2000), no. 9, 1219–1257.
19. G. Tarantello, *Multiple condensate solutions for the Chern–Simons–Higgs theory*, J. Math. Phys. **37** (1996), no. 8, 3769–3796.
20. Y. S. Yang, *Solitons in field theory and nonlinear analysis*, Springer Monogr Math., Springer, New York, NY, 2001.

Institut für Mathematik, Humboldt-Universität zu Berlin, 10099 Berlin, Germany
Current address: Institut für Angewandte Mathematik, Universität Bonn, Wegelerstraße 10, 53115 Bonn, Germany
E-mail address: kurzke@iam.uni-bonn.de

School of Mathematics, University of Minnesota, Minneapolis, MN 55455, USA
E-mail address: spirn@math.umn.edu

Centre de Recherches Mathématiques
CRM Proceedings and Lecture Notes
Volume **44**, 2008

Some Recent Results about a Class of Singularly Perturbed Elliptic Equations

Andrea Malchiodi

Abstract. We study a class singularly perturbed elliptic equations, with motivations arising from chemistry, biology or physics. We investigate the behavior of solutions as some parameter tends to zero, and we study in particular their location and profile. We report some recent progress in this direction.

1. Introduction

We consider a class of singularly perturbed elliptic equations, and our goal is to study the asymptotic behavior of solutions when some parameter tends to zero.

One of the main motivations for the study of these equations arises from reaction-diffusion systems, concerning in particular the so-called Turing's instability. More precisely, see [15, 45], it is known that scalar diffusion equations like

$$\begin{cases} u_t = \Delta u + f(u) & \text{in } \Omega; \\ \dfrac{\partial u}{\partial \nu} = 0 & \text{on } \partial\Omega, \end{cases}$$

for convex domains Ω admit only constant (linearly) stable steady state solutions. On the other hand, as noticed in [54], reaction-diffusion *systems* with different diffusivities might lead to nonhomogeneous stable steady states.

A well-know example is the following one

$$\text{(GM)} \qquad \begin{cases} \mathcal{U}_t = d_1 \Delta \mathcal{U} - \mathcal{U} + \dfrac{\mathcal{U}^p}{\mathcal{V}^q} & \text{in } \Omega \times (0, +\infty), \\ \mathcal{V}_t = d_2 \Delta \mathcal{V} - \mathcal{V} + \dfrac{\mathcal{U}^r}{\mathcal{V}^s} & \text{in } \Omega \times (0, +\infty), \\ \dfrac{\partial \mathcal{U}}{\partial \nu} = \dfrac{\partial \mathcal{V}}{\partial \nu} = 0 & \text{on } \partial\Omega \times (0, +\infty), \end{cases}$$

introduced in [26] to describe some biological experiment. The functions $\mathcal{U}$ and $\mathcal{V}$ represent the densities of some chemical substances, the numbers p, q, r, s are nonnegative and such that $0 < (p-1)/q < r/(s+1)$, and it is assumed that the diffusivities d_1 and d_2 satisfy $d_1 \ll 1 \ll d_2$. In the stationary case of (GM), as

2000 *Mathematics Subject Classification.* Primary 35B25, 35B40; Secondary 35B34, 35B38.

Key words and phrases. Singularly perturbed elliptic equations, concentration phenomena, geometric partial differential equations.

This is the final form of the paper.

explained in [47, 50], when $d_2 \to +\infty$ the function $\mathcal{V}$ is close to a constant (being nearly harmonic and with zero normal derivative at the boundary), and therefore the equation satisfied by $\mathcal{U}$ is similar to the following one

$$(\mathrm{P}_\varepsilon) \qquad \begin{cases} -\varepsilon^2 \Delta u + u = u^p & \text{in } \Omega, \\ \dfrac{\partial u}{\partial \nu} = 0 & \text{on } \partial\Omega, \\ u > 0 & \text{in } \Omega, \end{cases}$$

with $\varepsilon^2 = d_1$. One is interested in general in the profile of solution, their location and possibly in their stability properties.

The advantage of problem (P_ε) compared to (the stationary case of) (GM) is that, apart from being a scalar equation, its structure is variational and solutions can be found as critical points of the following Euler – Lagrange functional

$$I_\varepsilon(u) = \frac{1}{2} \int_\Omega (\varepsilon^2 |\nabla u|^2 + u^2) - \frac{1}{p+1} \int_\Omega |u|^{p+1}; \quad u \in H^1(\Omega).$$

The typical concentration behavior of solutions u_ε to (P_ε) is via a scaling of the variables in the form $u_\varepsilon(x) \sim u_0\big((x - Q)/\varepsilon\big)$, where Q is some point of $\overline{\Omega}$, and where u_0 is a solution of the problem

$$(1) \qquad -\Delta u_0 + u_0 = u_0^p \quad \text{in } \mathbb{R}^N \text{ (or in } \mathbb{R}^N_+ = \{(x_1, \dots, x_N) \in \mathbb{R}^N : x_N > 0\}),$$

the domain depending on whether Q lies in the interior of Ω or at the boundary; in the latter case Neumann conditions are imposed.

When $p < (N + 2)/(N - 2)$ (and indeed only if this inequality is satisfied), problem (1) admits positive radial solutions which decay to zero at infinity, see [12, 13, 53]. Solutions of (P_ε) which inherit this profile are called *spike-layers*, since they are highly concentrated near some point of $\overline{\Omega}$. There is an extensive literature regarding this type of solutions, beginning from the papers [37, 48, 49]. Indeed their structure is very rich, and we refer for example to the (far from complete) list of papers [18, 23, 28–31, 35, 36, 55, 56].

We list next some of the properties of the solutions constructed in the literature. One can first look at (nontrivial) solutions of (P_ε) with minimal possible energy: we notice that the functional I_ε, if p is subcritical, satisfies the assumptions of the mountain-pass theorem. It can be shown that the profile of the mountain-pass solution of (P_ε) is a soliton of (1) in the half-space, and it peaks at a point of $\partial\Omega$ with maximal mean curvature. Furthermore, there are solutions with interior and multiple peaks. Their energy (the corresponding value of I_ε) is of order ε^n, which is proportional to the volume of their *support*, heuristically identified with a ball of radius ε centered at the peak. Roughly speaking, concentration at the boundary occurs at critical points of the mean curvature, while concentration at the interior occurs at singular points of the distance function from the boundary. Methods of construction rely mainly on min-max arguments, finite-dimensional reductions and gluing techniques (for multiple spikes).

In some cases, see for example [50], it is possible to construct stationary solutions of (GM) starting from spike-layer solutions of (P_ε) and using perturbation arguments. In these references some stability conditions are also given: we notice that nontrivial solutions of (P_ε) are always unstable, but their counterparts in (GM) may gain stability through the coupling with a second equation.

There are other motivations for the study of the above kind of equations: (P_ε) arises also as limit of different reaction-diffusion systems (with chemotaxis for example, as shown in [47]). Another motivation comes from the nonlinear Schrödinger equation

$$\mathrm{i}\hbar\frac{\partial\psi}{\partial t} = -\hbar^2\Delta\psi + V(x)\psi + |\psi|^{p-1}\psi \quad \text{in } \mathbb{R}^n, \tag{2}$$

where ψ is a complex-valued function (the *wave function*), V is a potential and p is an exponent greater than 1. Indeed, if one looks for *standing waves*, namely solutions of the form $\psi(x,t) = \exp(-\mathrm{i}\omega t/\hbar)u(x)$, for some real function u, then the latter will satisfy

$$-\varepsilon^2\Delta u + V(x)u = u^p \quad \text{in } \mathbb{R}^n, \tag{3}$$

where we have set $\varepsilon = \hbar$ and we absorbed the constant ω into the potential V. Therefore, the problem very similar to (P_ε), apart from the addition of a potential term. Also for this problem we have highly peaked solutions, which in general concentrate at critical points of V: about this subject we refer the reader to the (still incomplete) list of papers [1–3, 6–8, 14, 21, 22, 25, 27, 32, 51] and to the bibliographies therein.

Apart from existence of solutions concentrating at points, one may ask whether there exist others which scale only in some of the variables, and not all of them, and which therefore concentrate at higher-dimensional sets, like curves or manifolds. Indeed under generic assumptions, see [47], if $\Omega \subseteq \mathbb{R}^n$ and $k = 1, \ldots, n-1$, it was conjectured that there exist solutions of (P_ε) concentrating at k-dimensional sets. The phenomenon was known for particular domains with some symmetry: for these and related issues see [4, 5, 9–11, 19, 20, 43, 46, 52].

2. A general result

We describe in this section a first general result (with no symmetry assumptions) concerning the above conjecture for $k = n-1$. This was obtained in [41] in the case $n = 2$, and then extended in [42] for arbitrary n. We first state the result and describe the new difficulties compared to the case of ordinary spike-layers. Then we sketch the main ideas of the proof.

2.1. Statement of the result and the new phenomena involved.

Theorem 2.1. *Let $\Omega \subseteq \mathbb{R}^n$ be a smooth bounded domain, and let $p > 1$. Then there exists a sequence $\varepsilon_j \to 0$ and a sequence of solutions u_{ε_j} of $(\mathrm{P}_{\varepsilon_j})$ with the following properties*

(i) *u_{ε_j} concentrates at $\partial\Omega$ as $j \to +\infty$, namely for every $r > 0$ one has $\int_{\Omega^r}(\varepsilon_j^2|\nabla u_{\varepsilon_j}|^2 + u_{\varepsilon_j}^2) \to 0$ as $j \to +\infty$, where $\Omega^r = \{x \in \Omega : \mathrm{dist}(x, \partial\Omega) \geq r\}$;*

(ii) *if $x_0 \in \partial\Omega$ and if ν_0 denotes the interior unit normal to $\partial\Omega$ at x_0, then for every $k \in \mathbb{N}$ there holds*

$$u_{\varepsilon_j}\big(\varepsilon_j(x - x_0)\big) \to w_0(x \cdot \nu_0) \quad \text{in } C^k_{\mathrm{loc}}(V_0),$$

where $V_0 = \{x \in \mathbb{R}^N : x \cdot \nu_0 > 0\}$, and where w_0 is the solution of

$$\begin{cases} -w_0'' + w_0 = w_0^p & \text{in } \mathbb{R}_+, \\ w_0'(0) = 0. \end{cases}$$

Roughly speaking, and we will use below this terminology, the result states that solutions concentrate at the whole boundary and, as expected, they scale in some of the variables only (one in this case), with profile given by a solution of (1) which does not decay o zero at infinity. To have a *genuine* solution of (1), we have then to reduce the dimension of the limit problem by the codimension of the set of concentration. Some comments are now in order.

Remark 2.2. (1) Solutions have large energy: their *support* (with the terminology used before) is a strip of width ε around $\partial\Omega$, so its volume is of order ε_j. From some scaling arguments it follows that also their energy is of order ε, namely (of order) $1/\varepsilon_j^{n-1}$ times bigger compared to the energy of the spikes. There are other types of solutions with large energy, constructed in [38], which consist of an increasing number of spike-layers, so with a profile different from ours.

(2) Solutions have large Morse index: the Morse index of u_{ε_j} is of order $1/\varepsilon_j^{n-1}$. This should be compared to a result in [17]: under some assumptions it is proved that if the Morse index of a family of solutions stays bounded as $\varepsilon \to 0$, these must concentrate at a finite number of points, namely they are ordinary spikes.

(3) The exponent p can also be supercritical: the profile is the solution of the *one-dimensional* equation $-\ddot{w}_0 + w_0 = w_0^p$, for which there is no critical exponent. We notice that, on the other hand, for p supercritical there are no spike-layers, see [16]. The method in the proof of Theorem 2.1 is rather robust: in fact one can also deal with other nonlinearities and nonhomogeneous equations as well.

(4) Existence is proved only along a sequence: it is due to *resonance phenomena*, which are described below, and is not related to compactness properties. It is peculiar of multi-dimensional spikes, not observed in the study of ordinary spike-layers.

The resonance phenomenon discussed in the above remark is the most distinctive aspect of concentration for (P_ε) at higher-dimensional sets, and below in this subsection we are going to describe how this occurs.

In one dimension the limit equation is

$$\begin{cases} -w_0'' + w_0 = w_0^p & \text{in } \mathbb{R}_+; \\ w_0'(0) = 0. \end{cases}$$

The function w_0 is a mountain-pass critical point of the functional

$$\overline{I}(u) = \frac{1}{2}\int_{\mathbb{R}_+} (|\nabla u|^2 + u^2) - \frac{1}{p+1}\int_{\mathbb{R}_+} |u|^{p+1}; \quad u \in H^1(\mathbb{R}_+).$$

It turns out that w_0 is nondegenerate, so the spectrum of $\overline{I}''(w_0)$ possesses one negative eigenvalue, while all the remaining ones are positive and bounded away from zero.

In higher dimensions, or when the limit equation is degenerate (due to translation invariance) then we have a finite-dimensional kernel, characterized in [34]. This is what allows (P_ε), or similar problems, to be studied via Lyapunov – Schmidt reductions.

To explain the spectral properties of the linearized equation at the boundary layers, we define some coordinates x' on $\partial\Omega$, and we let x_n denote the distance from $\partial\Omega$. Since the functions we are dealing with decay exponentially away from $\partial\Omega$, the main contribution of the Euler functional arises from a tubular neighborhood of $\partial\Omega$, where we have the above parametrization. Moreover, since at scale ε the boundary of Ω looks almost flat, the local coordinates (x', x_n) are nearly Euclidean, so we can write that

$$I_\varepsilon(u) \simeq \frac{1}{2}\int_{\partial\Omega\times\mathbb{R}_+} \varepsilon^2|\nabla u|^2 \, dx' \, dx_n + \int_{\partial\Omega\times\mathbb{R}_+} F(u) \, dx' \, dx_n.$$

Considering then the second differential, and splitting the Dirichlet energy into tangential and normal components we have that

$$(4) \qquad I''_\varepsilon(u_\varepsilon)[u,u] \simeq \int_{\partial\Omega\times\mathbb{R}_+} (\tfrac{1}{2}\varepsilon^2|\nabla u|^2 + F''(u_\varepsilon)u^2) \, dx' \, dx_n$$
$$\sim \frac{1}{2}\varepsilon^2 \int_{\partial\Omega\times\mathbb{R}_+} |\nabla_{x'} u|^2 \, dx' \, dx_n + \int_{\partial\Omega\times\mathbb{R}_+} \overline{I}''_\varepsilon(u_\varepsilon)[u,u] \, dx' \, dx_n.$$

The first term can be interpreted as the energy of an elastic membrane, while the second, since $u_\varepsilon(x', x_n) \simeq w_0(x_n/\varepsilon)$, is the integral over $\partial\Omega$ of the second differential at w_0 of $\overline{I}_\varepsilon$, the scaling in ε of the one-dimensional functional $\overline{I}$.

As a consequence we have that every eigenvalue μ of $\overline{I}''_\varepsilon(u_\varepsilon)$ (with eigenvector $v_\mu(x_n/\varepsilon)$) generates a sequence of eigenvalues of the form $\mu_j \sim \mu + \varepsilon^2 j^{2/(n-1)}$, $j \in \mathbb{N}$, for $I''_\varepsilon(u_\varepsilon)$, with eigenvectors $v_{\mu,j} \sim \phi_j(x') \times v_\mu(x_n/\varepsilon)$, where ϕ_j are the Fourier modes of $\partial\Omega$.

To see this in a concrete case, suppose for simplicity that $x' \in S^1$ (and that $x_n \in \mathbb{R}_+$). The eigenvalue equation for the linearized problem is

$$-\varepsilon^2\Delta v(x) + F''\big(w_0(x_n/\varepsilon)\big)v(x) = \lambda v(x).$$

Trough a separation of variables we write

$$v(x) = \sin(jx')v_j(x_n) \quad (x', x_n) \in S^1 \times \mathbb{R}_+ :$$

then the function v_j satisfies the ordinary differential equation

$$-\varepsilon^2\Delta_{x_n} v_j + \big(F''\big(w_0(x_n/\varepsilon)\big) + \varepsilon^2 j^2\big)v_j(x_n) = \lambda v_j(x_n).$$

From this formula one sees that the spectrum of this operator coincides with the one of $\overline{I}''_\varepsilon$, with the addition of the term $\varepsilon^2 j^2$. In the general case one has to replace the trigonometric functions with the Fourier modes of $\partial\Omega$, and the term $\varepsilon^2 j^2$ (qualitatively) with $\varepsilon^2 j^{2/(n-1)}$, by the Weyl's asymptotic formula.

As a consequence of these considerations one finds that the linearized operator has a number of negative eigenvalues which diverges as ε tends to zero with rate ε^{1-n}, as anticipated in Remark 2.2. Moreover, from the above asymptotic expansion of the eigenvalues, one sees that the average gap of two eigenvalues near zero is of order ε^{n-1}.

2.2. Idea of the proof. Since the Morse index of the solutions (or the number of negative eigenvalues of the linearization) depends on ε, it is not possible to employ standard min-max arguments, since they would lead to solutions with bounded Morse index. It is convenient instead to use local inversion arguments, also because the information about the profile of the solutions is automatic. The standard procedure consists in finding first some approximate solutions and then, when the linearized operator is invertible, one finds real solutions through the implicit function theorem: the main difficulty here is to understand the spectral properties of the linearized equation.

Approximate solutions. To begin, we consider the above coordinates (x', x_n) in a neighborhood of $\partial\Omega$, and we let w_0 be the solution of the problem

$$\begin{cases} -w_0'' + w_0 = w_0^p & \text{in } \mathbb{R}_+, \\ w_0'(0) = 0 & w_0 > 0. \end{cases}$$

Clearly the function $w_0(x_n/\varepsilon)$ constitutes an approximate solution (up to order ε) of (P_ε). However, since we expect the inverse operator to have a large norm, it is convenient to improve the accuracy of this solution with an iterative procedure. Indeed, setting

$$u_{k,\varepsilon} = w_0(x_n/\varepsilon) + \varepsilon w_1(x', x_n/\varepsilon) + \cdots + \varepsilon^k w_k(x', x_n/\varepsilon), \tag{5}$$

we can solve formally (P_ε) in powers of ε, so the functions $\{w_i\}$ are uniquely defined by recurrence by equations like

$$\begin{cases} -w_i'' + w_i - p w_0^{p-1} w_i = F_i(x', x_n/\varepsilon, w_0, w_1, \dots, w_{i-1}) & \text{in } \mathbb{R}_+, \\ w_i'(0) = 0. \end{cases}$$

This equation is always solvable by the invertibility of the linear operator acting on w_i. To get this recurrence formula, it is crucial to have a *slower* dependence in x' than in x_n (see (5)). This also reflects the fact that there is a *weak* interaction of the w_i's with the resonant Fourier modes (which are fast).

Invertibility of $I_\varepsilon''(u_{k,\varepsilon})$. We turn next to the invertibility issue, which is the most delicate in this problem. In the previous subsection we reasoned as if some neighborhood of $\partial\Omega$ inherited the product metric of $\partial\Omega$ and $\mathbb{R}_+$. Taking the curvature of the boundary into account, an expansion for the eigenvalues of $I_\varepsilon''(u_{k,\varepsilon})$ gives a behavior qualitatively like

$$\mu_j \sim \mu + \varepsilon^2 j^{2/(n-1)} + O(\varepsilon), \quad j \gg 1. \tag{6}$$

Since, as already remarked, the average gap between eigenvalues close to zero is of order ε^{n-1}, the above formula in general is not accurate enough to obtain invertibility. The idea then is to view the eigenvalues as functions of ε: an heuristic computation gives

$$\mu_j = \mu_j(\varepsilon) \implies \frac{\partial \mu_j}{\partial \varepsilon} \sim \varepsilon j^{2/(n-1)} \sim \frac{1}{\varepsilon} \quad (\text{for } j \sim \varepsilon^{-1/(n-1)}).$$

This idea can indeed be made rigorous using a classical perturbation result in linear theory due to T. Kato, see [33], which is the following.

Theorem 2.3 (T. Kato). *Let $s \mapsto T(s)$ be a smooth family of linear operators from an Hilbert space $\mathcal{H}$ into itself, of the form Identity$-$ compact. Let λ an isolated eigenvalue of $T(0)$. Then $\partial\lambda/\partial s$ exists (possibly as a multivalued function), and one has*

$$\frac{\partial \lambda}{\partial s}\Big|_{s=0} = \text{eigenvalues of} \left\{ P_\lambda \circ \frac{\partial T}{\partial s}\Big|_{s=0} \circ P_\lambda \right\},$$

where P_λ denotes the orthogonal projection onto the λ-eigenspace of $T(0)$ corresponding to the eigenvalue λ.

We notice that if $\lambda(0)$ has multiplicity greater than 1, then there could be a possible splitting. To use this theorem in the present case, we use ε as a parameter, and we take $\lambda \simeq 0$. Applying P_λ is equivalent to characterize the eigenfunctions with eigenvalue λ. This is done using Fourier analysis: in the *model* case described in the previous subsection, by the monotonicity of the eigenvalues in j, generically only one Fourier component of the eigenfunction will be nonzero. For the real situation one finds that indeed the Fourier modes of the eigenfunctions are highly peaked near some given ones, depending on ε. In this way, one obtains that eigenvalues which are close to zero vary with ε with approximately the same velocity.

At this point, using (6) and counting the number of eigenvalues in a fixed neighborhood of zero, one deduces that there will be (in the average) a certain number of spectral gaps in this interval. Then, using Kato's theorem and varying ε suitably, one can bring the spectral gaps near zero. In this way one obtains the following result.

Proposition 2.4. *There exists $C > 0$ and $\varepsilon_j \to 0$ such that the distance of the spectrum of $I''_{\varepsilon_j}(u_{k,\varepsilon_j})$ from 0 is greater or equal to $C^{-1}\varepsilon_j^{n-1}$.*

The proof of the theorem now just uses the contraction mapping theorem. When p is supercritical the problem is indeed not variational. However one can use in this case some a priori estimates in L^∞ and truncation arguments to obtain the same conclusion. Also, the method gives indeed existence not only for a sequence $\varepsilon_j \to 0$, but for a family of epsilon's whose density approaches 1 in measure as the parameter gets smaller and smaller.

3. Some extensions

In this section we discuss some extensions of the above result and some open problems and perspectives.

3.1. Higher codimensions. For the counterpart of Theorem 2.1 in higher codimension, the concentration set has to be *localized*. Heuristic considerations in terms of the energy functional suggest immediately that if a family of solutions concentrates near some submanifold of $\partial\Omega$, the limit set needs to be stationary with respect to the volume functional. If the dimension k is 1 for example, then the limit set should be a stationary curve for the length, namely a geodesic. The following result has been proved in [40] for $n = 3$, $k = 1$, and in [39] for the case of general n and k, settling the conjecture at the end of Section 1.

Theorem 3.1. *Let $\Omega \subseteq \mathbb{R}^n$ be a smooth bounded set, let k be an integer between 1 and $n-2$, and let $p \in \big(1, (n-k+2)/(n-k-2)\big)$. Let $K \subseteq \partial\Omega$ be a smooth, compact, embedded, nondegenerate minimal submanifold of dimension k. Then there exists $\varepsilon_j \to 0$ such that (P_{ε_j}) possesses solution u_j concentrating at K. The profile of u_{ε_j} converges to the (radial) solution of*

$$\begin{cases} -\Delta v + v = v^p & \text{in } \mathbb{R}^{n-k}_+, \\ \dfrac{\partial v}{\partial \nu} = 0 & \text{on } \partial\mathbb{R}^{n-k}_+ \\ v(x) \to 0 & \text{per } |x| \to +\infty. \end{cases} \tag{7}$$

By *minimal submanifold* we mean critical with respect to the variation of the volume (the mean curvature relative to $\partial\Omega$ is zero). *Non-degenerate* means that the second variation of the volume has no kernel: there is an explicit form involving the *Jacobi operator*, involving the *normal Laplacian* and the curvatures of K, $\partial\Omega$ (we refer to [39] for details).

Differently from Theorem 2.1, we require an upper bound on p: this condition is indeed natural, since the limit problem is solvable if and only if $p <$

$(n-k+2)/(n-k-2)$. In any case, supercritical exponents are allowed (with respect to the dimension n).

SKETCH OF THE PROOF. The idea of the proof is similar in spirit to that of Theorem 2.1, but more serious difficulties occur. Since the limit problem (7) is degenerate (we have now invariance by translation along the boundary of the half-space), its linearization has a nontrivial kernel, which is $n-k-1$ dimensional. In this case, by the comments after (4), one should expect the spectrum of the linearized equation roughly to look like

$$\underbrace{\{\mu+\varepsilon^2 j^{2/k}\}_j}_{\mu_j(\varepsilon)} \cup \underbrace{\{\varepsilon^2 l^{2/k}\}_l}_{\rho_l(\varepsilon)}, \quad j,l\in\mathbb{N},$$

together with another set of eigenvalues positive and bounded away from zero. The eigenvalues $\mu_j(\varepsilon)$ are as before, while the ρ_l's are *generated* by the zero eigenvalue of (7).

In the previous case there was a natural parametrization of a neighborhood of the limit set as a product $\partial\Omega\times[0,\delta]$, but in higher codimension this is not possible anymore, and one needs a more intrinsic geometric approach, based on the *normal bundle* of K in $\partial\Omega$, namely the set of vectors tangent to $\partial\Omega$ in K but orthogonal to K.

This has indeed a deep relation with the resonance modes of the linearized equation, and we need crucially to exploit this fact. To give an idea of the phenomenon, we choose some coordinates x' on K and we let x'' denote some coordinates in Ω transversal to K. The limit equation, after a dilation by ε, then takes the form

$$-\Delta_{x''}u+u=u^p \quad \text{in } \mathbb{R}^{n-k},$$

(we need $p<(n-k+2)/(n-k-2)$ for the existence of a solution w_0). Letting K_ε be the dilation of K by $1/\varepsilon$, the linearized operator at an approximate solution becomes $-\Delta_{K_\varepsilon}u+\underbrace{-\Delta_{x''}u+u-pw_0^{p-1}u}_{L_{x''}u}$. Separating variables, one can write

$$u(x',x'')=\sum_j u_j(x')\Psi_j(x''),$$

where $\Psi_j(x'')$ are eigenfunctions of $L_{x''}$. The first one turns out to be radially symmetric in x'', while the subsequent $n-k-1$ have the symmetry of the first spherical harmonics (in x''), which identify a *direction* in $\mathbb{R}^{m-k-1}$.

The first coefficient $u_0(x')$ can then be viewed as a *scalar function* on K_ε, while the subsequent m-tuple $u_1(x'),\dots,u_m(x')$ ($m=n-k-1$) can be seen as a section in the normal bundle to K_ε. Therefore, when considering the linearized operator on the x' variables, one sees first the Laplace-Beltrami operator on K, and then the *normal Laplacian* acting on the normal bundle.

An accurate analysis, taking the curvatures into account, shows that $\rho_l(\varepsilon)=\sigma_l\varepsilon^2+o(\varepsilon^2)$, where σ_l ($\neq 0$) is the l-th eigenvalue of the Jacobi operator (related to the second variation of the area). In finding approximate solutions, the component of the equation parallel to the kernel can be solved tilting the approximate solution perpendicularly to K, and using the nondegeneracy condition on K.

Since all the ρ_l's are nonzero, if some eigenvalue approaches zero faster than ε^2, then it should be of the form $\mu_j(\varepsilon)$ for some j. Notice also that

$$\mu_j(\varepsilon), \rho_l(\varepsilon) \simeq 0 \implies \frac{\partial \mu_j(\varepsilon)}{\partial \varepsilon} \simeq \frac{1}{\varepsilon}; \qquad \frac{\partial \rho_l(\varepsilon)}{\partial \varepsilon} \simeq \varepsilon.$$

Therefore, for having a good lower bound on the derivatives of the eigenvalues, we need to show the following: every u such that

$$I''_\varepsilon(u_{k,\varepsilon})[u] = \lambda u \quad \text{with } \lambda = o(\varepsilon^2),$$

is *approximately* a linear combination of eigenfunctions corresponding to $\mu_j(\varepsilon)$. Then Kato's theorem will yield the conclusion as for the proof of Theorem 2.1.

To show this, besides the approximate solutions, we construct a family of *approximate eigenfunctions* of the linearized operator, which allow us to characterize the resonant eigenfunctions with sufficient accuracy. □

3.2. Some results with symmetry and open problems. In this subsection we discuss some results in the presence of symmetries, which give us results different from before and suggest some new possible directions of investigation. The following result was obtained in [4] and extended in [43] for the case of multiple peaks.

Theorem 3.2. *Let $p > 1$, and consider problem (P_ε) in $B_1 \subseteq \mathbb{R}^n$. Then there exists a radial interior $(n-1)$-dimensional spike-layer solution at distance $\varepsilon|\log \varepsilon|$ from ∂B_1.*

We notice that the solutions found in Theorem 3.2 are of different kind from those in Theorem 2.1, since now the profile is that of an interior layer and not that of a boundary layer. Since we are in a radial context, we can use a simpler proof, based on a Lyapunov – Schmidt reduction, joint with some decay estimates (since solutions have large energy compared to ordinary spikes). Resonances in this case are ruled out by the symmetry if we work in the class of invariant functions.

This method also gives existence results for equation (3) if we consider radial potentials. As explained in the introduction, ordinary spikes concentrate at critical points of V. However, solutions which are concentrated near spheres have an *energy* which depends both on the potential and on the surface area. The whole energy can be expressed by introducing the auxiliary radial potential

$$M(r) = r^{n-1}\big(V(r)\big)^{\theta}; \quad \theta = \frac{p+1}{p-1} - \frac{1}{2}.$$

In [4] we obtained the following result, see also [9, 10, 19, 46] for related ones with reduced symmetry assumptions.

Theorem 3.3. *Let $p > 1$, and let $V(x) = V(r) \in C^2(\mathbb{R}^n)$ be bounded from above and from below by positive constants. Then the condition $M'(r) = 0$ gives necessary and sufficient conditions for concentration at spheres of radial solutions u_ε.*

The resonance however is still underlying, and this can be seen from the bifurcation of nonradial solutions for some values of ε.

Theorem 3.4 ([4]). *Let $\bar{r}$ be a critical point of M such that $M''(\bar{r}) \neq 0$. Let u_ε be the radial solutions given in Theorem 3.3. Then there exists $\varepsilon_0 > 0$ such that*

$$\Lambda_{\varepsilon_0} = \{(\varepsilon, u_\varepsilon) : \varepsilon \in (0, \varepsilon_0)\}$$

is a smooth curve. Moreover, there exists a sequence $\varepsilon_l \to 0$ *such that from every* $(\varepsilon_l, u_{\varepsilon_l}) \in \Lambda_{\varepsilon_0}$ *bifurcate nonradial solutions of* (3).

The above results suggest some possible counterparts when the symmetry assumptions are removed. Concerning Theorem 3.2, concentration at the boundary arises from a *combination* of the surface energy of the solution and the interaction with $\partial\Omega$, in the same spirit of Theorem 3.3. Since (as can be seen with energetic considerations) the boundary tends to *attract* the spikes, we expect the result to persist if the mean curvature of Ω is everywhere positive. Some results in this direction have been obtained in [44] for the Allen–Cahn equation.

Also for (3) there should be concentration at higher-dimensional sets with no symmetry condition on the potential. Still using energetic considerations, in [4] we conjectured that if concentration occurs at a k-dimensional manifold $\Sigma \subseteq \mathbb{R}^n$, then the following condition should hold

$$\theta_k \nabla^{\perp} V = V\mathbf{H},$$

where $\mathbf{H}$ is the mean curvature of Σ, and where $\theta_k = (p+1)/(p-1) - \frac{1}{2}(n-k)$. For $n = 2$ and $k = 1$, and under suitable nondegeneracy assumptions, in [24] this condition has been proved to be also sufficient for concentration.

It is also an open problem to understand the global structure of solutions, depending on ε, when concentration at a positive dimensional set occurs. When symmetry is present, this is known by Theorem 3.4. However, we do not expect this situation to be generic since it could be replaced by *imperfect bifurcation*. We wonder whether the pure bifurcation could persist under stronger regularity assumptions of the data, like analyticity or similar.

References

1. A. Ambrosetti, M. Badiale, and S. Cingolani, *Semiclassical states of nonlinear Schrödinger equations*, Arch. Ratiion. Mech. Anal. **140** (1997), no. 3, 285–300.
2. A. Ambrosetti, V. Felli, and A. Malchiodi, *Ground states of nonlinear Schrödinger equations with potentials vanishing at infinity*, J. Eur. Math. Soc. (JEMS) **7** (2005), no. 1, 117–144.
3. A. Ambrosetti and A. Malchiodi, *Perturbation methods and semilinear elliptic problems on* $\mathbb{R}^n$, Progr. in Math., vol. 240, Birkhäuser, Basel, 2006.
4. A. Ambrosetti, A. Malchiodi, and W.-M. Ni, *Singularly perturbed elliptic equations with symmetry: Existence of solutions concentrating on spheres.* I, Comm. Math. Phys. **235** (2003), no. 3, 427–466.
5. ______, *Singularly perturbed elliptic equations with symmetry: Existence of solutions concentrating on spheres.* II, Indiana Univ. Math. J. **53** (2004), no. 2, 297–329.
6. A. Ambrosetti, A. Malchiodi, and D. Ruiz, *Bound states of nonlinear Schrödinger equations with potentials vanishing at infinity*, J. Anal. Math. **98** (2006), 317–348.
7. A. Ambrosetti, A. Malchiodi, and S. Secchi, *Multiplicity results for some nonlinear Schrödinger equations with potentials*, Arch. Ratiion. Mech. Anal. **159** (2001), no. 3, 253–271.
8. G. Arioli and A. Szulkin, *A semilinear Schrödinger equation in the presence of a magnetic field*, Arch. Ratiion. Mech. Anal. **170** (2003), no. 4, 277–295.
9. M. Badiale and T. D'Aprile, *Concentration around a sphere for a singularly perturbed Schrödinger equation*, Nonlinear Anal. **49** (2002), no. 7, 947–985.
10. T Bartsch and S. Peng, *Semiclassical symmetric Schrödinger equations: existence of solutions concentrating simultaneously on several spheres.* preprint.
11. V. Benci and T. D'Aprile, *The semiclassical limit of the nonlinear Schrödinger equation in a radial potential*, J. Differential Equations **184** (2002), no. 1, 109–138.
12. H. Berestycki and P.-L. Lions, *Nonlinear scalar field equations.* I: *Existence of a ground state*, Arch. Rational Mech. Anal. **82** (1983), no. 4, 313–345.
13. ______, *Nonlinear scalar field equations.* II: *Existence of infinitely many solutions*, Arch. Rational Mech. Anal. **82** (1983), no. 4, 347–375.

14. J. Byeon and Z.-Q. Wang, *Standing waves with a critical frequency for nonlinear Schrödinger equations*, Arch. Ratiion. Mech. Anal. **165** (2002), no. 4, 295–316.
15. R. G. Casten and C. J. Holland, *Instability results for reaction diffusion equations with Neumann boundary conditions*, J. Differential Equations **27** (1978), no. 2, 266–273.
16. S. Cingolani and A. Pistoia, *Nonexistence of single blow-up solutions for a nonlinear Schrödinger equation involving critical Sobolev exponent*, Z. Angew. Math. Phys. **55** (2004), no. 2, 201–215.
17. E. N. Dancer, *Stable and finite Morse index solutions on $\mathbb{R}^n$ or on bounded domains with small diffusion*, Trans. Amer. Math. Soc. **357** (2005), no. 3, 1225–1243.
18. E. N. Dancer and S. Yan, *Multipeak solutions for a singularly perturbed Neumann problem*, Pacific J. Math. **189** (1999), no. 2, 241–262.
19. ———, *A new type of concentration solutions for a singularly perturbed elliptic problem*, Trans. Amer. Math. Soc. **359** (2007), no. 4, 1765–1790.
20. T. D'Aprile, *On a class of solutions with nonvanishing angular momentum for nonlinear Schrödinger equations*, Differential Integral Equations **16** (2003), no. 3, 349–384.
21. M. del Pino and P. L. Felmer, *Local mountain passes for semilinear elliptic problems in unbounded domains*, Calc. Var. Partial Differential Equations **4** (1996), no. 2, 121–137.
22. ———, *Semi-classcal states for nonlinear Schrödinger equations*, J. Funct. Anal. **149** (1997), no. 1, 245–265.
23. M. del Pino, P. L. Felmer, and J. Wei, *On the role of the mean curvature in some singularly perturbed Neumann problems*, SIAM J. Math. Anal. **31** (1999), no. 1, 63–79.
24. M. del Pino, M. Kowalczyk, and J. Wei, *Concentration at curves for nonlinear Schrödinger equations*, Comm. Pure Appl. Math. **60** (2007), no. 1, 113–146.
25. A. Floer and A. Weinstein, *Nonspreading wave packets for the cubic Schrödinger equation with a bounded potential*, J. Funct. Anal. **69** (1986), no. 3, 397–408.
26. A. Gierer and H. Meinhardt, *A theory of biological pattern formation*, Biological Cybernetics **12** (1972), no. 1, 30–39.
27. M. Grossi, *Some results on a class of nonlinear Schrödinger equations*, Math. Z. **235** (2000), no. 4, 687–705.
28. M. Grossi, A. Pistoia, and J. Wei, *Existence of multipeak solutions for a semilinear Neumann problem via nonsmooth critical point theory*, Calc. Var. Partial Differential Equations **11** (2000), no. 2, 143–175.
29. C. Gui, *Multipeak solutions for a semilinear Neumann problem*, Duke Math. J. **84** (1996), no. 3, 739–769.
30. C. Gui and J. Wei, *On multiple mixed interior and boundary peak solutions for some singularly perturbed Neumann problems*, Canad. J. Math. **52** (2000), no. 3, 522–538.
31. C. Gui, J. Wei, and M. Winter, *Multiple boundary peak solutions for some singularly perturbed Neumann problems*, Ann. Inst. H. Poincaré Anal. Non Linéaire **17** (2000), no. 1, 47–82.
32. L. Jeanjean and K. Tanaka, *A positive solution for a nonlinear Schrödinger equation on $\mathbb{R}^n$*, Indiana Univ. Math. J. **54** (2005), no. 2, 443–464.
33. T. Kato, *Perturbation theory for linear operators*, 2nd ed., Grundlehren Math. Wiss., vol. 132, Springer, New York, 1976.
34. M. K. Kwong, *Uniqueness of positive solutions of $\Delta u - u + u^p = 0$ in $\mathbb{R}^n$*, Arch. Rational Mech. Anal. **105** (1989), no. 3, 243–266.
35. Y. Y. Li, *On a singularly perturbed equation with Neumann boundary conditions*, Comm. Partial Differential Equations **23** (1998), no. 3-4, 487–545.
36. Y. Y. Li and L. Nirenberg, *The Dirichlet problem for singularly perturbed elliptic equations*, Comm. Pure Appl. Math. **51** (1998), no. 11-12, 1445–1490.
37. C.-S. Lin, W.-M. Ni, and I. Takagi, *Large amplitude stationary solutions to a chemotaxis systems*, J. Differential Equations **72** (1988), no. 1, 1–27.
38. F.-H. Lin, W.-M. Ni, and J. Wei, *On the number of interior peak solutions for a singularly perturbed Neumann problem*, Comm. Pure Appl. Math. **60** (2007), no. 2, 252–281.
39. F. Mahmoudi and A. Malchiodi, *Concentration on minimal submanifolds for a singularly perturbed Neumann problem*, Adv. Math. **209** (2007), no. 2, 460–525.
40. A. Malchiodi, *Concentration at curves for a singularly perturbed Neumann problem in three-dimensional domains*, Geom. Funct. Anal. **15** (2005), no. 6, 1162–1222.
41. A. Malchiodi and M. Montenegro, *Boundary concentration phenomena for a singularly perturbed elliptic problem*, Comm. Pure Appl. Math. **55** (2002), no. 12, 1507–1568.

42. ———, *Multidimensional boundary layers for a singularly perturbed Neumann problem*, Duke Math. J. **124** (2004), no. 1, 105–143.
43. A. Malchiodi, W.-M. Ni, and J. Wei, *Multiple clustered layer solutions for semilinear Neumann problems on a ball*, Ann. Inst. H. Poincaré Anal. Non Linéaire **22** (2005), no. 2, 143–163.
44. A Malchiodi and J. Wei, *Boundary interface for the Allen–Cahn equation*, J. Fixed Point Theory Appl. **1** (2007), no. 2, 305–336.
45. H. Matano, *Asymptotic behavior and stability of solutions of semilinear diffusion equations*, Publ. Res. Inst. Math. Sci. **15** (1979), no. 2, 401–454.
46. R. Molle and D. Passaseo, *Concentration phenomena for solutions of superlinear elliptic problems*, Ann. Inst. H. Poincaré Anal. Non Linéaire **23** (2006), no. 1, 63–84.
47. W.-M. Ni, *Diffusion, cross-diffusion, and their spike-layer steady states*, Notices Amer. Math. Soc. **45** (1998), no. 1, 9–18.
48. W.-M. Ni and I. Takagi, *On the shape of least-energy solution to a semilinear Neumann problem*, Comm. Pure Appl. Math. **41** (1991), no. 7, 819–851.
49. ———, *Locating the peaks of least-energy solutions to a semilinear Neumann problem*, Duke Math. J. **70** (1993), no. 2, 247–281.
50. W.-M. Ni, I. Takagi, and E. Yanagida, *Stability of least energy patterns of the shadow system for an activator-inhibitor model*, Japan J. Indust. Appl. Math. **18** (2001), no. 2, 259–272.
51. Y.-G. Oh, *On positive multi-lump bound states of nonlinear Schrödinger equations under multiple well potentials*, Comm. Math. Phys. **131** (1990), no. 2, 223–253.
52. J. Shi, *Semilinear Neumann boundary value problems on a rectangle*, Trans. Amer. Math. Soc. **354** (2002), no. 8, 3117–3154.
53. W. A. Strauss, *Existence of solitary waves in higher dimensions*, Comm. Math. Phys. **55** (1977), no. 2, 149–162.
54. A. M. Turing, *The chemical basis of morphogenesis*, Philosophical Transactions of the Royal Society of London. Series B, Biological Sciences **237** (1952), no. 641, 37–72.
55. Z. Q. Wang, *On the existence of multiple, single-peaked solutions for a semilinear Neumann problem*, Arch. Rational Mech. Anal. **120** (1992), no. 4, 375–399.
56. J. Wei, *On the boundary spike layer solutions of a singularly perturbed semilinear Neumann problem*, J. Differential Equations **134** (1997), no. 1, 104–133.

Scuola Internazionale Superiore di Studi Avanzati, via Beirut 2-4, 34014 Trieste, Italy

E-mail address: `malchiod@sissa.it`

Centre de Recherches Mathématiques
CRM Proceedings and Lecture Notes
Volume **44**, 2008

The Dipole Problem for $H^{1/2}(\mathbb{S}^2;\mathbb{S}^1)$-Maps and Application

Vincent Millot

ABSTRACT. In this note, we consider a dipole type problem related to the question of smooth approximation in the functional space $X = H^{1/2}(\mathbb{S}^2;\mathbb{S}^1)$. The results presented here are consequences of previous results obtained by A. Pisante and the author in [16]. Extensions to more general geometries for the starting manifold and open questions are also addressed.

1. Introduction

Recently there has been some interest in the study of maps between manifolds with $H^{1/2}$-regularity, especially in the context of the complex Ginzburg–Landau equations and their applications to superconductivity models (see [1, 2, 4, 7–10, 13, 16–20]). In this note, we shall consider one of the most common class of such maps, namely,

$$X = H^{1/2}(\mathbb{S}^2;\mathbb{S}^1) := \{g \in L^2(\mathbb{S}^2;\mathbb{R}^2) : |g| = 1 \text{ a.e. in } \mathbb{R}^2 \text{ and } |g|_{1/2} < +\infty\}, \tag{1}$$

where $|\cdot|_{1/2}$ denotes the standard Gagliardo fractional seminorm, i.e.,

$$|g|^2_{1/2} = \int_{\mathbb{S}^2}\int_{\mathbb{S}^2} \frac{|g(x)-g(y)|^2}{|x-y|^3}\,dx\,dy.$$

One easily sees that X is a closed subset of the Sobolev space $H^{1/2}(\mathbb{S}^2;\mathbb{R}^2)$ and that it defines a complete metric space for the distance induced by the $H^{1/2}$-norm $\|\cdot\|_{1/2} = \|\cdot\|_{L^2(\mathbb{S}^2)} + |\cdot|_{1/2}$.

The main feature of the space X is that the subspace of smooth maps $X \cap C^\infty(\mathbb{S}^2)$ fails to be dense in X with respect to the strong topology. However density holds for the sequential weak topology (see [19]). A typical example of a map in X which can not be strongly approximated by smooth maps is given by $g\colon \mathbb{S}^2 \to \mathbb{S}^1$,

$$g(x_1,x_2,x_3) = \frac{(x_1,x_2)}{\sqrt{x_1^2+x_2^2}}.$$

2000 *Mathematics Subject Classification.* 46E35, 49Q15, 49Q20.

Key words and phrases. fractional Sobolev spaces; topological singularities; minimal connections.

The author was partially supported by the Center for Nonlinear Analysis (CNA) under the National Science Foundation Grant No. 0405343.

This is the final form of the paper.

Indeed, it can be shown, using degree theory, that g cannot be smoothed near its singularities at the north and south poles without violating the constraint of being $\mathbb{S}^1$-valued. This simple example contains the essence of the obstruction and shows its topological nature.

For an arbitrary map $g \in X$, a characterization of the topologically relevant part of the singular set of g has been obtained in terms of a certain distribution $T(g)$ in [2, 11, 19] (we also refer to [8] for an alternative approach in terms of Cartesian currents and to [3, 11] for higher-dimensional analogues). Given $\varphi \in \mathrm{Lip}(\mathbb{S}^2;\mathbb{R})$, an arbitrary extension $u \in H^1(B^3;\mathbb{R}^2)$ of g and an arbitrary extension $\Phi \in \mathrm{Lip}(B^3;\mathbb{R})$ of φ, the distribution $T(g)$ is defined through its action on φ by

$$\langle T(g),\varphi\rangle = \int_{B^3} H(u)\cdot\nabla\Phi \tag{2}$$

where

$$H(u) = 2\big(\partial_2 u\wedge\partial_3 u, \partial_3 u\wedge\partial_1 u, \partial_1 u\wedge\partial_2 u\big) \in L^1(B^3). \tag{3}$$

One observes that $\operatorname{div} H(u) = 0$ in $\mathcal{D}'(B^3)$ so that the integral in (2) formally reduces to an integration on the boundary. In particular, this integral only depends on g and φ and so $T(g)$ is well defined (see [2]). As shown in [19] (see also [2]), g can be approximated strongly by smooth maps if and only if $T(g) = 0$ and in particular $T(g) = 0$ whenever g is smooth.

If $g \in X\cap W^{1,1}(\mathbb{S}^2)$, the integration by parts in (2) can be rigorously performed (see [2]) and yields

$$\langle T(g),\varphi\rangle = -\int_{\mathbb{S}^2} (g\wedge\nabla_T g)\cdot\nabla_T^{\perp}\varphi \tag{4}$$

where ∇_T denotes the tangential gradient in a direct orthonormal frame (τ_1,τ_2,ν) and ν is the outward normal on $\mathbb{S}^2 = \partial B^3$. If in addition g is smooth except at a finite number of points $\{a_j\}_{j=1}^N$ then (4) leads to (see, e.g., [6])

$$\langle T(g),\varphi\rangle = 2\pi\sum_{j=1}^N d_j\varphi(a_j) \tag{5}$$

where $d_j \in \mathbb{Z}$ is the topological degree of g around a_j and $\sum_{j=1}^N d_j = 0$ for obvious topological reasons.

Our main objective in this note is to measure the obstruction to smooth approximation in terms of energy. We shall rely on the method and techniques developed in [16] (see also [8, 19] for an approach in terms of Cartesian currents). We consider the following energy for $g \in H^{1/2}(\mathbb{S}^2;\mathbb{R}^2)$,

$$E(g) = \int_{B^3} |\nabla u_g|^2$$

where $u_g \in H^1(B^3,\mathbb{R}^2)$ is the harmonic extension of g to the tridimensional unit ball $B^3 \subset \mathbb{R}^3$, i.e.,

$$\begin{cases} \Delta u_g = 0 & \text{in } H^{-1}(B^3),\\ u_{g|\mathbb{S}^2} = g. \end{cases}$$

Equivalently,

$$E(g) = \mathrm{Inf}\bigg\{\int_{B^3} |\nabla u|^2\ ;\ u \in H^1_g(B^3;\mathbb{R}^2)\bigg\}.$$

It is well known that $|g|^2_{1/2} \sim E(g)$ and in particular, $g \mapsto \big(E(g)\big)^{1/2}$ provides a seminorm equivalent to $|\cdot|_{1/2}$.

As a first step in our analysis, we introduce a suitable "dipole problem" motivated by the following remark. Let $g \in X$ and $\{h_n\}_{n\in\mathbb{N}} \subset X \cap C^\infty$ be such that $h_n \to g$ a.e. in $\mathbb{S}^2$. If one consider the sequence $\{\bar{h}_n g\}_{n\in\mathbb{N}} \subset X$, where $\bar{h}_n g$ denotes the complex product between g and the complex conjugate of h_n (we identify $\mathbb{R}^2$ with the complex plane $\mathbb{C}$), it turns out that $T(\bar{h}_n g) = T(\bar{h}_n) + T(g) = T(g)$ for every $n \in \mathbb{N}$ since $T(\bar{h}_n) = 0$ (see [2]). And obviously, $\bar{h}_n g \to (1,0)$ a.e. in $\mathbb{S}^2$. Assuming that $T(g)$ takes the simple form $T(g) = 2\pi\big(\delta_P - \delta_Q\big)$ for some distinct points $P, Q \in \mathbb{S}^2$, it is therefore relevant to consider the quantity

$$\begin{aligned}(6)\quad & m\big(2\pi(\delta_P - \delta_Q)\big)\\ &= \operatorname{Inf}\Big\{\liminf_{n\to+\infty} E(g) : \{g_n\}_{n\in\mathbb{N}} \subset X, T(g_n) = 2\pi(\delta_P - \delta_Q), g_n \to (1,0) \text{ a.e.}\Big\}\end{aligned}$$

which heuristically gives the minimal $H^{1/2}$-energy necessary to remove the pair of singularities (P,Q). The explicit computation of $m\big(2\pi(\delta_P - \delta_Q)\big)$ is referred to as the "dipole problem" and is the object of our first result.

Theorem 1.1. *Let P and Q be two distinct points on $\mathbb{S}^2$. We have*

$$m\big(2\pi(\delta_P - \delta_Q)\big) = 2\pi d_{\mathbb{S}^2}(P,Q),$$

where $d_{\mathbb{S}^2}$ denotes the geodesic distance on the sphere $\mathbb{S}^2$.

The result of Theorem 1.1 can be generalized to an arbitrary $T(g)$ (see [16]) but for our purposes it suffices to concentrate on the special case (5). Indeed, the class of maps which belong to $X \cap W^{1,1}(\mathbb{S}^2)$ and are smooth except for a finite number of points, is dense in X for the strong topology. This important result was first obtained in [19] (see also [2, 16]). The statement of Theorem 1.1 for (5) should be modified as follows. Let $a_1, \ldots, a_N$ be N distinct points on $\mathbb{S}^2$, let $d_1, \ldots, d_N \in \mathbb{N}$ be such that $\sum_{j=1}^N d_j = 0$, and consider the quantity $m\big(2\pi\sum_j d_j\delta_{a_j}\big)$ defined as in (6) using the constraint $T(g_n) = 2\pi\sum_j d_j\delta_{a_j}$ (admissible sequences can be easily constructed, see, e.g., Section 3). Since the sum of the d_j's equals zero, we can relabel the a_j's taking into account their multiplicity $|d_j|$ as two lists $(p_1, \ldots, p_K)$ and $(q_1, \ldots, q_K)$ in such a way that

$$(7)\qquad 2\pi\sum_{j=1}^N d_j\delta_{a_j} = 2\pi\sum_{i=1}^K \delta_{p_i} - \delta_{q_i}.$$

Then we have

$$(8)\qquad m\left(2\pi\sum_{j=1}^N d_j\delta_{a_j}\right) = 2\pi \operatorname*{Min}_{\sigma\in\mathcal{S}_K} \sum_{i=1}^K d_{\mathbb{S}^2}(p_i, q_{\sigma(i)}),$$

where $\mathcal{S}_K$ denotes the set of all permutations of K indices. We shall omit the proof of (8) since the main ingredients are already contained in the proof of Theorem 1.1 (see also Section 3).

The quantity on the right-hand side of (8) is often referred to as the length of a minimal connection between the singularities relative to the distance $d_{\mathbb{S}^2}$. It has been first introduced in [5] for similar questions involving $\mathbb{S}^2$-valued maps from tridimensional domains with H^1-regularity, and it can be written in the dual form (see [5]):

$$(9)\quad \operatorname*{Min}_{\sigma\in\mathcal{S}_K}\sum_{i=1}^{K} d_{\mathbb{S}^2}(p_i,q_{\sigma(i)}) = \operatorname{Sup}\Big\{\sum_{i=1}^{K}\varphi(p_i)-\varphi(q_i) : \varphi\in \operatorname{Lip}(\mathbb{S}^2;\mathbb{R}),\ \varphi \text{ is 1-Lipschitz with respect to } d_{\mathbb{S}^2}\Big\}.$$

The formula above motivates the following definition for the length of a minimal connection relative to $d_{\mathbb{S}^2}$ for an arbitrary map $g \in X$,

$$L(g) = \frac{1}{2\pi}\operatorname{Sup}\big\{\langle T(g),\varphi\rangle : \varphi\in\operatorname{Lip}(\mathbb{S}^2;\mathbb{R}), \varphi \text{ is 1-Lipschitz with respect to } d_{\mathbb{S}^2}\big\}.$$

It turns out that $L(g)$ is the right quantity to quantify the obstruction to smooth approximation. If one wants to measure, for a given $g \in X$, the distance of g to the subspace of smooth maps, it is natural to consider the relaxed energy $\overline{E}\colon X \to \mathbb{R}$ defined by

$$\overline{E}(g) = \operatorname{Inf}\Big\{\liminf_{n\to+\infty} E(g_n) : \{g_n\}_{n\in\mathbb{N}} \subset X\cap C^\infty(\mathbb{S}^2),\ g_n \to g \text{ a.e.}\Big\}.$$

Obviously $\overline{E}(g) \geq E(g)$ and the gap between $\overline{E}(g)$ and $E(g)$ is precisely the quantity we are interested in. We have the following result.

Theorem 1.2. *For every $g \in X$, we have*

$$\overline{E}(g) = E(g) + 2\pi L(g).$$

A similar result has first been obtained by H. Brezis, J.-M. Coron, and E. H. Lieb [5] in the context of $\mathbb{S}^2$-valued maps. Since then many studies have been done on this type of relaxation problems for Sobolev maps between manifolds, see, e.g., [6, 12, 14, 15], see also the recent works by M. Giaquinta, G. Modica, J. Souček [8] and M. Giaquinta, D. Mucci [10] for $H^{1/2}$-maps in the context of Cartesian currents.

The plan of the paper is as follows. Sections 2 and 3 are devoted to the proofs of Theorem 1.1 and Theorem 1.2 respectively. In Section 4, we present some generalizations of the previous results when $\mathbb{S}^2$ is replaced by a more general two dimensional manifold $\mathcal{M}$. Then we compare our dipole problem with the more classical notion where one simply computes the infimum of the energy over the class of maps having two prescribed singularities of degree $+1$ and -1.

2. The dipole problem

Proposition 2.1. *Let P and Q be two distinct points on $\mathbb{S}^2$. There exists a sequence $\{g_n\}_{n\in\mathbb{N}} \subset X$ such that $T(g_n) = 2\pi(\delta_P - \delta_Q)$ for every n, $g_n \to (1,0)$ a.e. and*

$$\limsup_{n\to+\infty} E(g_n) \leq 2\pi\, d_{\mathbb{S}^2}(P,Q).$$

Proof. We split the proof into several steps.

Step 1. Due to the invariance of the energy E with respect to rotations, we may assume without loss of generality that the points P and Q satisfy

$$P = (s,0,-h) \quad\text{and}\quad Q = (-s,0,-h)$$

for some $s > 0$ and $h \geq 0$. In the sequel, we shall make use of the conformal transformation $\Phi\colon \overline{\mathbb{R}^3_+} = \mathbb{R}^2 \times \overline{\mathbb{R}_+} \to \overline{B^3} \subset \mathbb{R}^3$ given by

$$\Phi(y_1, y_2, y_3) = \left(\frac{2y_1}{y_1^2 + y_2^2 + (y_3+1)^2}, \frac{2y_2}{y_1^2 + y_2^2 + (y_3+1)^2}, \frac{y_1^2 + y_2^2 + y_3^2 - 1}{y_1^2 + y_2^2 + (y_3+1)^2}\right).$$

We observe that $\Phi(\cdot,\cdot,0) \in C^\infty(\mathbb{R}^2; \mathbb{S}^2 \setminus \{(0,0,1)\})$ is an inverse stereographic projection. Moreover

$$P = \Phi\big((l,0,0)\big) \quad \text{and} \quad Q = \Phi\big((-l,0,0)\big)$$

for some $0 \leq l \leq 1$ and the image of the segment $[(-l,0,0),(l,0,0)]$ through Φ is a minimizing geodesic on $\mathbb{S}^2$ connecting Q to P.

STEP 2. We introduce the second conformal transformation $\phi\colon \overline{\mathbb{R}^2_+} = \mathbb{R}\times\overline{\mathbb{R}_+} \to \mathbb{D}$ defined by

$$\phi(y_2, y_3) = \left(\frac{y_2^2 + y_3^2 - 1}{y_2^2 + (1+y_3)^2}, \frac{2y_2}{y_2^2 + (1+y_3)^2}\right),$$

so that $\phi(\cdot,0) \in C^\infty(\mathbb{R}; \mathbb{S}^1 \setminus \{(1,0)\})$ is an inverse stereographic projection. Using the complex notation, for $y_2 \neq 0$ we have

$$\phi(y_2, 0) = \mathrm{e}^{\mathrm{i}\theta(y_2)} \quad \text{with } \theta(y_2) = 2\arctan(1/y_2).$$

Next we define the sequence $\{\phi_n\}_{n\in\mathbb{N}}$ by

$$\phi_n(y_2,y_3) = \begin{cases} \phi(ny_2, ny_3) & \text{if } |y_2|^2 + |y_3|^2 \leq 1, \\ (1,0) & \text{if } |y_2|^2 + |y_3|^2 \geq 4, \\ \mathrm{e}^{\mathrm{i}(\theta(ny_2))(2-|y_2|)} & \text{if } 1 < |y_2| < 2 \text{ and } y_3 = 0, \\ \text{harmonic} & \text{if } 1 < |y_2|^2 + |y_3|^2 < 4 \text{ and } y_3 > 0. \end{cases}$$

We may easily check that $\phi_n \in \mathrm{Lip}(\mathbb{R}^2_+)$, $\phi_n(\cdot,0) \in \mathrm{Lip}(\mathbb{R};\mathbb{S}^1)$ and

$$\int_{\mathbb{R}^2_+} |\nabla\phi_n|^2 \,\mathrm{d}y_2\,\mathrm{d}y_3 \xrightarrow[n\to+\infty]{} \int_{\mathbb{R}^2_+} |\nabla\phi|^2 \,\mathrm{d}y_2\,\mathrm{d}y_3 = 2\pi. \tag{10}$$

Now we consider for $y = (y_1,y_2,y_3) \in \overline{\mathbb{R}^3_+}$,

$$v_n(y) = \begin{cases} \phi_n\left(\dfrac{y_2}{\eta_n(y_1)}, \dfrac{y_3}{\eta_n(y_1)}\right) & \text{if } |y_1| < l, \\ (1,0) & \text{if } |y_1| \geq l, \end{cases}$$

where $\eta_n(y_1) = (l - |y_1|)/n$. Once more, one may check that v_n is locally Lipschitz in $\overline{\mathbb{R}^3_+}$ away from $\{(-l,0,0),(l,0,0)\}$ and has finite energy. Moreover, v_n is identically equal to $(1,0)$ outside a small neighborhood $\mathcal{U} \subset \overline{\mathbb{R}^3_+}$ of $[(-l,0,0),(l,0,0)]$ independent of n. Finally, we define $u_n\colon \overline{B^3} \subset \mathbb{R}^3 \to \mathbb{R}^2$ by

$$u_n(x) = \begin{cases} v_n\big(\Phi^{-1}(x)\big) & \text{if } x \in \Phi(\mathcal{U}), \\ (1,0) & \text{otherwise.} \end{cases}$$

By construction $u_n \in H^1(B^3,\mathbb{R}^2)$, u_n is locally Lipschitz in $\overline{B^3}$ away from the points P and Q. In addition, $g_n := u_{n|\mathbb{S}^2} \in X \cap W^{1,1}(\mathbb{S}^2)$, $g_n \to (1,0)$ a.e. as $n \to +\infty$ and $T(g_n) = 2\pi(\delta_P - \delta_Q)$ for every n.

STEP 3. Using the conformal invariance of Φ, a straightforward computation yields

$$\int_{B^3} |\nabla u_n(x)|^2 \,\mathrm{d}x = \int_{\mathbb{R}^3_+} |\nabla v_n(y)|^2 a(y)\,\mathrm{d}y, \tag{11}$$

where the weight function $a(y)$ is given by

$$a(y) = \frac{2}{y_1^2 + y_2^2 + (y_3^2+1)^2}.$$

Then we estimate

$$\int_{\mathbb{R}^3_+} |\nabla v_n(y)|^2 a(y)\,\mathrm{d}y \le \int_{-l}^{l} \left(\int_{\mathbb{R}^2_+} |\nabla_{2,3}\, v_n|^2 a(y)\,\mathrm{d}y_2\,\mathrm{d}y_3 + 2\int_{\mathbb{R}^2_+} |\partial_1 v_n|^2 \,\mathrm{d}y_2\,\mathrm{d}y_3 \right) \mathrm{d}y_1.$$

Changing variables, $y_i = \eta_n(y_1) z_i$ for $i = 2,3$, we obtain

$$\begin{aligned}\int_{\mathbb{R}^2_+} |\nabla_{2,3} v_n|^2 a(y)\,\mathrm{d}y_2\,\mathrm{d}y_3 &= \int_{\mathbb{R}^2_+} |\nabla\phi_n(z_2,z_3)|^2 a(y_1, \eta_n(y_1)z_2, \eta_n(y_1)z_3)\,\mathrm{d}z_2\,\mathrm{d}z_3 \\ &= a(y_1,0,0)\int_{\mathbb{R}^2_+} |\nabla\phi_n(z_2,z_3)|^2 \,\mathrm{d}z_2\mathrm{d}z_3 + O(1/n)\end{aligned}$$

and

$$\begin{aligned}\int_{\mathbb{R}^2_+} |\partial_1 v_n|^2 \,\mathrm{d}y_2\,\mathrm{d}y_3 &\le |\dot\eta_n(y_1)|^2 \int_{\mathbb{R}^2_+} |\nabla\phi_n(z_2,z_3)|^2 (z_2^2+z_3^2)\,\mathrm{d}z_2\,\mathrm{d}z_3 \\ &\le 4|\dot\eta_n(y_1)|^2 \int_{\mathbb{R}^2_+} |\nabla\phi_n(z_2,z_3)|^2 \,\mathrm{d}z_2\,\mathrm{d}z_3.\end{aligned}$$

Since $\int_{-l}^{l} |\dot\eta_n(y_1)|^2\,\mathrm{d}y_1 = O(1/n^2)$, we deduce that

$$\int_{\mathbb{R}^3_+} |\nabla v_n(y)|^2 a(y)\,\mathrm{d}y \le \left(\int_{-l}^{l} a(y_1,0,0)\,\mathrm{d}y_1 \right) \int_{\mathbb{R}^2_+} |\nabla\phi_n|^2 + O(1/n).$$

which leads by (10) to

$$\limsup_{n\to+\infty} \int_{\mathbb{R}^3_+} |\nabla v_n(y)|^2 a(y)\,\mathrm{d}y \le 2\pi \int_{-l}^{l} a(y_1,0,0)\,\mathrm{d}y_1.$$

In view of (11) and the definition of E, we conclude

$$\limsup_{n\to+\infty} E(g_n) \le \limsup_{n\to+\infty} \int_{B^3} |\nabla u_n|^2 \,\mathrm{d}x \le 2\pi \int_{-l}^{l} a(y_1,0,0)\,\mathrm{d}y_1.$$

On the other hand, $\gamma \colon [-l,l] \to \mathbb{S}^2$ defined by $\gamma(t) = \Phi\big((t,0,0)\big)$ is a minimizing geodesic on $\mathbb{S}^2$ joining P and Q so that

$$d_{\mathbb{S}^2}(P,Q) = \int_{-l}^{l} |\dot\gamma(t)|\,\mathrm{d}t = \int_{-l}^{l} a\big((t,0,0)\big)\,\mathrm{d}t.$$

Therefore

$$\limsup_{n\to+\infty} E(g_n) \le 2\pi d_{\mathbb{S}^2}(P,Q)$$

and the proof is complete. □

Proposition 2.2. *Let P and Q be two distinct points on $\mathbb{S}^2$. Then, for any sequence $\{g_n\}_{n\in\mathbb{N}} \subset X$ such that $g_n \to (1,0)$ a.e. and $T(g_n) = 2\pi(\delta_P - \delta_Q)$ for every n, we have*

$$\liminf_{n\to+\infty} E(g_n) \geq 2\pi d_{\mathbb{S}^2}(P,Q).$$

PROOF. Without loss of generality, we may assume that

$$\liminf_{n\to+\infty} E(g_n) = \lim_{n\to+\infty} E(g_n) < +\infty.$$

Let $u_n \in H^1(B^3)$ be the harmonic extension of g_n to B^3. Then the sequence $\{u_n\}_{n\in\mathbb{N}}$ remains bounded in H^1. Since $g_n \to (1,0)$ a.e. on $\mathbb{S}^2$, extracting a subsequence, if necessary, we have $u_n \rightharpoonup (1,0)$ weakly in H^1. Next we observe that $|\nabla u_n|^2 \geq |H(u_n)|$ a.e. in B^3. We fix $0 < r \ll 1$ and we consider a cut-off function $\chi \in C^\infty(\mathbb{R};\mathbb{R})$ such that $0 \leq \chi \leq 1$, $\chi(t) \equiv 1$ if $|t| \geq 1-r$ and $\chi(t) \equiv 0$ if $|t| \leq 1-2r$. Then

$$E(g_n) = \int_{B^3} |\nabla u_n|^2 \,\mathrm{d}x \geq \int_{B^3} \chi(|x|)|H(u_n)| \,\mathrm{d}x = \int_{\mathcal{A}_{2r}} \chi(|x|)|H(u_n)| \,\mathrm{d}x,$$

where $\mathcal{A}_{2r} = \overline{B^3} \setminus B(0, 1-2r)$. Setting d_{2r} to be the geodesic distance in $\mathcal{A}_{2r}$, we define for $x \in \mathcal{A}_r$

$$\Psi(x) = d_{2r}(x, Q),$$

and for $x \in B^3$

$$\zeta(x) = \begin{cases} \chi(|x|)\Psi(x) & \text{if } x \in \mathcal{A}_{2r}, \\ 0 & \text{otherwise,} \end{cases}$$

so that $\zeta \in \mathrm{Lip}(B^3)$. Since $|\nabla\Psi| \leq 1$ a.e. in $\mathcal{A}_{2r}$, we may estimate

$$\begin{aligned}
(12) \quad \int_{\mathcal{A}_{2r}} \chi(|x|)|H(u_n)| \,\mathrm{d}x &\geq \int_{\mathcal{A}_{2r}} \chi(|x|)H(u_n)\cdot\nabla\Psi \,\mathrm{d} \\
&= \int_{B^3} H(u_n)\cdot\nabla\zeta \,\mathrm{d}x - \int_K \dot\chi(|x|)\Psi(x)H(u_n)\cdot\frac{x}{|x|}\,\mathrm{d}x \\
&= \langle T(g_n), \zeta_{|\mathbb{S}^2}\rangle - \int_K \dot\chi(|x|)\Psi(x)H(u_n)\cdot\frac{x}{|x|}\,\mathrm{d}x \\
&= 2\pi d_{2r}(P,Q) - \int_K \dot\chi(|x|)\Psi(x)H(u_n)\cdot\frac{x}{|x|}\,\mathrm{d}x
\end{aligned}$$

with $K = \mathcal{A}_{2r} \setminus \mathcal{A}_r$. By classical results, $\nabla u_n \to 0$ in $L^2_{\mathrm{loc}}(B^3)$ so that $H(u_n) \to 0$ in $L^1(K)$ and the integral on the right-hand side of (12) vanishes as $n \to +\infty$. Therefore

$$\lim_{n\to+\infty} E(g_n) \geq 2\pi d_{2r}(P,Q).$$

Now we let $r \to 0$ and since $d_{2r}(P,Q) \to d_{\mathbb{S}^2}(P,Q)$ as $r \to 0$, we recover the announced result. □

3. Relaxed energy and minimal connections

We begin with the proof of the upper bound $\overline{E} \leq E + 2\pi L$ in Theorem 1.2.

Proposition 3.1. *For every $g \in X$, there exists a sequence $\{g_n\}_{n\in\mathbb{N}} \subset X \cap C^\infty(\mathbb{S}^2)$ such that $g_n \to g$ a.e. in $\mathbb{S}^2$ and*

$$\limsup_{n\to+\infty} E(g_n) \leq E(g) + 2\pi L(g).$$

PROOF. STEP 1. First we assume that $g \in X \cap W^{1,1}(\mathbb{S}^2)$ and g is smooth except for a finite number of points. Then the distribution $T(g)$ may be written as $T(g) = 2\pi \sum_{i=1}^K \delta_{p_i} - \delta_{q_i}$ and without loss of generality, we may assume that

$$\sum_{i=1}^K d_{\mathbb{S}^2}(p_i, q_i) = \operatorname*{Min}_{\sigma \in \mathcal{S}_K} \sum_{i=1}^K d_{\mathbb{S}^2}(p_i, q_{\sigma(i)}) = L(g)$$

by (9). We consider for each $i = 1, \ldots, K$, the sequence $\{g_{i,n}\}_{n\in\mathbb{N}} \subset X$ associated to the pair (p_i, q_i) given by Proposition 2.1. We denote by u_g, resp. $u_{i,n}$, the harmonic extension of g, resp. of $g_{i,n}$, to the unit ball B^3. Next we consider the sequence

$$g_n = \left(\prod_{i=1}^K \bar{g}_{i,n}\right) g \in X,$$

using the complex product and complex conjugation. Obviously $g_n \to g$ a.e. since $g_{i,n} \to (1,0)$ a.e. for every i. By [2], we have

$$T(g_n) = T(g) - \sum_{i=1}^K T(g_{i,n}) = 0.$$

By an easy application of the maximum principle, one obtains $|u_g| \le 1$ and $|u_{i,n}| \le 1$ in B^3 so that a straightforward computation yields

$$E(g_n) \le \int_{B^3} |\nabla u_g|^2 + \sum_{i=1}^K \int_{B^3} |\nabla u_{i,n}|^2 = E(g) + \sum_{i=1}^K E(g_{i,n}).$$

In view of Proposition 2.1, we derive that

$$\limsup_{n\to+\infty} E(g_n) \le E(g) + 2\pi \sum_{i=1}^K d_{\mathbb{S}^2}(p_i, q_i) = E(g) + 2\pi L(g).$$

Since $T(g_n) = 0$, g_n can be strongly approximated by smooth maps in X (see [19]). Hence the required sequence may be obtained by approximating each g_n by smooth maps and then applying a standard diagonalization argument.

STEP 2. Now we consider an arbitrary map $g \in X$. First we observe that

$$|L(g_1) - L(g_2)| \le C\big(|g_1|_{1/2} + |g_2|_{1/2}\big)|g_1 - g_2|_{1/2} \quad \forall g_1, g_2 \in X, \tag{13}$$

for some universal constant $C > 0$ (see, e.g., [16] for more details). By the results in [19], we may find a sequence $\{g_n\}_{n\in\mathbb{N}} \subset X$ such that $g_n \to g$ strongly in $H^{1/2}$ and g_n satisfies the assumption of Step 1 for every n. Obviously, we may assume that $g_n \to g$ a.e. extracting a subsequence if necessary. By (13), we have

$$E(g_n) + 2\pi L(g_n) \xrightarrow[n\to+\infty]{} E(g) + 2\pi L(g).$$

Hence we may apply Step 1 to each g_n and then obtain the required sequence by a diagonalization argument. □

To obtain the lower bound $\overline{E} \ge E + 2\pi L$, we shall use an argument based on the lower semicontinuity result below. We introduce the expected lower bound $F \colon X \to \mathbb{R}_+$ defined by

$$F(g) = E(g) + 2\pi L(g).$$

We have the following.

Proposition 3.2. *The functional F is sequentially lower semicontinuous on X with respect to the a.e. convergence.*

PROOF. STEP 1. First we introduce for a given $0 < r < 1$, the functional $F_r \colon X \to \mathbb{R}_+$ defined by

$$F_r(g) = E(g) + 2\pi L_r(g),$$

where $L_r(g)$ denotes the length of a minimal connection relative to the geodesic distance d_r in the annulus $\mathcal{A}_r = \overline{B^3} \setminus B(0, 1-r)$, i.e.,

$$L_r(g) = \frac{1}{2\pi} \operatorname{Sup}\{\langle T(g), \Phi_{|\mathbb{S}^2}\rangle : \Phi \in \operatorname{Lip}(\mathcal{A}_r; \mathbb{R}),\ \Phi \text{ is 1-Lipschitz with respect to } d_r\}.$$

In the remaining of the proof we shall extend any $\Phi \in \operatorname{Lip}(\mathcal{A}_r;\mathbb{R})$ to the whole $\overline{B^3}$ by setting

$$\Phi(x) = \chi(|x|)\Phi\left(\frac{(1-r)x}{|x|}\right) \quad \forall x \in B(0, 1-r)$$

for a fixed function $\chi \in C^\infty(\mathbb{R})$ satisfying $0 \le \chi \le 1$, $\chi(t) = 1$ if $|t| \ge 1 - r$ and $\chi(t) = 0$ if $|t| \le 1 - 2r$. Then for any $g \in X$ and any $\Phi \in \operatorname{Lip}(\mathcal{A}_r;\mathbb{R})$, we have

$$\langle T(g), \Phi_{|\mathbb{S}^2}\rangle = \int_{B^3} H(u_g) \cdot \nabla\Phi$$

where u_g is the harmonic extension of g to B^3.

We claim that F_r is sequentially lower semicontinuous on X with respect to the a.e. convergence. Obviously, it suffices to prove that for any $\Phi \in \operatorname{Lip}(\mathcal{A}_r;\mathbb{R})$ with Φ 1-Lipschitz with respect to d_r,

$$g \in X \mapsto E(g) + \left|\int_{B^3} H(u_g)\cdot\nabla\Phi\right|$$

is sequentially lower semicontinuous on X with respect to the a.e. convergence since

$$F_r(g) = \operatorname{Sup}\left\{E(g) + \left|\int_{B^3} H(u_g)\cdot\nabla\Phi\right| : \Phi \in \operatorname{Lip}(\mathcal{A}_r;\mathbb{R}),\ \Phi \text{ is 1-Lipschitz with respect to } d_r\right\}.$$

Let $\{g_n\}_{n\in\mathbb{N}} \subset X$ be such that $g_n \to g \in X$ a.e. on $\mathbb{S}^2$. Without loss of generality, we may assume that

$$\liminf_{n\to+\infty} F_r(g_n) = \lim_{n\to+\infty} F_r(g_n) < +\infty.$$

Since $\sup_n E(g_n) \le \sup_n F_r(g_n) < +\infty$, we can extract a subsequence (not relabeled) such that $g_n \rightharpoonup g$ weakly in $H^{1/2}$. Setting u_n to be the harmonic extension of g_n to B^3, it is well known that u_n converges to u_g weakly in $H^1(B^3)$ and strongly in $H^1(K)$ for any compact set $K \subset B^3$. Writing $u_n = u + v_n$, v_n converges to 0 weakly in $H^1(B^3)$ and strongly in $H^1(K)$ for any compact set $K \subset B^3$. Therefore

$$E(g_n) = E(g) + \int_{B^3} |\nabla v_n|^2 + o(1).$$

Next we observe that

$$H(u_n) = H(v_n) + 2(\partial_2 v_n \wedge \partial_3 u_g, \partial_3 v_n \wedge \partial_1 u_g, \partial_1 v_n \wedge \partial_2 u_g) + 2(\partial_2 u_g \wedge \partial_3 u_n, \partial_3 u_g \wedge \partial_1 u_n, \partial_1 u_g \wedge \partial_2 u_n).$$

Hence $H(u_n) \to H(u)$ in $L^1(K)$ for any compact set $K \subset B^3$, so that

$$\int_{B^3} H(u_n) \cdot \nabla\Phi = \int_{B^3} H(u_g) \cdot \nabla\Phi + \int_{\mathcal{A}_r} H(v_n) \cdot \nabla\Phi + o(1).$$

Since $|\nabla\Phi| \leq 1$ a.e. in $\mathcal{A}_r$, we deduce

$$E(g_n)+\left|\int_{B^3} H(u_n)\cdot\nabla\Phi\right| \geq E(g)+\left|\int_{B^3} H(u_g)\cdot\nabla\Phi\right|+\int_{B^3} |\nabla v_n|^2-\int_{\mathcal{A}_r} |H(v_n)|+o(1)$$

and the conclusion follows because $|\nabla v_n|^2 \geq |H(v_n)|$ a.e. in $\mathcal{A}_r$.

STEP 2. We claim that

$$F(g) = \sup_{r>0} F_r(g) \quad \forall g \in X. \tag{14}$$

In view of Step 1, the proof of (14) will complete the proof of Proposition 3.2. Obviously it suffices to show that $L(g) = \sup_{r>0} L_r(g)$.

First we consider the case of a map $g \in X \cap W^{1,1}(\mathbb{S}^2)$ smooth except for a finite number of points. In this case, we may write $T(g) = 2\pi \sum_{i=1}^K \delta_{p_i} - \delta_{q_i}$ and by the results in [5],

$$L(g) = \operatorname*{Min}_{\sigma\in\mathcal{S}_K} \sum_{i=1}^K d_{\mathbb{S}^2}(p_i, q_{\sigma(i)}) \quad \text{and} \quad L_r(g) = \operatorname*{Min}_{\sigma\in\mathcal{S}_K} \sum_{i=1}^K d_r(p_i, q_{\sigma(i)}) \ \forall\, 0 < r < 1.$$

Now using the uniform convergence of d_r to $d_{\mathbb{S}^2}$ on $\mathbb{S}^2 \times \mathbb{S}^2$ as $r \to 0$, one easily obtains $L_r(g) \to L(g)$ (see [16] for details). Then we conclude by observing that $L_r(g) \leq L(g)$ since $d_r \leq d_{\mathbb{S}^2}$ as distances on $\mathbb{S}^2$.

Now we consider a general map $g \in X$. By the results in [19], given $\varepsilon > 0$ small, we may find $g_\varepsilon \in X \cap W^{1,1}(\mathbb{S}^2)$ smooth except for a finite number of points such that $\|g - g_\varepsilon\|_{1/2} \leq \varepsilon$. Using (13) and its equivalent form for L_r, i.e.,

$$|L_r(g_1) - L_r(g_2)| \leq C\big(|g_1|_{1/2} + |g_2|_{1/2}\big)|g_1 - g_2|_{1/2} \quad \forall g_1, g_2 \in X,$$

for a constant $C > 0$ independent of r (see [16] for details), we derive

$$\liminf_{r\to 0} L_r(g) \geq \lim_{r\to 0} L_r(g_\varepsilon) - C\varepsilon = L(g_\varepsilon) - C\varepsilon \geq L(g) - C\varepsilon$$

and

$$\limsup_{r\to 0} L_r(g) \leq \lim_{r\to 0} L_r(g_\varepsilon) + C\varepsilon = L(g_\varepsilon) + C\varepsilon \leq L(g) + C\varepsilon.$$

Letting $\varepsilon \to 0$, we recover $L_r(g) \to L(g)$ as $r \to 0$ and the conclusion follows since $L_r(g) \leq L(g)$. □

Corollary 3.1. *Let $g \in X$ and let $\{g_n\}_{n\in\mathbb{N}} \subset X \cap C^\infty(\mathbb{S}^2)$ be an arbitrary sequence such that $g_n \to g$ a.e. in $\mathbb{S}^2$. We have*

$$\liminf_{n\to+\infty} E(g_n) \geq E(g) + 2\pi L(g).$$

PROOF. Since $\{g_n\}_{n\in\mathbb{N}} \subset C^\infty(\mathbb{S}^2)$, we have $T(g_n) = 0$ and hence $L(g_n) = 0$. Next we infer from Proposition 3.2 that

$$\liminf_{n\to+\infty} E(g_n) = \liminf_{n\to+\infty} F(g_n) \geq F(g) = E(g) + 2\pi L(g)$$

which completes the proof. □

4. Some extensions and open problem

In this section, we extend the results of Theorems 1.1 and 1.2 to a more general starting manifold. We consider a smooth open bounded domain $\Omega \subset \mathbb{R}^3$ diffeomorphic to the unit ball B^3 and we set $\mathcal{M} = \partial\Omega$ so that the manifold $\mathcal{M}$ is diffeomorphic to the unit sphere $\mathbb{S}^2$. We are interested in the space of $\mathbb{S}^1$-valued maps

$$X_{\mathcal{M}} = H^{1/2}(\mathcal{M};\mathbb{S}^1) = \big\{g \in H^{1/2}(\mathcal{M};\mathbb{R}^2) : |g| = 1 \text{ a.e. on } \mathcal{M}\big\}$$

endowed with the metric induced by the $H^{1/2}$-norm.

For $g \in X_{\mathcal{M}}$, we define the distribution $T(g)$ as in (2) by integrating over Ω and taking extensions in Ω instead of B^3, and the energy of g is still given by the energy of its harmonic extension, i.e.,

$$E_{\mathcal{M}}(g) = \int_{\Omega} |\nabla u_g|^2,$$

with $u_g \in H^1_g(\Omega;\mathbb{R}^2)$ satisfying $\Delta u_g = 0$ in $H^{-1}(\Omega)$.

For two distinct points P and Q on $\mathcal{M}$, we define the dipole problem on $\mathcal{M}$ as

$$(15)\quad m_{\mathcal{M}}\big(2\pi(\delta_P - \delta_Q)\big) = \operatorname{Inf}\Big\{\liminf_{n\to+\infty} E_{\mathcal{M}}(g) : \{g_n\}_{n\in\mathbb{N}} \subset X_{\mathcal{M}},\\ T(g_n) = 2\pi\big(\delta_P - \delta_Q\big), g_n \to (1,0) \text{ a.e.}\Big\}.$$

Using a suitable diffeomorphism between Ω and B^3 and applying the method in Section 2 with minor modifications, we obtain the following variant of Theorem 1.1.

Theorem 1.1 bis. *Let P and Q be two distinct points on $\mathcal{M}$. We have*

$$m_{\mathcal{M}}\big(2\pi(\delta_P - \delta_Q)\big) = 2\pi d_{\mathcal{M}}(P,Q)$$

where $d_{\mathcal{M}}$ denotes the geodesic distance on the manifold $\mathcal{M}$.

To characterize the obstruction to smooth approximation in $X_{\mathcal{M}}$, we proceed as in the case $\mathcal{M} = \mathbb{S}^2$. We introduce the relaxed energy $\overline{E}_{\mathcal{M}} \colon X_{\mathcal{M}} \to \mathbb{R}_+$ defined by

$$\overline{E}_{\mathcal{M}}(g) = \operatorname{Inf}\Big\{\liminf_{n\to+\infty} E_{\mathcal{M}}(g_n) : \{g_n\}_{n\in\mathbb{N}} \subset X_{\mathcal{M}} \cap C^\infty(\mathcal{M}), g_n \to g \text{ a.e.}\Big\},$$

and we readily obtain from Section 3 and Theorem 1.1 bis the following result.

Theorem 1.2 bis. *For every $g \in X_{\mathcal{M}}$, we have*

$$\overline{E}_{\mathcal{M}}(g) = E_{\mathcal{M}}(g) + 2\pi L_{\mathcal{M}}(g),$$

where $L_{\mathcal{M}}(g)$ is the length of a minimal connection relative to $d_{\mathcal{M}}$, i.e.,

$$L_{\mathcal{M}}(g) = \frac{1}{2\pi} \operatorname{Sup}\{\langle T(g),\varphi\rangle : \varphi \in \operatorname{Lip}(\mathcal{M};\mathbb{R}),\ \varphi \text{ is 1-Lipschitz with respect to } d_{\mathcal{M}}\}.$$

To conclude this section, we present a more classical notion of "dipole problem" where one minimizes the energy over all maps in $X_{\mathcal{M}}$ having exactly two prescribed singularities (see, e.g., [6] in the context of $W^{1,1}$-maps with values into $\mathbb{S}^1$). More precisely, given two distinct points P and Q on $\mathcal{M}$, we consider

$$(16)\qquad \widetilde{m}_{\mathcal{M}}\big(2\pi(\delta_P - \delta_Q)\big) = \operatorname{Inf}\{E_{\mathcal{M}}(g) : g \in X_{\mathcal{M}}, T(g) = 2\pi(\delta_P - \delta_Q)\}.$$

The next result compares the dipole problems $m_{\mathcal{M}}$ and $\widetilde{m}_{\mathcal{M}}$.

Theorem 4.1. *Let P and Q be two distinct points on $\mathcal{M}$. We have*

$$2\pi d_\Omega(P,Q) \leq \widetilde{m}_{\mathcal{M}}\big(2\pi(\delta_P - \delta_Q)\big) \leq 2\pi d_{\mathcal{M}}(P,Q), \tag{17}$$

where d_Ω denotes the geodesic distance in $\overline{\Omega}$. Moreover, if $d_\Omega(P,Q) = d_{\mathcal{M}}(P,Q)$ and $d_\Omega(P,Q)$ is small enough, we can extract from any minimizing sequence $\{g_n\}_{n\in\mathbb{N}} \subset X_{\mathcal{M}}$, a subsequence $\{g_{n_k}\}_{k\in\mathbb{N}}$ such that $g_{n_k} \to \alpha$ a.e. on $\mathcal{M}$ for some constant $\alpha \in \mathbb{S}^1$. In particular, the infimum defining $\widetilde{m}_{\mathcal{M}}\big(2\pi(\delta_P - \delta_Q)\big)$ is not achieved whenever $d_\Omega(P,Q) = d_{\mathcal{M}}(P,Q)$ and $d_\Omega(P,Q)$ is small enough.

Observe that Theorem 4.1 gives a sufficient condition for which the two dipole problems coincide. We believe that in the case $d_\Omega(P,Q) < d_{\mathcal{M}}(P,Q)$, the two problems completely differ in nature. For instance, it seems that the lower bound in (17) is sharp but we do not have a proof of this fact. Beyond this quantitative problem, one can ask the much more interesting question:

Open problem. Let P and Q be two distinct points on the manifold $\mathcal{M}$ such that $d_\Omega(P,Q) < d_{\mathcal{M}}(P,Q)$. Is $\widetilde{m}_{\mathcal{M}}\big(2\pi(\delta_P - \delta_Q)\big)$ achieved?

We emphasize that the main difficulty of this problem comes from the constraint $T(g) = 2\pi(\delta_P - \delta_Q)$ which is obviously noncompact. The reader should certainly relate this question to recent results by P. Mironescu and A. Pisante [17] concerning $H^{1/2}$-maps from the circle into itself.

PROOF OF THEOREM 4.1. The upper bound in (17) is a trivial consequence of Theorem 1.1 bis and the lower bound can be easily deduced from the discussion below.

Let P and Q be two distinct points on $\mathcal{M}$ such that $d_\Omega(P,Q) = d_{\mathcal{M}}(P,Q)$, and let $\{g_n\}_{n\in\mathbb{N}} \subset X_{\mathcal{M}}$ be a minimizing sequence for $\widetilde{m}_{\mathcal{M}}\big(2\pi(\delta_P - \delta_Q)\big)$. Set u_n to be the harmonic extension of g_n to Ω. Since $\sup_n E_{\mathcal{M}}(g_n) < +\infty$, by classical results, we can extract a subsequence (not relabeled) such that $u_n \rightharpoonup u$ weakly in $H^1(\Omega)$ with $\Delta u = 0$ in $H^{-1}(\Omega)$, $g_n \rightharpoonup g$ weakly in $H^{1/2}(\mathcal{M})$ with $g = u_{|\mathcal{M}} \in X_{\mathcal{M}}$ and $g_n \to g$ a.e. on $\mathcal{M}$.

Next we consider the function $\Psi\colon \overline{\Omega} \to \mathbb{R}$ defined by

$$\Psi(x) = \min\{d_\Omega(x,Q), d_\Omega(P,Q)\}.$$

Obviously, Ψ is 1-Lipschitz in Ω with respect d_Ω so that $|\nabla\Psi| \leq 1$ a.e. in Ω. Since $|\nabla u_n|^2 \geq |H(u_n)|$ a.e. in Ω and $\nabla\Psi = 0$ a.e. in $K := \{x \in \overline{\Omega} : \Psi(x) = d_\Omega(P,Q)\}$, we easily estimate

$$E_{\mathcal{M}}(g_n) = \int_\Omega |\nabla u_n|^2 \geq \int_K |\nabla u_n|^2 + \int_{\Omega\setminus K} |H(u_n)| \geq \int_K |\nabla u_n|^2 + \int_\Omega H(u_n)\cdot\nabla\Psi.$$

Then we observe that $\int_\Omega H(u_n)\cdot\nabla\Psi = \langle T(g_n), \Psi_{|\mathcal{M}}\rangle = 2\pi d_\Omega(P,Q)$ and hence

$$E_{\mathcal{M}}(g_n) \geq \int_K |\nabla u_n|^2 + 2\pi d_\Omega(P,Q).$$

Now we infer from (17) and the assumption $d_\Omega(P,Q) = d_{\mathcal{M}}(P,Q)$ that $E_{\mathcal{M}}(g_n) \to 2\pi d_\Omega(P,Q)$. Therefore, $\int_K |\nabla u_n|^2 \to 0$. On the other hand, if $d_\Omega(P,Q)$ is small enough, we may find a ball $B(x_0,\varepsilon) \subset \mathbb{R}^3$ such that $x_0 \in \Omega$, $B(x_0,\varepsilon)\cap\mathcal{M} \neq \varnothing$ and $B(x_0,\varepsilon)\cap(\Omega\setminus K) = \varnothing$. By lower semicontinuity, one obtains

$$\int_{B(x_0,\varepsilon)\cap\Omega} |\nabla u|^2 \leq \int_K |\nabla u|^2 \leq \lim_{n\to+\infty} \int_K |\nabla u_n|^2 = 0,$$

so that u is constant in $B(x_0,\varepsilon)\cap\Omega$. Since $u_{|\mathcal{M}} = g \in X_{\mathcal{M}}$, we deduce that $|u| = 1$ in $B(x_0,\varepsilon)\cap\Omega$. Now, by an easy application of the maximum principle, one deduces that u is constant in Ω and hence $u \equiv \alpha$ for some constant $\alpha \in \mathbb{S}^1$. In particular, $g \equiv \alpha$ and the proof is complete. □

Acknowledgments. The author would like to thank S. Alama, L. Bronsard and P. Sternberg for giving the opportunity to deliver a lecture on the material presented in this note at a workshop that took place at the Centre de recherches mathématiques (CRM) in Montréal in July 2006.

References

1. F. Béthuel, *The approximation in the trace spaces defined between manifolds*, Nonlinear Anal. **24** (1995), no. 1, 121 – 130.
2. J. Bourgain, H. Brezis, and P. Mironescu, *$H^{1/2}$-maps with values into the circle: Minimal connections, lifting and Ginzburg – Landau equation*, Publ. Math. Inst. Hautes tudes Sci. **99** (2004), 1 – 115.
3. ———, *Lifting, degree and distributional Jacobian revisited*, Comm. Pure Appl. Math. **58** (2005), no. 4, 529 – 551.
4. A. Boutet de Monvel-Berthier, V. Georgescu, and R. Purice, *A boundary value problem related to the Ginzburg – Landau model*, Comm. Math. Phys. **142** (1991), no. 1, 1 – 23.
5. H. Brezis, J.-M. Coron, and E. H. Lieb, *Harmonic maps with defects*, Comm. Math. Phys. **107** (1986), no. 4, 649 – 705.
6. H. Brezis, P. Mironescu, and A. C. Ponce, *$W^{1,1}$-maps with values into $\mathbb{S}^1$*, Geometric Analysis of PDE and Several Complex Variables (S. Chanillo, P. D. Cordaro, N. Hanges, J. Hounie, and A. Meziani, eds.), Contemp. Math., vol. 368, Amer. Math. Soc., Providence, RI, 2005, pp. 69 – 100.
7. H. Brezis and P. Mironescu, *On some questions of topology for $\mathbb{S}^1$-valued fractional Sobolev spaces*, RACSAM Rev. R. Acad. Cienc. Exactas Fs. Nat. Ser. A Mat. **95** (2001), 121 – 143.
8. M. Giaquinta, G. Modica, and J. Souček, *On sequences of maps into $\mathbb{S}^1$ with equibounded $W^{1/2}$-energies*, Selecta Math. (N.S.) **10** (2004), no. 3, 359 – 375.
9. M. Giaquinta and D. Mucci, *Density results for the $W^{1/2}$ energy of maps into a manifold*, Math. Z. **251** (2005), no. 3, 535 – 549.
10. ———, *On sequences of maps into a manifold with equibounded energies*, J. Funct. Anal. **225** (2005), no. 1, 94 – 146.
11. F. Hang and F.-H. Lin, *A remark on the Jacobians*, Commun. Contemp. Math. **2** (2000), no. 1, 36 – 46.
12. R. Ignat, *The space* $\mathrm{BV}(\mathbb{S}^2,\mathbb{S}^1)$*: minimal connection and optimal lifting*, Ann. Inst. H. Poincaré Anal. Non Linéaire **22** (2005), no. 3, 283 – 302.
13. F.-H. Lin and T. Rivière, *Complex Ginzburg – Landau equations in high dimensions and codimension two area minimizing currents*, J. Eur. Math. Soc. (JEMS) **1** (1999), no. 3, 237 – 311.
14. V. Millot, *Energy with weight for $\mathbb{S}^2$-valued maps with prescribed singularities*, Calc. Var. Partial Differential Equations **24** (2005), no. 1, 83 – 109.
15. ———, *The relaxed energy for $\mathbb{S}^2$-valued maps and measurable weights*, Ann. Inst. H. Poincaré Anal. Non Linéaire **23** (2006), no. 2, 135 – 157.
16. V. Millot and A. Pisante, *Relaxed energies for $H^{1/2}$-maps with values into the circle and measurable weights*, Indiana Univ. Math. J., to appear.
17. P. Mironescu and A. Pisante, *A variational problem with lack of compactness for $H^{1/2}(\mathbb{S}^1;\mathbb{S}^1)$-maps of prescribed degree*, J. Funct. Anal. **217** (2004), no. 2, 249 – 279.
18. T. Rivière, *Line vortices in the* U(1)*-Higgs model*, ESAIM Contrôle Optim. Calc. Var. **1** (1995/1996), 77 – 167.
19. ———, *Dense subsets of $H^{1/2}(\mathbb{S}^2;\mathbb{S}^1)$*, Ann. Global Anal. Geom. **18** (2000), no. 5, 517 – 528.
20. É. Sandier, *Ginzburg – Landau minimizers from $\mathbb{R}^{n+1}$ to $\mathbb{R}^n$ and minimal connections*, Indiana Univ. Math. J. **50** (2001), no. 4, 1807 – 1844.

Center for Nonlinear Analysis, Department of Mathematical Sciences, Carnegie Mellon University, Pittsburgh, PA 15213, USA

Current address: Département de mathématiques, Université de Cergy-Pointoise, Site Saint-Martin, BP 222, 95302 Cergy-Pontoise, France

E-mail address: vmillot@andrew.cmu.edu

Centre de Recherches Mathématiques
CRM Proceedings and Lecture Notes
Volume **44**, 2008

Hodge Decompositions, Γ-Convergence and the Gross – Pitaevskii Energy

Alberto Montero

ABSTRACT. In this talk we presented a Γ-convergence result for the Gross–Pitaevskii energy. This result has been shown to hold in several particular cases, and our main contribution is to extend the data of the energy functional to which the result can be applied. The possibility of handling more general data hinges on being able to solve a weighted Hodge decomposition problem that does not seem to be in the literature at present time.

Let us consider the energy

$$\widetilde{E}_\varepsilon(u) = \int_{\mathbb{R}^n} \left\{ \frac{|(\nabla - \mathrm{i}V_\varepsilon)u|^2}{2} + \frac{a|u|^2}{2\varepsilon^2} + \frac{|u|^4}{4\varepsilon^2} \right\}. \tag{1}$$

under the constraint

$$\int_{\mathbb{R}^n} |u|^2 = 1. \tag{2}$$

Here $a \in C^\infty(\mathbb{R}^n)$ is a function, that we shall call the potential and that has

$$\lim_{|x|\to\infty} a(x) = +\infty, \tag{3}$$

$V_\varepsilon \in C^\infty(\mathbb{R}^n, \mathbb{R}^n)$ is a vector field and $\varepsilon > 0$ is a small parameter, all data of the functional. The variable is $u \in W^{1,2}(\mathbb{R}^n; \mathbb{C})$, and we consider this functional mainly for $n = 3$. When $n = 2$, very detailed descriptions of the global minimizers of $\widetilde{E}_\varepsilon$ (or an energy closely related to it) can be found in [2, 12, 13], and we refer the reader interested in this dimension to these articles. We concentrate from now on in the case $n = 3$.

This energy functional, usually referred to as the Gross – Pitaevskii energy, has been studied extensively by physicists, and more recently by mathematicians, in the context of Bose – Einstein condensation (see for instance [1, 20] for interesting surveys on this subject). A Bose – Einstein condensate is a peculiar kind of matter often characterized by the property that most of its particles share the same quantum state. In this situation, the particles act more as a single *super-particle*,

2000 *Mathematics Subject Classification.* Primary 35-06; Secondary 49S05.
This is the final form of the paper.

than as a collection of individual ones. The function u above is a kind of wave function for this super-particle, and $\widetilde{E}_\varepsilon(u)$ represents the energy of the condensate u, trapped in the potential a, and stirred by the vector field V_ε. One interesting aspect of these condensates is that, for V_ε in certain ranges, whirlpools appear in the condensate. This whirlpools are usually referred to as vortices, and their description and influence on $\widetilde{E}_\varepsilon$ is perhaps our main object of study.

We are interested in a Γ-convergence result that is somehow subjacent to several papers that study $\widetilde{E}_\varepsilon$ for particular potentials a and vector fields V_ε. Assuming certain bounds on $\widetilde{E}_\varepsilon$, this result gives a concrete expression for the relation between $\widetilde{E}_\varepsilon$ and the vortices in the condensate. Let us first introduce the notation we need to state this result.

Let $\lambda \in \mathbb{R}$ be a real number such that

$$\int_{\mathbb{R}^n} (\lambda - a)^+ = 1. \tag{4}$$

Such a number always exists, because of (3), and we assume it to be a regular value of a. Define the functions

$$\rho = (\lambda - 1)^+, \omega = (a - \lambda)^+. \tag{5}$$

The set

$$\Omega = \{x \in \mathbb{R}^n : \rho(x) > 0\}, \tag{6}$$

corresponds to the region of $\mathbb{R}^n$ where $a < \lambda$, and it will be important later. Define next

$$\mathcal{H}^* \tag{7}$$

to be the completion of $C_0^\infty(\mathbb{R}^n; \mathbb{C})$ with respect to the (norm induced by the) inner product

$$\langle f, g \rangle = \int_{\mathbb{R}^n} \{\nabla f \cdot \nabla \bar{g} + (1 + \omega) f \bar{g}\} \, \mathrm{d}x. \tag{8}$$

Here $\bar{z}$ denotes the complex conjugate of $z \in \mathbb{C}$. We will also use the notation

$$\mathcal{H} = \{f \in \mathcal{H}^* : \bar{f} = f\}, \tag{9}$$

the subspace of real valued functions in $\mathcal{H}^*$ (see [12]).

Notice now that, following Aftalion and Rivière [5], for ρ and ω from (5) one has

$$\widetilde{E}_\varepsilon(u) = \frac{1}{2} \int_{\mathbb{R}^n} \left\{ |(\nabla - \mathrm{i} V_\varepsilon) u|^2 + \frac{1}{2\varepsilon^2} (\rho - |u|^2)^2 + \frac{\omega}{\varepsilon^2} |u|^2 \right\} + \frac{1}{2\varepsilon^2} \int_{\mathbb{R}^n} \left\{ \lambda |u|^2 - \frac{\rho^2}{2} \right\}.$$

Under the constraint (2), the last integral in the above identity does not depend on u, so in this case the functional

$$E_\varepsilon(u) = \frac{1}{2} \int_{\mathbb{R}^n} \left\{ |(\nabla - \mathrm{i} V_\varepsilon) u|^2 + \frac{1}{2\varepsilon^2} (\rho - |u|^2)^2 + \frac{\omega}{\varepsilon^2} |u|^2 \right\} \tag{10}$$

is equivalent to $\widetilde{E}_\varepsilon$. We focus from here on in the study of E_ε under the constraint (2). We assume always the technical condition

$$|V_\varepsilon|^2 \leq C_\varepsilon \omega \tag{11}$$

for some constant $C_\varepsilon > 0$, so that for $u \in \mathcal{H}^*$, every term in $E_\varepsilon(u)$ is finite.

One can see from the above expression that there is a strong resemblance between E_ε and the Ginzburg–Landau energy of superconductivity, but also some differences. Perhaps the most important of these differences is the fact that the function ρ in E_ε above vanishes on $\partial\Omega$, where Ω is defined in (6), whereas the corresponding function in the Ginzburg-Landau case is identically equal to 1. The first step towards obtaining a Γ-limit for E_ε is a division argument, that goes back to [8, 18]. This division argument yields a splitting of the energy roughly in two terms: one that accounts for the cost that the condensate pays to accommodate its shape to the potential and for circulation of the condensate around possible holes in Ω, independent of u, and a second term that depends on u, and is a weighted version of the energy considered by Béthuel, Brezis and Hélein in [9], that we shall refer to as BBH energy. We describe this division argument presently.

Consider the global minimizer $f_1 \in \mathcal{H}$ of E_ε among real valued functions $f \in \mathcal{H}$, for $V_\varepsilon = 0$. This has been studied in [12], and in slightly different settings in [2, 23]. One can interpret f_1 as the density profile the condensate would have if it were trapped in the potential with no stirring. For $u \in \mathcal{H}^*$, one defines first

$$v_1 = u/f_1,$$

or $u = f_1 v_1$. Second, one considers a function $\theta\colon \Omega \to \mathbb{R}$, smooth, possibly multi-valued, with $\mathrm{e}^{\mathrm{i}\theta}$ smooth in Ω, to be chosen later, and defines

$$v = \mathrm{e}^{-\mathrm{i}\theta} v_1 \text{ in } \Omega.$$

We have the

Lemma 1 (Lassoued and Mironescu [18], Andre and Shafrir [8]). *With the notation described above, for every $u \in \mathcal{H}^*$ one has*

$$\begin{aligned}(12)\quad E_\varepsilon(u) = \int_\Omega \left\{ \frac{f_1^2}{2} \left| (\nabla - \mathrm{i}(V_\varepsilon - \nabla\theta))v \right|^2 + \frac{f_1^4}{4\varepsilon^2}(1 - |v|^2)^2 \right\} \\ + \int_{\mathbb{R}^n \setminus \Omega} \left\{ \frac{f_1^2}{2} |(\nabla - \mathrm{i}V_\varepsilon)v_1|^2 + \frac{f_1^4}{4\varepsilon^2}(1 - |v_1|^2)^2 \right\} \\ + \int_{\mathbb{R}^n} \left\{ \frac{1}{2}|\nabla f_1|^2 + \frac{1}{4\varepsilon^2}(\rho - f_1^2)^2 + \frac{\omega}{2\varepsilon^2} f_1^2 \right\}.\end{aligned}$$

One obtains two important facts from the identity above: first, from a Γ-convergence point of view, the domain Ω appears as a natural candidate for the support of some asymptotic lower bound for E_ε. The second important fact that we obtain from here comes from the following identity valid for any vector field $V\colon \Omega \to \mathbb{R}^n$, that to our knowledge was first pointed out by Béthuel and Rivière [10]:

$$|(\nabla - \mathrm{i}V)v|^2 = |\nabla v|^2 + |v|^2|V|^2 - 2V \cdot j(v).$$

Here the vector field $j(v)$ is given by

$$j(v) = \frac{1}{2\mathrm{i}}(\bar{v}\nabla v - v\nabla\bar{v}).$$

Using this fact, the integral over Ω in (12) above can be written as

$$(13)\quad E_\varepsilon(u) = \int_\Omega \left\{ \frac{f_1^2}{2}|\nabla v|^2 + \frac{f_1^4}{4\varepsilon^2}(1-|v|^2)^2 \right\} + \int_\Omega f_1^2 j(v)\cdot(\nabla\theta - V_\varepsilon) + \int_\Omega f_1^2|v|^2|\nabla\theta - V_\varepsilon|^2.$$

Here one recognizes

$$(14)\qquad H_\varepsilon(v_\varepsilon) := \int_\Omega \left\{ \frac{f_1^2}{2}|\nabla v_\varepsilon|^2 + \frac{f_1^4}{4\varepsilon^2}(1-|v_\varepsilon|^2)^2 \right\}$$

as the weighted version of the BBH energy, and also that the last term in (13) is almost independent of v, if $H_\varepsilon(v)$ is not too large.

Now we turn our attention to the choice of the function θ. We would like to choose θ so that

$$f_1^2(V_\varepsilon - \nabla\theta) \sim \nabla\times B \text{ in } \Omega$$

for some vector field B. If this were possible, we would have

$$(15)\qquad \int_\Omega f_1^2 j(v)\cdot(V_\varepsilon - \nabla\theta) \sim \int_\Omega j(v)\cdot\nabla\times B = \int_\Omega B\cdot\nabla\times j(v),$$

provided the boundary conditions of B on $\partial\Omega$ allow for the last integration by parts. By the properties of f_1 (cf. [12]), we know that $f_1^2 \to \rho$ as $\varepsilon\to 0$, so in fact it would suffice to obtain

$$(16)\qquad \rho(V_\varepsilon - \nabla\theta) \sim \nabla\times B \text{ in } \Omega.$$

We recognize the last integral in (15) as a Jacobian integral of the kind studied, in $n\geq 3$, by Jerrard and Soner [15], Alberti, Baldo and Orlandi [7], Sandier and Serfaty [21]. However, to apply these results to our problem one would need to relate the Jacobians

$$(17)\qquad J(v) = \tfrac{1}{2}\nabla\times j(v)$$

to H_ε defined in (14), rather than to the BBH energy.

The problem of finding θ such that (16) holds in some sense is of course a Hodge decomposition problem. There is an incredible amount of work in such problems, as can be seen for instance from Schwarz [22], but the decomposition one needs in this study has one important peculiarity: the function ρ that appears above in fact vanishes at a linear rate on $\partial\Omega$. Because of this, the problem (16) seems not to be covered by any of the usual theorems on Hodge decompositions one encounters in the literature. For example, Jerrard in [14] considers E_ε roughly in the model case

$$(18)\qquad \begin{aligned} V_0(x_1,x_2,x_3) &= (-y,x,0), \quad V_\varepsilon = |\ln(\varepsilon)|V_0,\\ \rho(x_1,x_2,x_3) &= 1-(\alpha_1x_1^2+\alpha_2x_2^2+\alpha_3x_3^2), \end{aligned}$$

As noted in [5],

$$\rho(\nabla s_0 - V_0) = \frac{-1}{2(1+(\alpha_1/\alpha_2)^2)}\nabla\times(\rho^2\hat{e}_3),$$

where

$$s_0 = \frac{\alpha_2^2-\alpha_1^2}{\alpha_2^2+\alpha_1^2}x_1x_2 \quad\text{and}\quad \hat{e}_3 = (0,0,1).$$

Slightly more general V and potential a were considered later by Alama, Bronsard and Montero in [6] when Ω is a torus of revolution, but making significant use of the symmetry of the torus. This of course restricts the kinds of potentials a one can use.

The following theorem allows for a reasonable extension of the kinds of vector fields V and potentials a that one can consider here. We use the notation

$$H_N(\Omega) = \{h \in W^{1,2}(\Omega;\mathbb{R}^3) : \nabla \times h = \nabla \cdot h = 0 \text{ in } \Omega,\ h \cdot \nu = 0 \text{ on } \partial\Omega\}. \tag{19}$$

Theorem 2 (Montero [19]). *Assume that $a \in C^\infty(\mathbb{R}^n)$ satisfies (3), that $\lambda \in \mathbb{R}$ in (4) is a regular value of a, and recall the definitions of ρ and Ω from (5) and (6). Let $\alpha \in]0,1[$ and $V_0 \in C^{2,\alpha}(\Omega;\mathbb{R}^3)$. One can always find $s_0 \in C^{2,\alpha}(\Omega)$, $B_0 \in C^{2,\alpha}(\Omega;\mathbb{R}^3)$ and $h_0 \in H_N(\Omega)$, such that*

$$\rho(\nabla s_0 + h_0 - V_0) = \nabla \times B_0 \text{ in } \Omega. \tag{20}$$

Furthermore, $\nabla s_0 \cdot \nu = V_0 \cdot \nu$ on $\partial\Omega$, and we can choose B_0 so that $B_0 = 0$ and $DB_0 = 0$ on $\partial\Omega$, where DB represents the Jacobian matrix of the vector field B. In particular, we can choose B_0 above to vanish quadratically on $\partial\Omega$, that is, there is a constant $C > 0$ such that

$$|B_0| \leq C\big(\operatorname{dist}(x;\partial\Omega)\big)^2.$$

The main difficulty in proving the theorem above is to establish the regularity for the solutions of a degenerate elliptic problem, in which the ellipticity constant vanishes at a linear rate on $\partial\Omega$. There are results in this regard in the literature (see for instance [11, 16]), but none of them seem to quite fit our needs. Concretely, the result we need is the following:

Lemma 3 (Montero [19]). *Denote by $W^{1,p}_m(\Omega)$ the completion of $C^\infty(\overline{\Omega})$ with respect to the norm*

$$\|s\| = \left\{\int_\Omega \rho^m(|\nabla s|^p + |s|^p)\right\}^{1/p}$$

for some integer $m \geq 1$, and assume that ρ and Ω are as above. For every $\phi_0 \in C^{0,\alpha}(\Omega)$ such that

$$\int_\Omega \rho^m \phi_0 \, \mathrm{d}x = 0$$

there is

$$s_0 \in W^{1,2}_m(\Omega)$$

unique up to a constant, that satisfies

$$-\operatorname{div}(\rho^m A \nabla s_0) = \rho^m \phi_0 \tag{21}$$

weakly in Ω, that is,

$$\int_\Omega \rho^m \nabla s \cdot A \nabla s_0 = \int_\Omega \rho^m \phi_0 s \quad \textit{for all } s \in W^{1,2}_m(\Omega). \tag{22}$$

Furthermore, $s_0 \in C^{2,\alpha}(\Omega)$ and $\nabla s_0 \cdot A\nu = 0$ on $\partial\Omega$.

With this Hodge decomposition in hand, one can choose θ so that

$$\nabla(\theta - s_0) \sim h_0.$$

On the other hand, the problem of relating the Jacobians $J(v)$ to $H_\varepsilon(v)$ has been addressed by Jerrard in [14]. As we mention above, this last paper is written for rather particular choices of potential a and stirring V_ε, but most of the argument can be adapted to a more general V_ε that satisfies the condition (11). The main difficulty is that the weight f_1^2 in H_ε is very close to ρ, which vanishes linearly on $\partial\Omega$. This requires the field B_ε in the decomposition of V_ε given by Theorem 2 to vanish quadratically on the boundary, which is exactly what the theorem gives. Roughly speaking, one obtains the following:

Theorem 4. *Assume $V_\varepsilon = \mu_o|\ln(\varepsilon)|V_0$, where $V_0 \in C^\infty(\mathbb{R}^n;\mathbb{R}^n)$ is a given vector field, and apply the Hodge decomposition theorem above to obtain*

$$\rho(\nabla s_0 + h_0 - V_0) = \nabla \times B_0. \tag{23}$$

With the notation from Lemma 1, assume that $u_\varepsilon \in \mathcal{H}^*$ is a sequence of functions that have

$$H_\varepsilon(v_\varepsilon) \le C|\ln(\varepsilon)|. \tag{24}$$

Here we denote by v_ε the result of the (two steps of the) division in Lemma 1 applied to u_ε, for θ so that

$$\nabla(\theta - s_0) \sim h_0.$$

In this situation, the Jacobians $J(v_\varepsilon)$ of the v_ε are weakly convergent to a current J, where $(1/\pi)J$ is a rectifiable, integer multiplicity 1-current. Furthermore, if we denote by $M_\rho(J)$ the weighted mass of J, that is

$$M_\rho(J) = \sup\{J(B) : \|B/\rho\|_\infty \le 1\},$$

we have $M_\rho(J) < \infty$, and

$$\liminf_{\varepsilon\to 0} \frac{E_\varepsilon(u_\varepsilon) - D_\varepsilon}{|\ln(\varepsilon)|} \ge E_0(J) := M_\rho(J) + 2\mu_0 J(B_0), \tag{25}$$

where

$$D_\varepsilon = \int_{\mathbb{R}^n} \left\{ \frac{1}{2}|\nabla f_1|^2 + \frac{1}{4\varepsilon^2}(\rho - f_1^2)^2 + \frac{\omega}{2\varepsilon^2} f_1^2 \right\} + \int_\Omega \rho|\nabla\theta - V_\varepsilon|^2.$$

Note that D_ε depends on V_ε and the potential a, but not on u. Finally, the line energy E_0 above is also the Γ-limit of E_ε under the energy bound (24).

This theorem is not new. It appears in [14] for V_ε and a very close to the model case, and also in [6] when Ω is a torus of revolution, for V and a having the same symmetry as the torus. Earlier, but only formally, the energy E_0 was also identified in [5].

To conclude, we mention that a standard application for this type of result is the construction of local minimizers of E_ε. Indeed, it is well known that the Kohn–Sternberg argument [17] leads to the existence of local minimizers of E_ε if one can find isolated local minimizers of E_0 in an appropriate sense. Such local minimizers of E_0 have indeed been found in [6, 14], although earlier studies of this line energy were carried out in [3, 4].

References

1. A. Aftalion, *Vortices in Bose–Einstein condensates*, Progr. Nonlinear Differential Equations Appl., vol. 67, Birkhäuser, Boston, 2006.
2. A. Aftalion, S. Alama, and L. Bronsard, *Giant vortex and the breakdown of strong pinning in a rotating Bose–Einstein condensate*, Arch. Ration. Mech. Anal. **178** (2005), no. 2, 247–286.
3. A. Aftalion and R. L. Jerrard, *Shape of vortices for a rotating Bose–Einstein condensate*, Phys. Rev. A (3) **66** (2002), no. 2, 023611.
4. ______, *Properties of a single vortex solution in a rotating Bose–Einstein condensate*, C. R. Math. Acad. Sci. Paris **336** (2003), no. 9, 713–718.
5. A. Aftalion and T. Rivière, *Vortex energy and vortex bending for a rotating Bose–Einstein condensate*, Phys. Rev. A (3) **64**, no. 4, 043611.

6. S. Alama, L. Bronsard, and A. Montero, *Vortices for a rotating toroidal Bose–Einstein condensate*. submitted.
7. G. Alberti, S. Baldo, and G. Orlandi, *Variational convergence for functionals of Ginzburg–Landau type*, Indiana Univ. Math. J. **54** (2005), no. 5, 1411–1472.
8. N. André and I. Shafrir, *Minimization of a Ginzburg–Landau type functional with nonvanishing Dirichlet boundary condition*, Calc. Var. Partial Differential Equations **7** (1998), no. 3, 191–217.
9. F. Béthuel, H. Brezis, and F. Hélein, *Ginzburg–Landau vortices*, Progr. Nonlinear Differential Equations Appl., vol. 13, Birkhäuser, Boston, 1994.
10. F. Béthuel and T. Rivière, *Vortices for a variational problem related to superconductivity*, Ann. Inst. H. Poincaré Anal. Non Linéaire **12** (1995), no. 3, 243–303.
11. P. Bolley, J. Camus, and G. Métivier, *Estimations de Schauder et régularité höldérienne pour une classe de problèmes aux limites singuliers*, Comm. Partial Differential Equations **11** (1986), no. 11, 1135–1203.
12. R. Ignat and V. Millot, *The critical velocity for vortex existence in a two-dimensional rotating Bose–Einstein condensate*, J. Funct. Anal. **233** (2006), no. 1, 260–306.
13. ———, *Energy expansion and vortex location for a two-dimensional rotating Bose–Einstein condensate*, Rev. Math. Phys. **18** (2006), no. 2, 119–162.
14. R. L. Jerrard, *Local minimizers with vortex filaments for a Gross–Pitaevsky functional*, ESAIM Control Optim. Calc. Var. **13** (2007), no. 1, 35–71.
15. R. L. Jerrard and H. M. Soner, *The Jacobian and the Ginzburg–Landau energy*, Calc. Var. Partial Differential Equations **14** (2002), no. 2, 151–191.
16. H. Koch, *Non-Euclidean singular integrals and the porous medium equation*, Habilitation thesis, University of Heidelberg, 1999.
17. R. V. Kohn and P. Sternberg, *Local minimisers and singular perturbations*, Proc. Roy. Soc. Edinburgh Sect. A **111** (1989), no. 1-2, 69–84.
18. L. Lassoued and P. Mironescu, *Ginzburg–Landau type energy with discontinuous constraint*, J. Anal. Math. **77** (1999), 1–26.
19. A. Montero, *Hodge decompositions with degenerate weights and the Gross–Pitaevskii energy*, J. Funct. Anal. to appear.
20. L. Pitaevskii and S. Stringari, *Bose–Einstein condensation*, Internat. Ser. Monogr. Phys., vol. 116, Oxford Univ. Press, Oxford, 2003.
21. É. Sandier and S. Serfaty, *A product-estimate for Ginzburg–Landau and corollaries*, J. Funct. Anal. **211** (2004), no. 1, 219–244.
22. G. Schwarz, *Hodge decomposition — a method for solving boundary value problems*, Lecture Notes in Math., vol. 1607, Springer, Berlin, 1995.
23. S. Serfaty, *On a model of rotating superfluids*, ESAIM Control Optim. Calc. Var. **6** (2001), 201–238.

Department of Mathematics, University of Toronto, Toronto, ON M5S 3G3, Canada

Current address: Department of Theoretical and Applied Mathematics, The University of Akron, Akron, OH 44325, USA

E-mail address: jm95@akron.edu

Centre de Recherches Mathématiques
CRM Proceedings and Lecture Notes
Volume **44**, 2008

Bifurcation of Vortex Solutions to a Ginzburg–Landau Equation in an Annulus

Yoshihisa Morita

Abstract. We consider a simplified Ginzburg-Landau model of superconductivity in an annulus domain. This model equation has two physical parameters λ and h which are related to the Ginzburg–Landau parameter and strength of an applied magnetic field respectively. Then a solution with k-mode in the polar angle bifurcates from the trivial solution at appropriate parameter values λ and h. This solution is vortexless, that is, it has no zeros. We study the bifurcation near the critical point on which bifurcation curves corresponding to two different modes intersect in the parameter space (h, λ). We then investigate the local bifurcation structure around the critical point and prove the existence of a vortex solution under a generic condition. In particular we show that the solution has zeros on a boundary if the parameters is on some curve emanating from the critical point. The stability of the vortex solution is also discussed by applying the center manifold theorem.

1. Introduction

We consider a Ginzburg–Landau energy in a thin domain occupied by a superconducting sample. Here the thin domain is defined by

$$\Omega(\varepsilon) = \{(x, z) = (x_1, x_2, z) \in \mathbb{R}^3 : x = (x_1, x_2) \in D, 0 < z < \varepsilon\},$$

where ε is a small parameter and D is a bounded domain in $\mathbb{R}^2$. Under a constant applied magnetic field $H_a = (0, 0, h)$ vertical to the horizontal plane the Ginzburg–Landau energy is given by

$$(1.1)\quad E_\varepsilon(\Psi, A) = \int_{\Omega(\varepsilon)} \frac{1}{2}\left\{|(\nabla - \mathrm{i}A)\psi|^2 + \frac{\kappa^2}{4}(1 - |\Psi|^2)^2\right\} \mathrm{d}x\, \mathrm{d}z + \frac{1}{2}\int_{\mathbb{R}^3} |\operatorname{curl} A - H_a|^2 \,\mathrm{d}x\, \mathrm{d}z$$

where $\Psi = \Psi(x, z)$ is complex-valued, $A \in \mathbb{R}^3$ is a vector potential of the magnetic field and κ is a positive parameter (called the Ginzburg–Landau parameter). We assume that the material is surrounded by the vacuum, hence the integral for the magnetic field is taken over the whole space $\mathbb{R}^3$.

2000 *Mathematics Subject Classification.* 35B32, 35J60, 35Q99, 37L10.
Key words and phrases. Ginzburg–Landau equation, bifurcation, vortex solution.
This is the final form of the paper.

A formal reduction of the energy as $\varepsilon \to 0$ tells

$$\Psi(x,z) = \psi(x) + o(1), \quad A = (A_a(x),0) + o(1), \ \mathrm{curl}(A_a,0) = H_a$$

and

$$E_\varepsilon(\Psi,A)/\varepsilon \to E(\psi) := \int_D \frac{1}{2}|(\nabla_x - \mathrm{i}A_0)\psi|^2 + \frac{\kappa^2}{4}(1-|\psi|^2)^2\,\mathrm{d}x$$

($\varepsilon \to 0$). This reduction is partially justified in a mathematically rigorous way under a certain condition (see [2, 7, 8]).

Henceforth we deal with this reduced energy and the corresponding Euler–Lagrange equation in an annulus domain

$$D = \{x = (x_1,x_2) \in \mathbb{R}^2 : 1 < |x| < R\}, \tag{1.2}$$

with the given vector potential

$$A_0 := \frac{h}{2}(-x_2,x_1). \tag{1.3}$$

Namely we consider the simplified Ginzburg–Landau energy

$$E(\psi) = \int_D \frac{1}{2}|(\nabla - \mathrm{i}A_0)\psi|^2 + \frac{\lambda}{4}(1-|\psi|^2)^2\,\mathrm{d}x, \tag{1.4}$$

and the Euler–Lagrange equation

$$(\nabla - \mathrm{i}A_0)^2\psi + \lambda(1-|\psi|^2)\psi = 0, \quad x \in D, \tag{1.5}$$

$$\frac{\partial\psi}{\partial \boldsymbol{n}} = 0, \qquad x \in \partial D, \tag{1.6}$$

where A_0 is defined as (1.3), $\lambda := \kappa^2$ and $\boldsymbol{n}$ denotes the outer normal vector on the boundaries. Note that $A_0 \cdot \boldsymbol{n} = 0$ on the boundaries of the annulus domain (1.2). Equation (1.5)–(1.6) can be regarded as a simplified Ginzburg–Landau model on a thin film with a hole. Using the polar coordinates, we easily see that (1.5)–(1.6) is written as

$$\frac{\partial^2\psi}{\partial r^2} + \frac{1}{r}\frac{\partial\psi}{\partial r} + \left(\frac{1}{r}\frac{\partial}{\partial\theta} - \mathrm{i}\frac{hr}{2}\right)^2\psi + \lambda(1-|\psi|^2)\psi, \quad (r,\theta) \in (1,R)\times\mathbb{R}/2\pi \tag{1.7}$$

$$\frac{\partial\psi}{\partial r} = 0, \qquad r = 1, R. \tag{1.8}$$

Henceforth we often abbreviate the notation as $\psi_r = \partial\psi/\partial r$, $\psi_{rr} = \partial^2\psi/\partial r^2$.

We study a bifurcation of a solution from the trivial solution $\psi = 0$ in the parameter space (h,λ). By the ansatz

$$\psi = V(r)\exp(\mathrm{i}k\theta), \quad V \in \mathbb{R},$$

we obtain the equation for V as

$$V_{rr} + \frac{1}{r}V_r - \left(\frac{k}{r} - \frac{hr}{2}\right)^2 V + \lambda(1-V^2)V = 0, \quad 1 < r < R, \tag{1.9}$$

$$V_r(1) = V_r(R) = 0. \tag{1.10}$$

For each $k \in \mathbb{Z}$ there exists a bifurcation curve in (h,λ)-space at which a unique positive solution to (1.9)–(1.10) bifurcates from the trivial solution $V = 0$. Then the corresponding bifurcating solution of (1.7)–(1.8) is a vortexless solution. Namely the solution does not have (isolated) zeros with a topological degree. However we suspect that mixing of two modes induces a bifurcation of a vortex solution. In fact the recent study [3] tells that such a vortex solution certainly takes place in a different but closely related model equation (see also [1, 9] as related works). We will apply the arguments in [3] to show the existence of a vortex solution to the present equation. More precisely, we show that a vortex solution certainly bifurcates in a neighborhood of an intersection point of bifurcation curves corresponding to two different modes. For the proof we derive bifurcation equations from (1.5)–(1.8) around the intersection point by using the Lyapunov–Schmidt reduction and investigate the bifurcation equations to obtain the vortex solution. Moreover, we show the transient phenomenon from a vortexless solution to a vortex solution, that is, as the parameter varies suitably, a solution with zeros on each boundary arises. In consequence, we obtain similar results to those in [3].

Before stating the main result, we prepare some notations. Let $\sigma^{(k)} = \sigma^{(k)}(h) > 0$ be the first eigenvalue of the operator

$$\mathcal{L}_h^{(k)}[v] := -v_{rr} - \frac{1}{r}v_r + (k/r - hr/2)^2 v, \tag{1.11}$$

defined on the space $H_N^2(1, R) := \{v \in H^2((1, R); \mathbb{R}) : v_r(1) = v_r(R) = 0\}$. We let $v^{(k)}(r; h)$ be a positive eigenfunction corresponding to $\sigma^{(k)}(h)$. Note that $\varphi_k := v^{(k)} \exp(\mathrm{i}k\theta)$ is an eigenfunction corresponding to the eigenvalue $\sigma^{(k)}(h) - \lambda$ of the linearized operator

$$\mathcal{L}[\varphi] := -(\nabla - \mathrm{i}A_0)^2\varphi - \lambda\varphi = -\varphi_{rr} - \frac{1}{r}\varphi_r - \left(\frac{1}{r}\frac{\partial}{\partial\theta} - \mathrm{i}\frac{hr}{2}\right)^2 \varphi - \lambda\varphi \tag{1.12}$$

with the domain

$$\mathrm{Dom}(\mathcal{L}) = \{\varphi \in H^2(D; \mathbb{C}) : \partial\varphi/\partial\boldsymbol{n} = 0 \ (x \in \partial D)\}.$$

We always identify $\mathbb{C}$ with $\mathbb{R}^2$. We, however, often use the complex expression for simplicity of notation. Note that the eigenspace for φ_k is of 2-dimensions in the real space.

We normalized the eigenfunction as

$$\|\varphi_k\| = \left(\int_D |\varphi_k|^2 \,\mathrm{d}x\right)^{1/2} = \left(2\pi \int_1^R |v^{(k)}(r; h)|^2 r \,\mathrm{d}r\right)^{1/2} = 1. \tag{1.13}$$

Define a curve and a region in (h, λ)-space for each $k \in \mathbb{Z}$ by

$$\begin{aligned} C_k &:= \{(h, \lambda) = (h, \sigma^{(k)}(h)), h \in \mathbb{R}\}, \\ \Sigma_k &:= \{(h, \lambda) : 0 \le h, \ \sigma^{(k)}(h) < \lambda\}. \end{aligned} \tag{1.14}$$

Also, define

$$\Sigma_{k,m} := \mathrm{Int}(\Sigma_k \cap \Sigma_m), \quad k < m. \tag{1.15}$$

We have the following lemma:

Lemma 1.1. *Let $0 \le k < m$. Then the two curves C_k and C_m intersect in $\{(h, \lambda) : (m + k)/R^2 < h < m + k, \lambda > 0\}$. If $k = 0$ and $R \in (1, \sqrt{2}]$, the curves transversally intersect at a unique point.*

Note that sign-changing eigenfunctions of $\mathcal{L}_h^{(k)}$ arises if R is larger. Given an integer $n \ge 1$, let $\sigma_n^{(k)}(h)$ be the $(n + 1)$th eigenvalue of $\mathcal{L}_h^{(k)}$ and let $v_n^{(k)}$ be a corresponding eigenfunction with n zeros. Then one of the parameters crosses the curve $\lambda = \sigma_n(h)$ a bifurcation takes place and there exists a solution $V = V_n^{(k)}(r)$ to (1.9)–(1.10) having n zeros in a neighborhood of the bifurcation point. When the solution $V_n^{(k)}$ bifurcates, the trivial solution is already unstable because of $\sigma^{(k)}(h) < \sigma_n^{(k)}(h)$. Thus every bifurcating solution $\psi = V_n^{(k)}\exp(ik\theta)$ is unstable near the bifurcation point, which is easily verified by a standard bifurcation theory. Although there is a possibility that such a bifurcating solution becomes stable far away from the bifurcation point, we are not able to discuss the stabilization through the present local bifurcation analysis.

We let (h_c, λ_c) be a intersection point of C_k and C_m. At $(h, c) = (h_c, \lambda_c)$ define the following quantities:

$$\begin{aligned} b_1 &:= 2\pi\int_0^1 (v^{(k)})^2 r^3\,\mathrm{d}r, & b_2 &:= 2\pi\int_0^1 (v^{(m)})^2 r^3\,\mathrm{d}r,\\ c_1 &:= 2\pi\lambda_c\int_0^1 (v^{(k)})^4 r\,\mathrm{d}r, & c_2 &:= 2\pi\lambda_c\int_0^1 (v^{(m)})^4 r\,\mathrm{d}r,\\ d &:= 2\pi\lambda_c\int_0^1 (v^{(k)}v^{(m)})^2 r\,\mathrm{d}r. \end{aligned} \tag{1.16}$$

Moreover, set

$$a_1 := k - \tfrac{1}{2}h_c b_1, \quad a_2 := m - \tfrac{1}{2}h_c b_2. \tag{1.17}$$

Proposition 1.2. *Let k and m be integers satisfying $0 \le k < m$. Suppose that only the two curves C_k and C_m intersect at $(h, \lambda) = (h_c, \lambda_c)$ and that $a_1 < a_2$ holds. Then there exist a positive parameter $\varepsilon_0 = \varepsilon_0(k, m)$, real-valued smooth functions $\mu(\alpha, \beta)$, $\nu(\alpha, \beta)$ and a smooth map $(\alpha, \beta) \mapsto \psi(\cdot, \cdot; \alpha, \beta) \in \mathrm{Dom}(\mathcal{L})$ defined for $(\alpha, \beta) \in \mathbb{R}^2$ with $|(\alpha, \beta)| := (|\alpha|^2 + |\beta|^2)^{1/2} < \varepsilon_0$ such that $\psi(\cdot, \cdot; \alpha, \beta)$ gives a nontrivial solution to (1.7)–(1.8) for $(h, \lambda) = \big(h_c + \mu(\alpha, \beta), \lambda_c + \nu(\alpha, \beta)\big)$ with $(\alpha, \beta) \neq (0, 0)$, and*

$$\lim_{(\alpha,\beta)\to(0,0)} \big(\mu(\alpha, \beta), \nu(\alpha, \beta)\big) = 0, \qquad \lim_{(\alpha,\beta)\to(0,0)} \|\psi(\cdot, \cdot; \alpha, \beta)\|_{H^2(D)} = 0.$$

Moreover, by scaling $(\alpha, \beta, \mu, \nu) = (\varepsilon\rho_1, \varepsilon\rho_2, \varepsilon^2\hat{\mu}, \varepsilon^2\hat{\nu})$, those functions are expressed as

$$\psi = \varepsilon\{\rho_1 v^{(k)}(r; h_c)\mathrm{e}^{\mathrm{i}k\theta} + \rho_2 v^{(m)}(r; h_c)\mathrm{e}^{\mathrm{i}m\theta}\} + O(\varepsilon^2), \tag{1.18}$$

$$\mu = \varepsilon^2\hat{\mu}(\rho_1, \rho_2, \varepsilon) = \varepsilon^2\tilde{\mu}(\rho_1, \rho_2) + O(\varepsilon^3), \tag{1.19}$$

$$\nu = \varepsilon^2\hat{\nu}(\rho_1, \rho_2, \varepsilon) = \varepsilon^2\tilde{\nu}(\rho_1, \rho_2) + O(\varepsilon^3), \tag{1.20}$$

where $(\tilde{\mu}, \tilde{\nu})$ is a solution of

$$\begin{cases} a_1\tilde{\mu} + \tilde{\nu} - c_1\rho_1^2 - 2d\rho_2^2 = 0,\\ a_2\tilde{\mu} + \tilde{\nu} - c_2\rho_2^2 - 2d\rho_1^2 = 0. \end{cases} \tag{1.21}$$

Since (1.5) is invariant under the transformations

$$\psi \mapsto \psi\exp(\mathrm{i}c), \ \psi(\cdot, \theta) \mapsto \psi(\cdot, \theta + c) \quad (\forall c \in \mathbb{R}), \tag{1.22}$$

we have a family of the solutions $\psi(r, \theta + \tau)\exp(\mathrm{i}c)$ $(\tau, c \in \mathbb{R})$ given a nontrivial solution $\psi(r, \theta)$ to (1.5). In the sequel we can show that for the solution $\psi(\cdot, \cdot; \alpha, \beta)$ in Proposition 1.2 $\psi = \psi(\cdot, \cdot; \alpha\mathrm{e}^{\mathrm{i}c_1}, \beta\mathrm{e}^{\mathrm{i}c_2})$ is also a solution for $\forall c_1, c_2 \in \mathbb{R}$ due to the invariance (1.22).

The next lemma is crucial when we show the existence of a vortex solution.

Lemma 1.3. *Given integers k and m with $0 \le k < m$, the positive function*

$$Q(r) := v^{(k)}(r; h_c)/v^{(m)}(r; h_c), \quad r \in [1, R]$$

is strictly decreasing. Moreover, in $[1, R] \times [0, 2\pi)$ the equation

$$\rho_1 v^{(k)}(r; h_c) + \rho_2 v^{(m)}(r; h_c) \exp(\mathrm{i}(m-k)\theta) = 0 \quad (\rho_1 . \rho_2 > 0)$$

has solutions $(r, \theta) = \big(r_0, (2j-1)\pi/(m-k)\big)$ $(j = 1, \dots, m-k)$ if and only if $\rho_2/\rho_1 \in [\gamma_R, \gamma_1]$, where

$$\gamma_1 := v^{(k)}(1; h_c)/v^{(m)}(1; h_c), \quad \gamma_R := v^{(k)}(R; h_c)/v^{(m)}(R; h_c). \tag{1.23}$$

Theorem 1.4. *In addition to the assumptions of Proposition 1.2 suppose that $c_1 c_2 \neq 4d^2$ holds. Let $\psi(\cdot, \cdot; \alpha, \beta)$ be the solution obtained in Proposition 1.2 and let γ_1 and γ_R be the numbers defined by (1.23). Then the solution has zeros in the following cases:*

(i) *Given (ρ_1, ρ_2) satisfying $\rho_2/\rho_1 \in (\gamma_R, \gamma_1)$, there exist $\varepsilon_1 \in (0, \varepsilon_0]$ and smooth functions $\big(r_j(\varepsilon), \theta_j(\varepsilon)\big)$, $j = 1, \dots, m-k$, defined for $\varepsilon \in [0, \varepsilon_1)$ such that $\big(r_j(0), \theta_j(0)\big) = \big(r_0, (2j-1)\pi/(m-k)\big)$, and*

$$\psi(r_j(\varepsilon), \theta_j(\varepsilon); \varepsilon\rho_1, \varepsilon\rho_2) = 0 \quad (j = 1, \dots, m-k) \tag{1.24}$$

holds for $(h, \lambda) = \big(h_c + \varepsilon^2 \hat{\mu}(\rho_1, \rho_2, \varepsilon), \lambda_c + \varepsilon^2 \hat{\nu}(\rho_1, \rho_2, \varepsilon)\big)$, $0 < \varepsilon < \varepsilon_1$.

(ii) *Given (ρ_1, ρ_2) satisfying $\rho_2/\rho_1 = \gamma_1$, there exist $\varepsilon_2 \in (0, \varepsilon_0]$ and smooth functions $\hat{\theta}(\varepsilon), \hat{\gamma}_1(\varepsilon)$ such that $\hat{\theta}(0) = (2j-1)\pi/(m-k), \hat{\gamma}_1(0) = \gamma_1$, and*

$$\psi\big(1, \hat{\theta}(\varepsilon); \varepsilon\rho_1, \varepsilon\rho_1\hat{\gamma}_1(\varepsilon)\big) = 0 \tag{1.25}$$

holds for $(h, \lambda) = \big(h_c + \varepsilon^2 \hat{\mu}(\rho_1, \rho_1\hat{\gamma}_1(\varepsilon), \varepsilon), \lambda_c + \varepsilon^2 \hat{\nu}(\rho_1, \rho_1\hat{\gamma}_1(\varepsilon), \varepsilon)\big)$, $0 < \varepsilon < \varepsilon_2$.

(iii) *Given (ρ_1, ρ_2) satisfying $\rho_2/\rho_1 = \gamma_R$, there exist $\varepsilon_3 \in (0, \varepsilon_0]$ and smooth functions $\tilde{\theta}(\varepsilon), \tilde{\gamma}_R(\varepsilon)$ such that $\tilde{\theta}(0) = (2j-1)\pi/(m-k), \tilde{\gamma}_R(0) = \gamma_R$, and*

$$\psi\big(R, \tilde{\theta}(\varepsilon); \varepsilon\rho_1, \varepsilon\rho_1\tilde{\gamma}_R(\varepsilon)\big) = 0 \tag{1.26}$$

holds for $(h, \lambda) = \big(h_c + \varepsilon^2 \hat{\mu}(\rho_1, \rho_1\hat{\gamma}_R(\varepsilon), \varepsilon), \lambda_c + \varepsilon^2 \hat{\nu}(\rho_1, \rho_1\hat{\gamma}_R(\varepsilon), \varepsilon)\big)$, $0 < \varepsilon < \varepsilon_3$.

Remark 1.1. Theorem 1.4(i) tells that the solution $\psi(r, \theta; \varepsilon\rho_1, \varepsilon\rho_2)$ to (1.7)–(1.8) has j zeros in the 2-dimensional region

$$(h, \lambda) = \big(h_c + \varepsilon^2 \hat{\mu}(\rho_1, \rho_1\gamma, \varepsilon), \lambda_c + \varepsilon^2 \hat{\nu}(\rho_1, \rho_1\gamma, \varepsilon)\big), \quad (\gamma, \varepsilon) \in (\gamma_R, \gamma_1) \times (0, \varepsilon_1).$$

Note that ε_1 depends on γ. On the other hand Theorem 1.4(ii) asserts that the solution has a zero on the boundary $\{r = 1\}$ if the parameters lie on the curve

$$(h, \lambda) = (h_c, \lambda_c) + \varepsilon^2 \big(\hat{\mu}(\rho_1, \rho_2\hat{\gamma}_1(\varepsilon), \varepsilon), \hat{\nu}(\rho_1, \rho_2\hat{\gamma}_1(\varepsilon), \varepsilon)\big), \quad \varepsilon \in (0, \varepsilon_2)$$

emanating from (h_c, λ_c). Unfortunately in this case, if $m - k \ge 2$, it is not clear that all the curves for $j = 1, \dots, m-k$ coincide or not. Similarly, for the case (iii) the same problem is left.

Remark 1.2. Since $\psi^\star(r, \theta) = \psi(r, \theta + c; \alpha, \beta)$ is also a solution, $\psi^\star(r, \theta)$ has zeros $(r_j(\varepsilon), \theta_j(\varepsilon) - c)$ $(j = 1, \dots, m-k)$, for instance, in case (i) of Theorem 1.4.

Finally we discuss the stability. (1.5)–(1.6) is the stationary equation of the time dependent equation

$$\begin{cases} \psi_t = (\nabla - \mathrm{i}A_0)^2\psi + \lambda(1 - |\psi|^2)\psi, & (x, t) \in D \times (0, \infty), \\ \partial\psi/\partial\boldsymbol{n} = 0, & (x, t) \in \partial D \times (0, \infty), \\ \psi(x, 0) = \psi_0(x, 0), & x \in D. \end{cases} \tag{1.27}$$

This also gives the gradient flow of (1.4). Thus we can discuss the stability of a solution to (1.5)–(1.6) (or (1.7)–(1.8)) as an equilibrium solution to (1.27) or as a critical point of (1.4). A solution to (1.5) is stable (or unstable) if it is a stable (or unstable) equilibrium of (1.27). If the equilibrium solution is stable, it is also a local minimizer of (1.4). We note that the stability is meant by the sense of Lyapunov stability since a nontrivial equilibrium is not isolated by (1.22) (see [6] for the definition of the Lyapunov stability).

Corollary 1.5. *Let $\psi(\cdot,\cdot;\alpha,\beta)$ be the solution in Proposition* 1.2 *for $k = 0, m = 1$. Take the same scaling as in Theorem* 1.4. *Then for a sufficiently small $\varepsilon > 0$ the solution is unstable if $c_1c_2 < 4d^2$ while stable if $c_1c_2 > 4d^2$.*

2. Proof of Lemmas 1.1 and 1.3

By the Fourier expansion

$$\varphi = \sum_{k=-\infty}^{\infty} v_k(x)\exp(ik\theta)$$

to (1.12) gives the decomposition of the linearized operator $\mathcal{L}$ as follows:

$$\mathcal{L}_h^{(k)}[v_k] - \lambda v_k, \quad k = 0, \pm 1, \dots,$$

where $\mathcal{L}_h^{(k)}$ is defined by (1.11). Therefore when $\lambda = \mu^{(k)}(h)$, $\mathcal{L}$ has a zero eigenvalue and an eigenfunction $v^{(k)}(x;h,d)\exp(ik\theta)$ corresponding to it, where $\mu^{(k)}$ and $v^{(k)}$ are the first eigenvalue and the first eigenfunction of $\mathcal{L}_h^{(k)}$ respectively.

Proof of Lemma 1.1. First note that

$$0 = \sigma^{(0)}(0) < \sigma^{(k)}(0) < \sigma^{(m)}(0) \quad (0 < k < m).$$

Define the energy functional associated with the eigenvalue problem of $\mathcal{L}_h^{(k)}$:

$$E_k(v) := \int_1^R \{v_r^2 + (k/r - hr/2)^2 v^2\} r\, \mathrm{d}r.$$

Using

$$(m/r - hr/2)^2 - (k/r - hr/2)^2 = \{(m+k)/r - hr\}(m-k)/r,$$

we easily see that for any $v \in H^2\big((1,R);\mathbb{R}\big)$, $v \not\equiv 0$

$$\begin{cases} E_m(v) > E_k(v), & h \le (m+k)/R^2, \\ E_m(v) < E_k(v), & m+k \le h \end{cases}$$

because of $m > k$. This implies

$$\begin{cases} \sigma^{(m)}(h) > \sigma^{(k)}(h), & h \le (m+k)/R^2, \\ \sigma^{(m)}(h) < \sigma^{(k)}(h), & m+k \le h. \end{cases}$$

Thus the two curves C_k and C_m has an intersection point in the region $\{(h,\lambda) : (m+k)/R^2 < h < m+k, \lambda > 0\}$.

Since the first eigenvalue of $\mathcal{L}_h^{(k)}$ is simple, we can verify that $\sigma^{(k)}(h)$ and $v^{(k)}(\cdot;h)$ are differentiable in h. Differentiating $\mathcal{L}_h^{(k)}[v^{(k)}(\cdot;h)] = \sigma^{(k)}(h)v^{(k)}(\cdot;h)$ with respect to h yields

$$\mathcal{L}_h^{(k)}[\partial v^{(k)}/\partial h] - r(k/r - hr/2)v^{(k)} = (\mathrm{d}\sigma^{(k)}/\mathrm{d}h)v^{(k)} + \sigma^{(k)}\partial v^{(k)}/\partial h.$$

From the normalization of the eigenfunction (1.13),

$$\frac{\mathrm{d}\sigma^{(k)}}{\mathrm{d}h} = -k + \frac{h}{2}\int_1^R (v^{(k)})^2 r^3\,\mathrm{d}r \tag{2.1}$$

follows.

Noticing $\mathrm{d}\sigma^{(0)}/\mathrm{d}h > 0$ $(h > 0)$,

$$\frac{\mathrm{d}\sigma^{(m)}}{\mathrm{d}h} > 0\ (h > 2m), \quad \frac{\mathrm{d}\sigma^{(m)}}{\mathrm{d}h} < 0\ (h < 2m/R^2)$$

and

$$\begin{cases} \sigma^{(m)}(h) > \sigma^{(0)}(h), & h \le m/R^2, \\ \sigma^{(m)}(h) < \sigma^{(0)}(h), & m \ge h. \end{cases}$$

we easily see that C_0 and C_m intersect uniquely in the region $\{(h,\lambda) : m/R^2 < h < m, \lambda > 0\}$ if $R \le \sqrt{2}$. □

Proof of Lemma 1.3. We prove $v^{(k)}(r;h_c)/v^{(m)}(r;hc)$ $(m > k)$ is monotone decreasing in $r \in [1,R]$. Multiply $v^{(m)}$ and $v^{(k)}$ with the equations

$$\mathcal{L}^{(k)}_{h_c}[v^{(k)}(\cdot,h_c)] = \lambda_c v^{(k)}(\cdot,h_c),$$
$$\mathcal{L}^{(m)}_{h_c}[v^{(m)}(\cdot,h_c)] = \lambda_c v^{(m)}(\cdot,h_c)$$

respectively and take the difference. For $Q = v_r^{(k)}v^{(m)} - v^{(k)}v_r^{(m)}$ we obtain

$$Q_r + \frac{1}{r}Q = (1/r)(rQ)_r = -\{(k+m)/r - h_c r\}\{(m-k)/r\}v^{(m)}v^{(k)}. \tag{2.2}$$

Then Q satisfies

$$Q(1) = Q(R) = 0,\ Q_r(1) < 0,\ Q_r(r) \ne 0 \quad \text{unless } r = \sqrt{(m+k)/h_c}.$$

This implies $Q(r) < 0$ $(1 < x < R)$, that is, $(v^{(k)}/v^{(m)})_r < 0$.

Next consider the equation

$$\begin{cases} v^{(k)}(r;h_c) + \gamma v^{(m)}(r;h_c)\cos\big((m-k)\theta\big) = 0, \\ v^{(m)}(r)\sin\big((m-k)\theta\big) = 0. \end{cases}$$

From the second equation we obtain $(m-k)\theta = N\pi$, $N \in \mathbb{Z}$. Substituting this into the first equation yields

$$v^{(k)}(r;h_c)/v^{(m)}(r;h_c) = -\gamma\cos(N\pi) = -(-1)^N\gamma.$$

Thus it is easy to complete the proof of the lemma. □

3. Lyapunov – Schmidt reduction

Define

$$\langle \Psi_1, \Psi_2 \rangle := \int_D \Psi_1 \overline{\Psi_2}\,\mathrm{d}x \tag{3.1}$$

and

$$\|\Psi\| = \sqrt{\langle \Psi, \Psi \rangle}.$$

We denote

$$X := L^2(D;\mathbb{C}), \quad Y := \mathrm{Dom}(\mathcal{L}) \subset X.$$

Then we see

$$Y = \{u \in H^2(D) : u_r(1) = u_r(R) = 0\}.$$

Denoted by

$$\|u\|_Y = \|u\|_{H^2(D)} \quad (u \in Y).$$

For the operator

$$\mathcal{L}_c := -\big(\nabla - \mathrm{i}h_c(-x_2, x_1)\big)^2 - \lambda_c, \tag{3.2}$$

with the domain Y, it is easy to verify that

$$\langle \mathcal{L}_c[\Psi_1], \Psi_2\rangle = \langle \Psi_1, \mathcal{L}_c[\Psi_2]\rangle.$$

We define the projections P and Q from X into itself by

$$\begin{cases} P\Psi := \langle \Psi, \varphi_k\rangle\varphi_k + \langle \Psi, \varphi_m\rangle\varphi_m, \\ Q\Psi := \Psi - P\Psi, \end{cases} \tag{3.3}$$

where

$$\varphi_k(x) := v^{(k)}(r; h_c)\exp(\mathrm{i}k\theta), \quad \varphi_m(x) := v^{(m)}(r; h_c)\exp(\mathrm{i}m\theta). \tag{3.4}$$

Note that φ_k and φ_m are normalized as (1.16). We also recall that $\mathbb{C}$ is identified with $\mathbb{R}^2$. Therefore the subspace through the projection P is of real 4 dimensions. By the assumption of Proposition 1.2,

$$\mathcal{L}_c[\Psi] = f$$

has solutions $\Psi = \alpha\varphi_k + \beta\varphi_m + \hat{\Psi}$ ($(\alpha, \beta) \in \mathbb{C}^2$) if and only if $f \in QX$, where $\hat{\Psi}$ is the unique solution of

$$\mathcal{L}_c[\hat{\Psi}] = f \ (f \in QX), \quad P\hat{\Psi} = 0.$$

We define $\mathcal{L}_c^{-1} : QX \to QY$ by the correspondence $f \mapsto \hat{\Psi}$, which is a bounded operator.

Now we consider (1.7) – (1.8). We write (1.7) as

$$\mathcal{L}_c[\Psi] = F(\Psi, h, \lambda), \tag{3.5}$$

where

$$[F(\Psi, h, \lambda)](x, y) := -\mathrm{i}(h - h_c)\Psi_\theta - \frac{h^2 - h_c^2}{4}r^2\Psi + (\lambda - \lambda_c)\Psi - \lambda|\Psi|^2\Psi, \tag{3.6}$$

Set

$$\mu = h - h_c, \quad \nu = \lambda - \lambda_c, \tag{3.7}$$

and

$$\Psi = \Phi(\alpha, \beta) + \hat{\Psi}, \quad [\Phi(\alpha, \beta)](x, y) := \alpha\varphi_k(x, y) + \beta\varphi_m(x, y). \tag{3.8}$$

Then (3.5) is equivalent to

$$\hat{\Psi} = \mathcal{L}_c^{-1}QG(\alpha, \beta, \mu, \nu, \hat{\Psi}), \tag{3.9}$$

$$PG(\alpha, \beta, \mu, \nu, \hat{\Psi}) = 0, \tag{3.10}$$

where

$$\begin{aligned} G(\alpha, \beta, \mu, \nu, \hat{\Psi}) &:= F(\Phi(\alpha, \beta) + \hat{\Psi}, h_c + \mu, \lambda_c + \nu) \\ &= -\mathrm{i}\mu(\mathrm{i}k\alpha\varphi_k + \mathrm{i}m\beta\varphi_m) - \{(2h_c\mu + \mu^2)r^2/4\}\Phi(\alpha, \beta) \\ &\quad + \nu\Phi(\alpha, \beta) - \mathrm{i}\mu\hat{\Psi}_\theta - \{(2h_c\mu + \mu^2)r^2/4\}\hat{\Psi} \\ &\quad + \nu\hat{\Psi} - (\lambda_c + \nu)|\Phi(\alpha, \beta) + \hat{\Psi}|^2\{\Phi(\alpha, \beta) + \hat{\Psi}\}. \end{aligned} \tag{3.11}$$

Note that $G(0,0,0,0,0) = 0$ and

$$(\partial G/\partial\alpha)(0,0,0,0,0) = 0, \quad (\partial G/\partial\beta)(0,0,0,0,0) = 0, \quad (\partial G/\partial\hat{\Psi})(0,0,0,0,0) = 0.$$

Since $\mathbb{C}$ is identified with $\mathbb{R}^2$, the map $(\Psi, h, \lambda) \mapsto F(\Psi, h, \lambda) \in X$ is of C^∞. Applying the implicit function theorem to (3.9) yields positive numbers ε_0, δ_0 and ζ_0 such that for $|(\alpha,\beta)| < \varepsilon_0, |(\mu,\nu)| < \delta_0$ and $\|\hat{\Psi}\|_Y < \zeta_0$, there exists a unique solution $\hat{\Psi} = \hat{\Psi}(\alpha,\beta,\mu,\nu) \in Y$ to (3.9). Moreover $\hat{\Psi}$ is of C^∞ and satisfies

$$\hat{\Psi}(0,0,0,0) = 0, \quad \partial_J\hat{\Psi}(0,0,0,0) = 0 \ (J = \alpha,\beta,\mu,\nu). \tag{3.12}$$

Substituting $\hat{\Psi}(\alpha,\beta,\mu,\nu)$ into (3.10), we obtain equations for the parameters, called bifurcation equations:

$$H_1(\alpha,\beta,\mu,\nu) := \langle G(\alpha,\beta,\mu,\nu,\hat{\Psi}(\alpha,\beta,\mu,\nu),\varphi_k\rangle = 0, \tag{3.13}$$

$$H_2(\alpha,\beta,\mu,\nu) := \langle G(\alpha,\beta,\mu,\nu,\hat{\Psi}(\alpha,\beta,\mu,\nu),\varphi_m\rangle = 0. \tag{3.14}$$

Recall that the original equation is invariant under the transformations of (1.22). Applying the same argument in [3, Section 3], we obtain the following lemma.

Lemma 3.1. *The following properties hold:*

$$\begin{cases} H_1(\alpha e^{ic},\beta,\mu,\nu) = e^{ic}H_1(\alpha,\beta,\mu,\nu), \\ H_2(\alpha e^{ic},\beta,\mu,\nu) = H_2(\alpha,\beta,\mu,\nu), \end{cases} \qquad \forall c \in \mathbb{R}, \tag{3.15}$$

and

$$\begin{cases} H_1(\alpha,\beta e^{ic},\mu,\nu) = H_1(\alpha,\beta,\mu,\nu), \\ H_2(\alpha,\beta e^{ic},\mu,\nu) = e^{ic}H_2(\alpha,\beta,\mu,\nu), \end{cases} \qquad \forall c \in \mathbb{R}. \tag{3.16}$$

In addition

$$\overline{H_j(\alpha,\beta,\mu,\nu)} = H_j(\overline{\alpha},\overline{\beta},\mu,\nu), \quad j = 1,2. \tag{3.17}$$

The lemma tells that we can regard the map (H_1, H_2) as

$$(H_1,H_2)\colon (\alpha,\beta,\mu,\nu) \in \mathbb{R}^4 \mapsto \big(H_1(\alpha,\beta,\mu,\nu), H_2(\alpha,\beta,\mu,\nu)\big) \in \mathbb{R}^2.$$

Moreover, by (3.15) and (3.16) we can decompose the equations into

$$H_1(\alpha,0,\mu,\nu) = 0, \quad \beta = 0, \tag{3.18}$$

$$H_2(0,\beta,\mu,\nu) = 0, \quad \alpha = 0, \tag{3.19}$$

and

$$\big(\hat{H}_1(\alpha,\beta,\mu,\nu), \hat{H}_2(\alpha,\beta,\mu,\nu)\big) = (0,0), \tag{3.20}$$

where

$$\hat{H}_1(\alpha,\beta,\mu,\nu) := H_1(\alpha,\beta,\mu,\nu)/\alpha, \quad \hat{H}_2(\alpha,\beta,\mu,\nu) := H_2(\alpha,\beta,\mu,\nu)/\beta. \tag{3.21}$$

We notice that a bifurcating solution $\Psi \approx \alpha\varphi_k(x,y)$ is obtained by solving (3.18) while $\Psi \approx \beta\varphi_m(x,y)$ by (3.19). A solution with $\alpha\beta \neq 0$ of (3.20) corresponds to the desired solution with the form $\Psi \approx \alpha\varphi_k(x,y) + \beta\varphi_m(x,y)$.

4. Sketch of the proof for the main results

We use the scaling $\alpha = O(\varepsilon)$, $\beta = O(\varepsilon)$, $\mu = O(\varepsilon^2)$ and $\nu = O(\varepsilon^2)$ in the bifurcation equations (3.20). Recall (3.11). A straightforward computation leads to

$$H_1 = k\mu\alpha - (h_c\mu\alpha/2)\cdot 2\pi\int_1^R (v^{(k)})^2 r^3\,\mathrm{d}x + \nu\alpha$$
$$- \alpha^3\lambda_c\cdot 2\pi\int_0^1 (v^{(k)})^4 dr - 2\alpha\beta^2\lambda_c\cdot 2\pi\int_1^R (v^{(k)}v^{(m)})^2\,\mathrm{d}r + \text{h.o.t.} = 0,$$

and

$$H_2 = m\mu\beta - (h_c\mu\beta/2)\cdot 2\pi\int_1^R (v^{(m)})^2 r^3\,\mathrm{d}r + \nu\beta$$
$$- 2\alpha^2\beta\lambda_c\cdot 2\pi\int_0^1 (v^{(k)}w^{(m)})^2\,\mathrm{d}x - \beta^3\lambda_c\cdot 2\pi\int_0^1 (v^{(m)})^4\,\mathrm{d}x + \text{h.o.t.} = 0,$$

hence

$$\tag{4.1} \begin{cases} H_1(\alpha,\beta,\mu,\nu) = \{a_1\mu + \nu - c_1\alpha^2 - 2d\beta^2 + R_1(\alpha,\beta,\mu,\nu)\}\alpha, \\ H_2(\alpha,\beta,\mu,\nu) = \{a_2\mu + \nu - 2d\alpha^2 - c_2\beta^2 + R_2(\alpha,\beta,\mu,\nu)\}\beta, \end{cases}$$

where c_1, c_2 and d are defined by (1.16), and

$$R_j = O(\varepsilon^3), \quad j = 1,2.$$

In consequence we obtain the next expressions for (3.20)

$$\tag{4.2} \begin{cases} \hat{H}_1(\alpha,\beta,\mu,\nu) = a_1\mu + \nu - c_1\alpha^2 - 2d\beta^2 + R_1(\alpha,\beta,\mu,\nu) = 0, \\ \hat{H}_2(\alpha,\beta,\mu,\nu) = a_2\mu + \nu - 2d\alpha^2 - c_2\beta^2 + R_2(\alpha,\beta,\mu,\nu) = 0. \end{cases}$$

By the condition $a_1 < a_2$, (4.2) can be expressed as

$$\tag{4.3} \begin{pmatrix}\mu\\ \nu\end{pmatrix} = \mathbb{F}(\mu,\nu,\alpha,\beta) := \begin{pmatrix}a_1 & 1\\ a_2 & 1\end{pmatrix}^{-1}\begin{pmatrix}c_1\alpha^2 + 2d\beta^2 - R_1(\alpha,\beta,\mu,\nu)\\ 2d\alpha^2 + c_2\beta^2 - R_2(\alpha,\beta,\mu,\nu)\end{pmatrix}.$$

Applying the contraction mapping theorem to $\mathbb{F}(\cdot,\cdot,\alpha,\beta) : \mathbb{R}^2 \mapsto \mathbb{R}^2$ yields a solution $(\mu,\nu) = \big(\mu(\alpha,\beta),\nu(\alpha,\beta)\big)$ of (4.3) for given (α,β) with $|(\alpha,\beta)| \ll 1$ and the solution is unique in a small neighborhood of $(\mu,\nu,\alpha,\beta) = (0,0,0,0)$. We see that $\big(\mu(\alpha,\beta),\nu(\alpha,\beta)\big)$ is of C^∞ since so is $\mathbb{F}(\mu,\nu,\alpha,\beta)$.

Now we set

$$\tag{4.4} \alpha = \varepsilon\rho_1, \quad \beta = \varepsilon\rho_2, \quad \mu = \varepsilon^2\hat{\mu}, \quad \nu = \varepsilon^2\hat{\nu}$$

and define

$$\tag{4.5} \begin{cases} \hat{K}_1(\hat{\mu},\hat{\nu},\rho_1,\rho_2,\varepsilon) := \hat{H}_1(\varepsilon\rho_1,\varepsilon\rho_2,\varepsilon^2\hat{\mu},\varepsilon^2\hat{\nu})/\varepsilon^2, \\ \hat{K}_2(\hat{\mu},\hat{\nu},\rho_1,\rho_2,\varepsilon) := \hat{H}_2(\varepsilon\rho_1,\varepsilon\rho_2,\varepsilon^2\hat{\mu},\varepsilon^2\hat{\nu})/\varepsilon^2. \end{cases}$$

Let $\big(\tilde{\mu}(\rho_1,\rho_2),\tilde{\nu}(\rho_1,\rho_2)\big)$ be a solution of

$$\tag{4.6} \hat{K}_1(\tilde{\mu},\tilde{\nu},\rho_1,\rho_2,0) = 0, \quad \hat{K}_2(\tilde{\mu},\tilde{\nu},\rho_1,\rho_2,0) = 0.$$

Note that the Jacobi matrix

$$\begin{pmatrix}\partial_{\hat{\mu}}\hat{K}_1 & \partial_{\hat{\nu}}\hat{K}_1\\ \partial_{\hat{\mu}}\hat{K}_2 & \partial_{\hat{\nu}}\hat{K}_2\end{pmatrix}_{|(\hat{\mu},\hat{\nu},\rho_1,\rho_2,\varepsilon)=(\tilde{\mu},\tilde{\nu},\rho_1,\rho_2,0)} = \begin{pmatrix}a_1 & 1\\ a_2 & 1\end{pmatrix}$$

is invertible, so the implicit function theorem can be applied to the equations to obtain a solution $(\hat{\mu}(\rho_1,\rho_2,\varepsilon),\hat{\nu}(\rho_1,\rho_2,\varepsilon))$ of $(\hat{K}_1,\hat{K}_2)=(0,0)$ with the properties that

$$\hat{\mu}(\rho_1,\rho_2,\varepsilon)=\tilde{\mu}(\rho_1,\rho_2)+O(\varepsilon),\quad \hat{\nu}(\rho_1,\rho_2,\varepsilon)=\tilde{\nu}(\rho_1,\rho_2)+O(\varepsilon). \tag{4.7}$$

Consequently, we complete the proof of Proposition 1.2.

Next investigate how the vortex solution takes place. Set

$$\begin{cases} V_0:=\{(\mu,\nu):\big(\hat{K}_1(\mu,\nu,\rho_1,\rho_2,0),\hat{K}_2(\mu,\nu,\rho_1,\rho_2,0)\big)=(0,0), \\ \qquad\qquad\qquad\qquad\qquad\qquad\qquad \gamma_R<\rho_2/\rho_1<\gamma_1\}, \\ \ell_1:=\{(\mu,\nu):\big(\hat{K}_1(\mu,\nu,\rho_1,,\rho_2,0),\hat{K}_2(\mu,\nu,\rho_1,\rho_2,0)\big)=(0,0), \\ \qquad\qquad\qquad\qquad\qquad\qquad\qquad \rho_2/\rho_1=\gamma_1\}, \\ \ell_R:=\{(\mu,\nu):\big(\hat{K}_1(\mu,\nu,\rho_1,\rho_2,0),\hat{K}_2(\mu,\nu,\rho_1,\rho_2,0)\big)=(0,0), \\ \qquad\qquad\qquad\qquad\qquad\qquad\qquad \rho_2/\rho_1=\gamma_R\}, \end{cases} \tag{4.8}$$

where γ_1 and γ_R were defined in (1.23). For $\gamma\in[\gamma_R,\gamma_1]$, letting

$$\tilde{\Phi}(r,\theta;\gamma,\varepsilon):=\exp(-\mathrm{i}k\theta)\Psi(r,\theta;\varepsilon\rho_1,\varepsilon\rho_1\gamma)/\varepsilon\rho_1,$$

we see from Lemma 1.1 that

$$\tilde{\Phi}(r_0,\theta_j;\gamma,0)=v^{(k)}(r_0;h_c)+\gamma v^{(m)}(r_0;h_c)\exp(\mathrm{i}(m-k)\theta_j)=0,$$

where $\theta_j:=(2j-1)\pi/(m-k)$, Moreover, if $\gamma\in(\gamma_R,\gamma_1)$,

$$\partial_r\tilde{\Phi}(r_0,\theta_j;\gamma,0)=v_r^{(k)}(r_0)+\gamma v_r^{(m)}(r_0)\exp(\mathrm{i}(m-k)\theta_j)=v_r^{(k)}(r_0)-\gamma v_r^{(m)}(r_0),$$
$$\partial_\theta\tilde{\Phi}(r_0,\theta_j;\gamma,0)=\mathrm{i}(m-k)\gamma v^{(m)}(r_0)\exp(\mathrm{i}(m-k)\theta_j)=-\mathrm{i}(m-k)\gamma v^{(m)}(r_0).$$

In view of $\gamma=v^{(k)}(r_0)/v^{(m)}(r_0)$,

$$\partial_r\tilde{\Phi}(r_0,\theta_j;\rho_1,\rho_1\gamma)=\{v_r^{(k)}(r_0)v^{(m)}(r_0)-v^{(k)}(r_0)v_r^{(m)}(r_0)\}/v^{(m)}(r_0)\neq 0.$$

Thus the implicit function theorem yields (i) of Theorem 1.4.

In case $\gamma=\gamma_1$,

$$\tilde{\Phi}(1,\theta_j;\gamma_1,0)=v^{(k)}(1;h_c)-\gamma_1 v^{(m)}(1;h_c)=0.$$

Since

$$\partial_\theta\tilde{\Phi}(1,\theta_j;\gamma_1,0)=-\mathrm{i}(m-k)\gamma_1 v^{(m)}(1;h_c)$$
$$\partial_\gamma\tilde{\Phi}(1,\theta_j;\gamma_1,0)=-v^{(m)}(1;h_c),$$

we again apply the implicit function theorem to obtain (ii) of the theorem. The case of (iii) can be treated similarly. This completes the proof of Theorem 1.4.

5. Sketch of the proof for Corollary 1.5

In this section we discuss the stability of the bifurcating solution in a neighborhood of the intersection point of C_0 and C_1 by applying a standard center manifold theory.

Recall $\mathbb{C}$ is identified with $\mathbb{R}^2$. Solutions of (1.27) generate a smooth semiflow in $X_1:=H^1(D:\mathbb{C})$ (see [6]). It is easy to verify that

$$\frac{\mathrm{d}}{\mathrm{d}t}E(\psi(\cdot,\cdot,t))=-\int_D|\psi_t|^2\,\mathrm{d}x$$

and an equilibrium state of the flow is a solution of (1.7)–(1.8).

Write (1.27) as

$$\psi_t = -\mathcal{L}_c[\psi] + F(\psi, h, \lambda), \tag{5.1}$$

where $\mathcal{L}_c$ and F are defined by (3.2) and (3.6) respectively. We decompose (5.1) through the projection (3.3). Put

$$\psi = P\psi + Q\psi = z_0\varphi_0(x,y) + z_1\varphi_1(x,y) + \hat{\psi},$$
$$z_j = \langle \psi, \varphi_j \rangle, \quad j = 0, 1.$$

With (3.7) and (3.11), we see (5.1) is decomposed into

$$\dot{z}_0 = \langle G(z_0, z_1, \mu, \nu, \hat{\psi}), \varphi_0 \rangle, \tag{5.2}$$

$$\dot{z}_1 = \langle G(z_0, z_1, \mu, \nu, \hat{\psi}), \varphi_1 \rangle, \tag{5.3}$$

$$\hat{\psi}_t = -\mathcal{L}_c[\hat{\psi}] + QG(z_0, z_1, \mu, \nu, \hat{\psi}), \tag{5.4}$$

where the dot stands for the time derivative. Since the least eigenvalue of $\mathcal{L}_c$ restricted on QX is positive, there exist

$$\mathcal{U} := \{(z_0, z_1, \mu, \nu, \hat{\psi}) : |(z_0, z_1, \mu, \nu)| < \delta_1, \|\hat{\psi}\|_{X_1} < \delta_2\}$$

and a smooth map $K(z_0, z_1, \mu, \nu)$ such that the graph of K is contained in $\mathcal{U}$ and the flow reduced on the manifold is given by

$$\begin{cases} \dot{z}_0 = f_0(z_0, z_1, \mu, \nu) := \langle G\big(z_0, z_1, \mu, \nu, K(z_0, z_1, \mu, \nu)\big), \varphi_0 \rangle, \\ \dot{z}_1 = f_1(z_0, z_1, \mu, \nu) := \langle G\big(z_0, z_1, \mu, \nu, K(z_0, z_1, \mu, \nu)\big), \varphi_1 \rangle. \end{cases} \tag{5.5}$$

By the attractivity of the manifold and the invariant foliation along the manifold, if (5.5) has a nondegenerate stable (or an unstable) equilibrium (z_0^*, z_1^*) such that

$$\big(z_0^*, z_1^*, \mu, \nu, K(z_0^*, z_1^*, \mu, \nu)\big) \in \mathcal{U},$$

then $\psi^* := z_0^*\varphi_0 + z_1^*\varphi_1 + K(z_0^*, z_1^*, \mu, \nu)$ gives a stable (or an unstable) equilibrium of (1.27) (see [4–6]).

We may consider only the case that z_0, z_1 are real. Indeed, as for the complex case, we obtain corresponding solutions by simply putting $z_0 e^{ic_0}, z_1 e^{ic1}$ $(c_0, c_1 \in \mathbb{R})$. We also obtain, by the same computation as done for the bifurcation equations and the scaling $z_j = O(\varepsilon), \mu, \nu = O(\varepsilon^2)$, the explicit expression of (5.5) as follows:

$$\begin{cases} \dot{z}_0 = \{a_1\mu + \nu - c_1 z_0^2 - 2dz_1^2 + R_0(z_0, z_1, \mu, \nu)\}z_0, \\ \dot{z}_1 = \{a_2\mu + \nu - 2dz_0^2 - c_2 z_1^2 + R_1(z_0, z_1, \mu, \nu)\}z_1, \end{cases} \tag{5.6}$$

where $R_j = O(\varepsilon^3)$, $j = 0, 1$.

Introducing the new variables $u = z_0^2$, $v = z_1^2$ and neglecting the higher order terms $R_j, j = 0, 1$ in (5.6) yield the following Lotka–Voltera system

$$\begin{cases} \dot{u} = 2(\kappa_0 - c_1 u - 2dv)u, \\ \dot{v} = 2(\kappa_1 - 2du - c_2 v)v, \end{cases} \tag{5.7}$$

where

$$\kappa_0 := a_1\mu + \nu, \quad \kappa_1 = a_2\mu + \nu.$$

The dynamical structure of the system (5.7) in the region $\{(u, v) : 0 \leq u, 0 \leq v\}$ is well known. As a matter of fact the equations have equilibria

$$\begin{cases} P_{00} := (0,0), \\ P_0 := (\kappa_1/c_1, 0) \quad \text{if } \kappa_0 > 0, \\ P_1 := (0, \kappa_2/c_2) \quad \text{if } \kappa_1 > 0, \end{cases} \tag{5.8}$$

and

$$P_c := \left(\frac{c_2\kappa_0 - 2d\kappa_1}{c_1c_2 - 4d^2}, \frac{-2d\kappa_0 + c_1\kappa_1}{c_1c_2 - 4d^2} \right) \tag{5.9}$$

if

$$c_1c_2 < 4d^2, \quad c_1/2d < \kappa_1/\kappa_0 < 2d/c_2, \tag{5.10}$$

or if

$$c_1c_2 > 4d^2, \quad 2d/c_1 < \kappa_1/\kappa_0 < c_2/2d \tag{5.11}$$

in the region $\{(u, v) : 0 \leq u, 0 \leq v\}$.

We remark that the equilibria P_0 and P_1 correspond to the solutions bifurcating from the trivial solution at the curves C_0 and C_1 respectively. On the other hand P_c corresponds to the solution obtained in Proposition 1.2.

The following stability results are easily verified by the phase plane analysis of (5.7):

(1) $\kappa_0, \kappa_1 < 0$: only P_{00} exists and it is asymptotically stable.
(2) $\kappa_1 < 0 < \kappa_0$: P_{00} and P_0 exist. P_{00} is unstable while P_0 is asymptotically stable.
(3) $\kappa_0 < 0 < \kappa_1$: P_{00} and P_1 exist. P_{00} is unstable while P_1 is asymptotically stable.
(4) $\kappa_0, \kappa_1 > 0$:
  (A) $c_1c_2 < 4d^2$:
    (i) $0 < 2d/c_1 < \kappa_1/\kappa_0$: P_{00}, P_0 and P_1 exist. P_{00} and P_0 are unstable while P_1 is asymptotically stable.
    (ii) $0 < \kappa_1/\kappa_0 < c_1/2d$: P_{00}, P_0 and P_1 exist. P_{00} and P_1 are unstable while P_0 is asymptotically stable.
    (iii) $c_2/2d < \kappa_1/\kappa_0 < 2d/c_1$: P_{00}, P_0, P_1 and P_2 exist. P_{00} and P_c unstable while P_0 and P_1 are asymptotically stable.
  (B) $c_1c_2 > 4d^2$:
    (i) $0 < c_2/2d < \kappa_1/\kappa_0$: P_{00}, P_0 and P_1 exist. P_{00} and P are unstable while P_1 is asymptotically stable.
    (ii) $0 < \kappa_1/\kappa_0 < 2d/c_2$: P_{00}, P_0 and P_1 exist. P_{00} and P_1 are unstable while P_0 is asymptotically stable.
    (iii) $2d/c_1 < \kappa_1/\kappa_0 < c_2/2d$: P_{00}, P_0, P_1 and P_2 exist. P_{00}, P_0 and P_1 are unstable while P_c is asymptotically stable.

Using the results in (4)(A)(iii) and (4)(B)(iii), we can obtain the desired stability in Corollary 1.5.

References

1. T. Boeck and S. J. Chapman, *Bifurcation to vortex solutions in superconducting films*, European J. Appl. Math. **8** (1997), no. 2, 125–148.

2. S. J. Chapman, Q. Du, and M. D. Gunzberger, *A model for variable thickness superconducting thin films*, Z. Angew. Math. Phys. **47** (1996), no. 3, 410–431.
3. C.-N. Chen and Y. Morita, *Bifurcation of vortex and boundary-vortex solutions in a Ginzburg–Landau model*, Nonlinearity **20** (2007), 943–964.
4. S.-N. Chow, X.-B. Lin, and K. Lu, *Smooth invariant foliations in infinite-dimensional spaces*, J. Differential Equations **94** (1991), no. 2, 266–291.
5. S.-N. Chow and K. Lu, *Invariant manifolds for flows in Banach spaces*, J. Differential Equations **74** (1988), no. 2, 285–317.
6. D. Henry, *Geometric theory of semilinear parabolic equations*, Lecture Notes in Math., vol. 840, Springer, New York, 1981.
7. S. Jimbo and Y. Morita, *Ginzburg–Landau equation with magnetic effect in a thin domain*, Calc. Var. Partial Differential Equations **15** (2002), no. 3, 325–352.
8. Y. Morita, *Stable solutions to the Ginzburg–Landau equation with magnetic effect in a thin domain*, Japan J. Indust. Appl. Math. **21** (2004), no. 2, 129–147.
9. P. Takáč, *Bifurcations and vortex formation in the Ginzburg–Landau equations*, ZAMM Z. Angew. Math. Mech. **81** (2001), 523–539.

Department of Applied Mathematics and Informatics, Ryukoku University, Seta, Otsu 520-2194, Japan

E-mail address: morita@rins.ryukoku.ac.jp

Centre de Recherches Mathématiques
CRM Proceedings and Lecture Notes
Volume **44**, 2008

An Allen – Cahn Type Problem with Curvature Modification

Xiaofeng Ren

ABSTRACT. A two component system driven by both interface area and interface curvature is studied with a new phase field model. The Euler – Lagrange equation derived from the free energy functional of the system is a fourth order nonlinear partial differential equation. Formal asymptotic analysis shows that if the curvature impact in the system is strong, there exists a bubble profile in each space dimension. A bubble profile describes a pattern of an inner core of one component surround by an outer membrane of the other component. There are four distinct cases for dimension equal to 1, 2, 3, or 4 and larger. Outlines of the rigorous proofs of the existence theorems are given, based on the formal asymptotic analysis.

1. Introduction

In a two component system of a condensed matter, when the conformation of the two components is solely determined by the area of the interfaces separating the two components, the free energy of the system is often described by the Allen – Cahn model [1]

$$I_{\mathrm{ac}}(u) = \int_D \left[\frac{\varepsilon^2}{2}|\nabla u|^2 + F(u)\right] \mathrm{d}x, \tag{1.1}$$

where F is a balanced double well potential such as $F(u) = \frac{1}{4}u^2(1-u)^2$, and D is a region in $\mathbb{R}^n$. The equilibria of I_{ac} are solutions of the second-order nonlinear PDE

$$-\varepsilon^2\Delta u(x) + f\big(u(x)\big) = 0,\ x \in D, \quad \partial_\nu u(x) = 0,\ x \in \partial D \tag{1.2}$$

where ∂_ν is the outward normal derivative on ∂D. The parameter ε is positive and small. The nonlinear function f is the derivative of F, such as $f(u) = u(u-\frac{1}{2})(u-1)$ if F is the example given above.

If $u(x)$ is close to 0, then the first component occupies x; if $u(x)$ is close to 1, then the second component occupies x. These two regions are separated by a transitional region where $u(x)$ is somewhat greater than 0 and less than 1. We may call the set $\{x \in D : u(x) = \frac{1}{2}\}$, which is usually an $(n-1)$-dimensional surface, or a union of several surfaces, in $\mathbb{R}^n$, the interface. Given a configuration $u(x)$ with x

2000 *Mathematics Subject Classification.* Primary ??; Secondary ??
The work of the author is supported in part by NSF grant DMS-0509725.
This is the final form of the paper.

on the interface, we may roughly interpret $-\varepsilon^2\Delta u + f(u)$ as the mean curvature of the interface at x. The equation (1.2) then states that at an equilibrium state, the mean curvature of the interface must be everywhere equal to 0.

In such an interface area driven system, it is difficult for the two components to co-exist. Casten and Holland [3] (and Matano [10] independently) showed that when D is bounded and convex, any nonconstant solution of (1.2) must be unstable.

More recently in the study of polymer blends (see Tang and Freed [18]) an additional molecular weight dependent curvature term is found to contribute to the free energy. In this case one observes two immiscible homopolymers, one forming an outer membrane and the other constituting an inner core.

This morphology pattern may be explained phenomenologically by a very simple model. Suppose that the two homopolymers are separated by a closed curve Γ in $\mathbb{R}^2$. We propose that the free energy I_c of the system is given by

$$I_c(\Gamma) = \int_\Gamma \kappa^2 \, ds + \gamma \int_\Gamma ds, \tag{1.3}$$

where s is the length element, κ is the curvature and $\gamma > 0$ is a parameter. If we assume that Γ is a circle of radius ρ, then the curvature is everywhere $1/\rho$ and (1.3) becomes

$$\frac{2\pi}{\rho} + 2\pi\gamma\rho, \tag{1.4}$$

A stable configuration is obtained by minimizing (1.4) with respect to ρ. One finds that

$$\rho = \frac{1}{\sqrt{\gamma}}. \tag{1.5}$$

In this paper we study a more sophisticated phase field version of (1.3). As in the Allen-Cahn approach we let u be the phase field variable of a two component system. Again $u(x) \approx 0$ means that x is taken by one component; $u(x) \approx 1$ means that x is taken by the other component. The free energy of the system is now

$$I(u) = \frac{1}{2}\int_D |\Delta u - f(u)|^2 \, dx + \gamma \int_D [\tfrac{1}{2}|\nabla u|^2 + F(u)] \, dx. \tag{1.6}$$

Here $\Delta u - f(u)$ plays the role of curvature and $\frac{1}{2}|\nabla u|^2 + F(u)$ plays the role of length element. The constant $1/2$ in front of the first integral is put there for simplicity later. Also note that γ is the only parameter in I. It gives the relative weights of the two parts in I. There is no ε in (1.6) because one can always scale out ε by changing the x variable.

The connection between I and I_c is not very clear at the moment. There is a well-known relationship between I_{ac} and $\int_\Gamma ds$, i.e., I_c without the curvature part: $\int_\Gamma ds$ is the Gamma-limit of I_{ac} to as $\varepsilon \to 0$. See De Giorgi [6], Modica and Mortola [12], Modica [11], Kohn and Sternberg [9], etc., for this theory. We do not know if a Gamma-convergence theory between I and I_c is available. The curvature part of I_c, i.e., $\int_\Gamma \kappa^2 \, ds$, is known as the Willmore functional [19] (also see Simon [17]). There are some partial results regarding the convergence of

$$\int_D [-\varepsilon^2\Delta u + f(u)]^2 \, dx \tag{1.7}$$

to the Willmore functional (see Moser [13]).

We will study (1.6) in the general case of n dimensions, i.e., $D \subset \mathbb{R}^n$ with n being a positive integer. Although in (1.3) we have assumed that γ is positive when $n = 2$, here we allow γ to be negative if $n \geq 3$. We are only interested in the situation where (1.6) is sufficiently different from (1.1), so we assume that $|\gamma|$ is small.

The Euler–Lagrange equation of (1.6) is a fourth-order partial differential equation

$$\Delta(\Delta u - f(u)) - f'(u)(\Delta u - f(u)) - \gamma(\Delta u - f(u)) = 0. \tag{1.8}$$

If D has a boundary, then we impose the Neumann boundary conditions

$$\frac{\partial u}{\partial \nu} = \frac{\partial(\Delta u - f(u))}{\partial \nu} = 0 \quad \text{on } \partial D. \tag{1.9}$$

If we introduce a new variable $v = \Delta u - f(u)$, then (1.8) may be written as a system

$$\Delta u - f(u) - v = 0, \ \Delta v - f'(u)v - \gamma v = 0 \quad \text{in } D. \tag{1.10}$$

If D has a boundary, then u and v both should satisfy the Neumann boundary condition there.

We are interested in the outer membrane/inner core pattern mentioned earlier, in the phase field model (1.6). More specifically we seek a radially symmetric solution of (1.8). The domain D is the entire space $\mathbb{R}^n$. We require that the solution $u = u(|x|) = u(r)$ satisfy the conditions

$$u(0) > 0, \quad \lim_{r \to \infty} u(r) = 0. \tag{1.11}$$

We often call such a solution a bubble profile. Recall that γ is either positive or negative, but $|\gamma|$ is sufficiently small. This means that the curvature term in the free energy (1.6) is significant and the problem is very different from the Allen-Cahn problem (1.1). Note that the Allen–Cahn problem does not have a bubble profile solution.

In this paper we give a detailed description of the profile in each dimension using formal asymptotic analysis. The reader will see from our intuitive argument how the radius of a bubble profile is determined by the parameter γ in (1.6).

It turns out that our asymptotic analysis is very sensitive to the dimension n of the underlying space. The $n = 2$ and $n \geq 4$ cases are easier because one needs only to track quantities that are of algebraically small orders, measured by the bubble radius ρ. The results one obtains can be guessed from I_c. However in the $n = 1$ and $n = 3$ cases, one must study quantities that are exponentially small in terms of ρ. One can not derive any useful information from I_c. We study $n = 2$ case in Section 2, $n \geq 4$ case in Section 3, $n = 1$ case in Section 4, and $n = 3$ case in Section 5.

The formal asymptotic analysis leads to rigorous proofs of the existence of bubble profiles by Ren and Wei [16]. Bubble profiles exist in 1 dimension and 2 dimensions if γ is positive and sufficiently small. Bubble profiles exist, in $n = 3$ and $n \geq 4$ dimensions if γ is negative and sufficiently close to 0. In Section 6 we outline some key steps used in [16]. Many of the intricate constructions there are motivated by the formal study given here.

2. $n = 2$

Let (u, v) be a solution of (1.10). With u and v being radial in $\mathbb{R}^n$ the system (1.10) becomes

$$u'' + \frac{(n-1)u'}{r} - f(u) = v \tag{2.1}$$

$$v'' + \frac{(n-1)v'}{r} - f'(u)v = \gamma v \tag{2.2}$$

where u and v are now functions of $r = |x| \in (0, \infty)$. There are boundary conditions at $r = 0$:

$$u'(0) = 0 \tag{2.3}$$

$$v'(0) = 0. \tag{2.4}$$

One important quantity to analyze is the location of the interface of u. Define the interface location $\rho \in (0, \infty)$ to be the point where

$$u(\rho) = \tfrac{1}{2}. \tag{2.5}$$

In this section we consider the case $n = 2$. It turns out that it suffices to study the behavior of u and v near the interface point ρ. The leading order approximation of u near ρ is the function H which is the solution of

$$-H'' + f(H) = 0 \text{ in } \mathbb{R}; \quad H(-\infty) = 1, \; H(\infty) = 0, \; H(0) = \tfrac{1}{2}. \tag{2.6}$$

In the special case $f(H) = H(H - \frac{1}{2})(H - 1)$, H is known explicitly,

$$H(y) = \frac{1}{2}\left[\tanh\left(-\frac{y}{2\sqrt{2}}\right) + 1\right]. \tag{2.7}$$

We define a constant τ which is often called the surface tension from H:

$$\tau = \int_{\mathbb{R}} \left(H'(y)\right)^2 \mathrm{d}y. \tag{2.8}$$

Therefore the first approximation of u is found:

$$u(\rho + y) = H(y) + \cdots. \tag{2.9}$$

In the next step we use H for u in (2.2) and look for the leading order approximation of v which we denote by v_1:

$$v(\rho + y) = \frac{v_1(y)}{\rho} + \cdots \tag{2.10}$$

Then v_1 must satisfy the following equation.

$$v_1'' - f'(H)v_1 = 0. \tag{2.11}$$

This implies that

$$v_1(y) = c_1 H'(y) \tag{2.12}$$

where c_1 is to be determined.

Now the next order approximation of u is sought. Write

$$u(\rho + y) = H(y) + \frac{u_1(y)}{\rho} + \cdots. \tag{2.13}$$

Use $v_1 = c_1 H'$ for v in (2.1) to find the equation for u_1:

$$u_1'' - f'(H)u_1 + H' = c_1 H', \qquad u_1(0) = 0. \tag{2.14}$$

Here $u_1(0) = 0$ since $u(\rho) = H(0) = \frac{1}{2}$. If one multiplies the last equation by H' and integrates over $(-\infty, \infty)$, then integration by parts yields that

$$c_1 = 1, \tag{2.15}$$

and $u_1 = c_2 H'$. Therefore because $u_1(0) = 0$ and $H'(0) \neq 0$,

$$u_1 = 0. \tag{2.16}$$

With $c_1 = 1$ one also derives that

$$v_1(y) = H'(y). \tag{2.17}$$

We move to the next order by setting

$$v = \frac{H'}{\rho} + \frac{v_2}{\rho^2} + \cdots . \tag{2.18}$$

With u set to be H in (2.2) one finds the equation for v_2:

$$v_2'' - f'(H)v_2 + H'' = \gamma_1 H' \tag{2.19}$$

where γ_1 is the leading order approximation of γ:

$$\gamma = \frac{\gamma_1}{\rho} + \cdots . \tag{2.20}$$

Multiplying (2.19) by H' and integrating over $(-\infty, \infty)$ show that

$$\gamma_1 = 0 \tag{2.21}$$

and v_2 satisfies

$$v_2'' - f'(H)v_2 + H'' = 0. \tag{2.22}$$

The last equation can be solved to give

$$v_2 = -\frac{yH'}{2} + c_3 H' \tag{2.23}$$

where c_3 is to be determined. The expansion for u is now

$$u = H' + \frac{u_2}{\rho^2} + \cdots . \tag{2.24}$$

The equation for u_2 is

$$u_2'' - f'(H)u_2 - yH' = v_2. \tag{2.25}$$

Multiplying (2.25) by H' and integrating show that the constant c_2 in (2.23) must vanish and

$$v_2 = -\frac{yH'}{2}. \tag{2.26}$$

Then (2.25) becomes

$$u_2'' - f'(H)u_2 = \frac{yH'}{2}. \tag{2.27}$$

We now come to the last order. Let

$$v = \frac{H'}{\rho} - \frac{yH'}{2\rho^2} + \frac{v_3}{\rho^3} + \cdots . \tag{2.28}$$

The equation for v_3 is

$$v_3'' - f'(H)v_3 - yH'' + v_2' - f''(H)u_2 H' = \gamma_2 H' \tag{2.29}$$

where γ_2 is the second-order approximation of γ:

$$\gamma = \frac{\gamma_2}{\rho^2} + \cdots . \tag{2.30}$$

Multiplying by H' and integrating yield that

$$-\int_{\mathbb{R}} yH''H' + \int_{\mathbb{R}} v_2' H' - \int_{\mathbb{R}} f''(H)(H')^2 u_2 = \gamma_2 \int_{\mathbb{R}} (H')^2 \tag{2.31}$$

Differentiating the equation (2.27) for u_2 leads to

$$u_2''' - f'(H)u_2' - f''(H)H'u_2 = \tfrac{1}{2}(H' + yH''). \tag{2.32}$$

Multiplying by H' and integrating show that

$$-\int_{\mathbb{R}} f''(H)(H')^2 u_2 = \int_{\mathbb{R}} \frac{1}{2}(H' + yH'')H'. \tag{2.33}$$

Substituting (2.33) back to (2.31) yields

$$\gamma_2 \int_{\mathbb{R}} (H')^2 = -\int_{\mathbb{R}} yH''H' + \int_{\mathbb{R}} v_2' H' + \frac{1}{2}\left(\int_{\mathbb{R}} (H')^2 + \int_{\mathbb{R}} yH'H'' \right). \tag{2.34}$$

Note that

$$\int_{\mathbb{R}} yH''H' = -\frac{1}{2}\int_{\mathbb{R}} (H')^2 = -\frac{\tau}{2} \tag{2.35}$$

for integration by parts shows that

$$\int_{\mathbb{R}} yH'H'' = \int_{\mathbb{R}} yH' \, \mathrm{d}H' = -\int_{\mathbb{R}} H'(H' + yH'') = -\int_{\mathbb{R}} (H')^2 - \int_{\mathbb{R}} yH'H''. \tag{2.36}$$

Moreover

$$\int_{\mathbb{R}} v_2' H' = \int_{\mathbb{R}} -\left(\frac{yH'}{2}\right)' H' = -\frac{1}{2}\int_{\mathbb{R}} (H' + yH'')H' = -\frac{1}{4}\int_{\mathbb{R}} (H')^2 = -\frac{\tau}{4}. \tag{2.37}$$

We find from (2.34) that

$$\gamma_2 = \tfrac{1}{2}. \tag{2.38}$$

There is enough information now about the presumed solution. Particularly we know that

$$\gamma = \frac{1}{2\rho^2} + \cdots . \tag{2.39}$$

The last equation should be read backwards so that

$$\rho = \frac{1}{\sqrt{2\gamma}} + \cdots . \tag{2.40}$$

This tells us where the interface of the solution should be. Note that γ must be a positive number.

In two dimensions, the phase field problem (1.6) is consistent with the simple model (1.3). We have the following existence theorem.

Theorem 2.1 ([16]). *When γ is positive and sufficiently small, there exists a bubble profile. The radius of the bubble is $1/\sqrt{2\gamma} + o(\gamma^{-1/2})$.*

3. $n \geq 4$

The asymptotic analysis of a radial solution of (1.8) in the case $n \geq 4$ is similar to that in the case $n = 2$. Again near ρ we can expand u, v and γ to

$$u(\rho + y) = H(y) + \frac{u_1}{\rho} + \frac{u_2}{\rho^2} + \cdots \tag{3.1}$$

$$v(\rho + y) = \frac{v_1(y)}{\rho} + \frac{v_2(y)}{\rho^2} + \frac{v_3(y)}{\rho^3} + \cdots \tag{3.2}$$

$$\gamma = \frac{\gamma_1}{\rho} + \frac{\gamma_2}{\rho^2} \cdots . \tag{3.3}$$

Similar calculations show that

$$v_1 = (n-1)H'; \quad u_1 = 0; \quad \gamma_1 = 0; \quad v_2 = -\frac{(n-1)^2 y H'}{2}; \tag{3.4}$$

the equation for u_2 is

$$u_2'' - f'(H)u_2 = \frac{(n-1)(3-n)}{2} y H', \quad u_2(0) = 0; \tag{3.5}$$

and the equation for v_3 is

$$v_3'' - f'(H)v_3 - (n-1)^2 y H'' + (n-1)v_2' - (n-1)f''(H)H'u_2 = (n-1)\gamma_2 H'. \tag{3.6}$$

From the last equation and the equation for u_2' derived by differentiating (3.5) one finds

$$\gamma_2 = \frac{(n-1)(3-n)}{2}. \tag{3.7}$$

Note that in this case γ_2 is negative. Reading the equation (3.7) backwards we deduce that

$$\rho = \sqrt{\frac{(n-1)(n-3)}{-2\gamma}} + \cdots . \tag{3.8}$$

The outcome of our analysis is consistent with (1.3) for $n \geq 4$. With Γ being a $n-1$ dimensional sphere,

$$I_c(\Gamma) = \omega_{n-1}\rho^{n-3} + \omega_{n-1}\gamma\rho^{n-1} \tag{3.9}$$

where ω_{n-1} is the area of the $(n-1)$-dimensional unit sphere. It is clear that if $\gamma > 0$, the right side is increasing in ρ. Only if $\gamma < 0$, there exists a critical point, but this critical point is a maximum. It is (3.8) up to a multiplicative constant.

The existence theorem in the $n \geq 4$ case for the phase field model is the following.

Theorem 3.1 ([16]). *When γ is negative and sufficiently close to 0, there exists a bubble profile. The radius of the bubble is $\sqrt{(n-1)(n-3)/(-2\gamma)} + o(\gamma^{-1/2})$.*

4. $n = 1$

When $n = 1$, the phase field problem is far more complex than (1.3). A zero-dimensional sphere is just the union of two points in $\mathbb{R}$, and ρ is half the distance between the two points. This sphere has no curvature. Hence

$$I_c(\Gamma) = 2\gamma, \tag{4.1}$$

a constant independent of ρ. No conclusion can be drawn from (4.1). For the phase field problem, we can still carry out an asymptotic analysis. However we must analyze the effect of the point $x = 0$ very carefully.

A little more information from H (2.6) is needed. The derivative of H, $H'(y)$, decays to 0 exponentially fast as $|y| \to \infty$. More precisely we have $a > 0$ and $k < 0$ such that

$$H'(y) = k\mathrm{e}^{-a|y|} + O(\mathrm{e}^{-2a|y|}). \tag{4.2}$$

Necessarily

$$a^2 = f'(1) = f'(1). \tag{4.3}$$

In the special case $f(H) = H(H-1/2)(H-1)$,

$$a = \frac{1}{\sqrt{2}}, \quad k = -\frac{1}{\sqrt{2}}.$$

The leading order approximation of u is still H near ρ. But we need to include a correction due to the boundary condition $u'(0) = 0$. To have this boundary condition satisfied let

$$u(x) = H(x-\rho) + \frac{H'(-\rho)}{a}\mathrm{e}^{-ax} + \cdots. \tag{4.4}$$

The leading order approximation of v near ρ is v_1,

$$v(\rho + y) = v_1(y) + \cdots \tag{4.5}$$

where v_1 satisfied

$$v_1'' - f'(H)v_1 = 0. \tag{4.6}$$

Hence

$$v_1(y) = dH'(y) \tag{4.7}$$

for some d to be determined. For v_1 we must also add a correction so that it satisfies the boundary condition $v'(0) = 0$:

$$v(\rho+y) = d\Big(H'(y) + \frac{H''(-\rho)}{a}\mathrm{e}^{-a(\rho+y)}\Big) + \cdots. \tag{4.8}$$

We turn to the next order approximation of u near ρ:

$$u(\rho+y) = H(y) + \frac{H'(-\rho)}{a}\mathrm{e}^{-a(\rho+y)} + u_1(y) + \cdots \tag{4.9}$$

where u_1 satisfies the equation

(4.10)

$$u_1'' - f'(H)u_1 + \frac{H'(-\rho)}{a}\mathrm{e}^{-a(\rho+y)}\big(f'(1) - f'(H)\big) = dH', \quad u_1(0) = -\frac{H'(-\rho)}{a}\mathrm{e}^{-a\rho}.$$

Multiplying by H' and integrating over $(-\infty, \infty)$ yield that

$$d = \frac{1}{\tau}\int_{\mathbb{R}} \frac{H'(-\rho)}{a}\mathrm{e}^{-a(\rho+y)}\big(f'(1) - f'(H)\big)H'\,\mathrm{d}y. \tag{4.11}$$

This constant d is of order $O(\mathrm{e}^{-2a\rho})$. The integral in (4.11) may be calculated as in (4.18). One sees that $u_1 = O(\mathrm{e}^{-2a\rho})$, and $v_1 = O(\mathrm{e}^{-2a\rho})$.

The next order approximation of v near ρ is denoted v_2:

$$v(\rho+y) = d\Big(H'(y) + \frac{H''(-\rho)}{a}\mathrm{e}^{-a(\rho+y)}\Big) + v_2(y) + \cdots \tag{4.12}$$

where $v_2(y)$ satisfies

$$(4.13)\quad v_2'' - f'(H)v_2 + d\frac{H''(-\rho)}{a}\mathrm{e}^{-a(\rho+y)}\big(f'(1) - f'(H)\big) - df''(H)H'\Big(\frac{H'(-\rho)}{a}ee^{-a(\rho+y)} + u_1\Big) = \gamma_1 dH'$$

where γ_1 is the leading order approximation of γ:

$$(4.14)\qquad \gamma = \gamma_1 + \cdots .$$

Differentiation of (4.10) implies

$$(4.15)\quad u_1''' - f'(H)u_1' - f''(H)H'u_1 - H'(-\rho)\mathrm{e}^{-a(\rho+y)}\big(f'(1) - f'(H)\big) - \frac{H'(-\rho)}{a}\mathrm{e}^{-a(\rho+y)}f''(H)H' = dH''.$$

Let $w = v_2 - du_1'$. Then w satisfies

$$(4.16)\quad w'' - f'(H)w + d\Big(\frac{H''(-\rho)}{a} + H'(-\rho)\Big)\mathrm{e}^{-a(\rho+y)}\big(f'(1) - f'(H)\big) = \gamma_1 dH' - d^2H''.$$

Multiplying by H' and integrating find that

$$(4.17)\qquad \gamma_1\tau = \Big(\frac{H''(-\rho)}{a} + H'(-\rho)\Big)\int_{\mathbb{R}} \mathrm{e}^{-a(y+\rho)}\big(f'(1) - f'(H)\big)H'\,\mathrm{d}y.$$

To calculate the integral in (4.17) note that $H''' - f'(H)H' = 0$ by differentiating (2.6). Then

$$(4.18)\qquad \begin{aligned}\int_{\mathbb{R}} \mathrm{e}^{-ay}\big(f'(1) - f'(H)\big)H'\,\mathrm{d}y &= \int_{\mathbb{R}} (f'(1)H' - H''')\mathrm{e}^{-ay}\,\mathrm{d}y\\ &= \big[-H''(y)\mathrm{e}^{-ay} - aH'(y)\mathrm{e}^{-ay}\big]_{-\infty}^{\infty}\\ &= 2ak.\end{aligned}$$

This implies that

$$(4.19)\qquad \gamma_1 = \frac{1}{\tau}\Big(\frac{H''(-\rho)}{a} + H'(-\rho)\Big)\mathrm{e}^{-a\rho}2ak \approx \frac{4ak^2\mathrm{e}^{-2a\rho}}{\tau}.$$

One can see that $v_2 = O(\mathrm{e}^{-4a\rho})$.

With the leading order term of γ determined the structure of the solution is rather clear. The formula (4.19) is read backwards to give the location of the interface:

$$(4.20)\qquad \rho = \frac{1}{2a}\log\frac{1}{\gamma} + \cdots .$$

Note in this case γ must be positive. As γ tends to 0, ρ approaches ∞, but very slowly. The following existence theorem is proved in [16].

Theorem 4.1 ([16]). *When γ is positive and sufficiently small, there exists a bubble profile. The radius of the bubble is $(1/2a)\log(1/\gamma) + o\big(\log(1/\gamma)\big)$.*

5. $n = 3$

When $n = 3$, for a sphere $\kappa = 1/\rho$ and $\int_\Gamma \kappa^2 \, \mathrm{d}s = (1/\rho^2)4\pi\rho^2 = 4\pi$. Hence

$$I_c(\Gamma) = 4\pi + 4\pi\gamma\rho^2, \tag{5.1}$$

which has no critical point for positive ρ. The phase field problem is again very different.

In the three-dimensional case it is often more convenient to consider

$$\tilde{v}(r) = rv(r) \tag{5.2}$$

instead of v. Then $\tilde{v}$ satisfies the equation

$$\tilde{v}'' - f'(u)\tilde{v} = \gamma\tilde{v}, \quad \tilde{v}(0) = 0. \tag{5.3}$$

The first approximation of u, taking the condition $u'(0) = 0$ into account, is

$$u(r) = H(r-\rho) + \frac{H'(-\rho)}{a}\mathrm{e}^{-ar} \cdots . \tag{5.4}$$

The leading order approximation of $\tilde{v}$ which we denote by $\tilde{v}_1$ satisfies the equation

$$\tilde{v}_1'' - f'(H)\tilde{v}_1 = 0 \tag{5.5}$$

near ρ. This implies that

$$\tilde{v}_1(y) = dH'(y) \tag{5.6}$$

where d is to be determined. To have the boundary condition $\tilde{v}(0) = 0$ satisfied we add a correction term so that

$$\tilde{v}(r) = d(H'(r-\rho) - H'(-\rho)\mathrm{e}^{-ar}) + \cdots \tag{5.7}$$

Let

$$u(r) = H(r-\rho) + \frac{H'(-\rho)}{a}\mathrm{e}^{-ar} + u_1(r-\rho) + \cdots \tag{5.8}$$

where $u_1(y)$ satisfies

$$\begin{aligned} u_1'' - f'(H)u_1 + \frac{H'(-\rho)}{a}\mathrm{e}^{-a(\rho+y)}(f'(1) - f'(H)) &= \frac{d-2}{\rho}H', \\ u_1(0) &= -\frac{H'(-\rho)}{a}\mathrm{e}^{-a\rho}. \end{aligned} \tag{5.9}$$

Multiplying the equation by H' and integrating yield

$$\begin{aligned} \frac{(d-2)\tau}{\rho} &= \frac{H'(-\rho)}{a}\int_{\mathbb{R}} (f'(1) - f'(H))H'\mathrm{e}^{-a(\rho+y)} \, \mathrm{d}y \\ &= \frac{H'(-\rho)}{a}\mathrm{e}^{-a\rho}2ak = 2kH'(-\rho)\mathrm{e}^{-a\rho}, \end{aligned} \tag{5.10}$$

where we have used (4.18). One sees that $u_1 = O(\mathrm{e}^{-2a\rho})$. This also implies that $\tilde{v}_1 = O(1)$.

The next order approximation of $\tilde{v}$ is denoted by $\tilde{v}_2(y)$ so that

$$\tilde{v}(r) = d(H'(r-\rho) - H'(-\rho)\mathrm{e}^{-ar}) + \tilde{v}_2(r-\rho) + \cdots . \tag{5.11}$$

The equation for $\tilde{v}_2$ is

$$\tilde{v}_2'' - f'(H)\tilde{v}_2 - 2H'(-\rho)\mathrm{e}^{-ar}\big(f'(1) - f'(H)\big) - 2f''(H)H'\Big(\frac{H'(-\rho)}{a}\mathrm{e}^{-ar} + u_1\Big) = 0. \tag{5.12}$$

Comparing this with the equation for u_1', which is

$$u_1''' - f'(H)u_1' - f''(H)H'u_1 - H'(-\rho)\mathrm{e}^{-ar}\big(f'(1) - f'(H)\big) - \frac{H'(-\rho)}{a}\mathrm{e}^{-ar}f''(H)H' = \frac{d-2}{\rho}H'', \tag{5.13}$$

we find that the difference $w = \tilde{v}_2 - 2u_1'$ satisfies

$$w'' - f'(H)w = -\frac{2(d-2)}{\rho}H''. \tag{5.14}$$

The equation (5.14) can be solve to yield

$$w = -\frac{d-2}{\rho}yH' + eH', \tag{5.15}$$

where the constant e is to be determined. Hence

$$\tilde{v}_2 = 2u_1' - \frac{d-2}{\rho}yH' + eH'. \tag{5.16}$$

In this case we have to expand to one more order. Let

$$u = H + \frac{H'(-\rho)}{a}\mathrm{e}^{-ar} + u_1 + u_2 \cdots \tag{5.17}$$

where u_2 satisfies

$$u_2'' - f'(H)u_2 = -\frac{d-2}{\rho^2}yH' + \frac{\tilde{v}_2 - 2u_1'}{\rho}, \quad u_2(0) = 0. \tag{5.18}$$

Multiplying by H' and integrating show that

$$0 = \int_{\mathbb{R}} \frac{\tilde{v}_2 - 2u_1'}{\rho}H' = \frac{e}{\rho}\int_{\mathbb{R}}(H')^2. \tag{5.19}$$

Therefore

$$e = 0 \tag{5.20}$$

and

$$\tilde{v}_2 = 2u_1' - \frac{d-2}{\rho}yH'. \tag{5.21}$$

Here we see that $u_2 = O(\mathrm{e}^{-2a\rho}/\rho)$ and $\tilde{v}_2 = O(\mathrm{e}^{-2a\rho})$.

Introduce $\tilde{v}_3$ so that

$$v(\rho + y) = d(H'(y) - H'(-\rho)\mathrm{e}^{-a(\rho+y)}) + 2u_1' - \frac{d-2}{\rho}yH' + \tilde{v}_3 + \cdots. \tag{5.22}$$

The new function $\tilde{v}_3(y)$ satisfies

$$\tilde{v}_3'' - f'(H)\tilde{v}_3 - 2f''(H)H'u_2 = 2\gamma_1 H' \tag{5.23}$$

where γ_1 is the leading order approximation of γ. Multiplying by H' and integrating yield

$$\gamma_1\tau = -\int_{\mathbb{R}} f''(H)(H')^2 u_2 \tag{5.24}$$

To compute the integral in (5.24) we differentiate the equation for u_2 to obtain

$$u_2''' - f'(H)u_2' - f''(H)H'u_2 = -\frac{d-2}{\rho^2}(yH')' + \left(\frac{\tilde{v}_2 - 2u_1'}{\rho}\right)'. \tag{5.25}$$

Multiplying by H' and integrating show that

$$\begin{aligned} -\int_{\mathbb{R}} f''(H)(H')^2 u_2 &= -\frac{d-2}{\rho^2}\int_{\mathbb{R}}(yH')'H' + \frac{1}{\rho}\int_{\mathbb{R}} w'H' \\ &= -\frac{d-2}{\rho^2}\int_{\mathbb{R}}(yH')'H' + \frac{1}{\rho}\int_{\mathbb{R}}\left(-\frac{d-2}{\rho}yH'\right)'H' \\ &= -\frac{2(d-2)}{\rho^2}\int_{\mathbb{R}}(yH')'H' \\ &= -\frac{(d-2)\tau}{\rho^2} \end{aligned}$$

where we have used (2.35). This and (5.10) show that

$$\gamma_1 = -\frac{d-2}{\rho^2} = -\frac{2kH'(-\rho)e^{-a\rho}}{\tau\rho} \approx -\frac{2\big(H'(-\rho)\big)^2}{\tau\rho} \approx -\frac{2k^2 e^{-2a\rho}}{\tau\rho}. \tag{5.26}$$

One also sees that $\tilde{v}_3 = O(e^{-2a\rho}/\rho)$.

Note that this γ_1 is negative. To read (5.26) backwards, let $l(s)$ be the inverse function of

$$\rho \to -\frac{2k^2 e^{-2a\rho}}{\tau\rho}. \tag{5.27}$$

Here $l\colon (-\infty, 0) \to (0, \infty)$. As s tends to 0, $l(s)$ grows to ∞, but more slowly than $-(1/2a)\log(-s)$ does, i.e., more slowly than in the $n = 1$ case. The interface ρ is at

$$\rho = l(\gamma) + \cdots. \tag{5.28}$$

In this case the existence theorem is the following.

Theorem 5.1 ([16]). *When γ is negative and sufficiently close to 0, there exists a bubble profile. The radius of the bubble is $l(\gamma) + o\big(l(\gamma)\big)$.*

6. Localized energy method

The rigorous proofs of Theorems 2.1, 3.1, 4.1 and 5.1 are given in [16]. We used the so-called localized energy method which is a combination of the Lyapunov–Schmidt reduction argument and variational techniques. In this section we list some key steps in this method.

Let ρ be the location of the interface of bubble profile u in the sense that $u(\rho) = \frac{1}{2}$. Near ρ, u has a rather particular shape. This shape is mostly described by the function H given in (2.6). When r is much less than ρ, $u(r)$ is close to 1; when r is much larger than ρ, $u(r)$ is close to 0. The proofs of the existence theorems are very much about locating ρ. As $\gamma \to 0$, $\rho \to \infty$. The construction is divided into two steps: in the first step, we fix ρ large and solve a nonlinear problem with an orthogonal condition. In the next step, we locate ρ by finding a critical point for a reduced energy function involving ρ only. For the localized energy method used in other problems, see [2, 4, 5, 7, 8, 14, 15].

$n = 2$. Let us first consider the case $n = 2$. The operator on the left side of (1.8) is denoted by S, i.e.,

$$S(u) = (\Delta - f'(u) - \gamma)[\Delta u - f(u)] \tag{6.1}$$

For each $\rho \in (1/(2\sqrt{2\gamma}), \frac{2}{\sqrt{2\gamma}})$, we construct an approximate solution w to (1.8) of the form

$$w(r;\rho) = H(r-\rho) + \beta(r;\rho) \tag{6.2}$$

where

$$\beta(r;\rho) = c_{1,\rho}\mathrm{e}^{-ar} + c_{2,\rho}r\mathrm{e}^{-ar}. \tag{6.3}$$

The constants $c_{1,\rho}$ and $c_{2,\rho}$ are so chosen that $w'(0) = w'''(0) = 0$, which ensures that $S(w)$ is regular at $r = 0$. More explicitly

$$c_{1,\rho} = \frac{H'(-\rho)}{a} + \frac{1}{2a}\left(H'(-\rho) - \frac{H'''(-\rho)}{a^2}\right), \tag{6.4}$$

$$c_{2,\rho} = \frac{1}{2}\left(H'(-\rho) - \frac{H'''(-\rho)}{a^2}\right). \tag{6.5}$$

In the case $n = 2$, the exact form of β is not too important. The main role played by β is to have the boundary conditions at $r = 0$ satisfied.

We define an approximate kernel

$$h(r;\rho) = H'(r-\rho) + b_{1,\rho}\mathrm{e}^{-ar} + b_{2,\rho}r\mathrm{e}^{-ar} \tag{6.6}$$

where $b_{1,\rho} = O(\mathrm{e}^{-a\rho})$ and $b_{2,\rho} = O(\mathrm{e}^{-a\rho})$ are constants so chosen that $h'(0) = h'''(0) = 0$.

Let π_ρ be the projection operator to the subspace perpendicular to h:

$$\pi_\rho g = g - \frac{\langle g, h\rangle}{\|h\|_2^2}h. \tag{6.7}$$

Here $\langle g, h\rangle$ is the $L^2(\mathbb{R}^2)$ inner product of g and h, and $\|h\|_2$ is the $L^2(\mathbb{R}^2)$ norm of h. We view

$$M = \left\{w(\cdot;\rho) : \frac{1}{2\sqrt{2\gamma}} < \rho < \frac{2}{\sqrt{2\gamma}}\right\} \tag{6.8}$$

as a one-dimensional submanifold in $H_r^4(\mathbb{R}^2)$. The space $H_r^4(\mathbb{R}^2)$ is the collection of radial H^4 functions on $\mathbb{R}^2$. At each $w(\cdot;\rho)$ we define an approximate normal subspace

$$F_\rho = \{\phi \in H_r^4(\mathbb{R}^2) : \phi \perp h(\cdot;\rho)\}. \tag{6.9}$$

It turns out that in each F_ρ there exists $\phi(\cdot;\rho)$ so that

$$\pi_\rho S\big(w(\cdot;\rho) + \phi(\cdot;\rho)\big) = 0. \tag{6.10}$$

The free energy of $w + \phi$ is close to the free energy of w. One can show that

$$I(w+\phi) = I(w) + O(\gamma^{1.5}). \tag{6.11}$$

The free energy of w can be calculated:

$$I(w) = 2\pi\tau\left(\frac{1}{2\rho} + \gamma\rho\right) + O(\gamma^{1.5}). \tag{6.12}$$

Therefore we conclude that

$$I\big(w(\cdot;\rho)+\phi(\cdot;\rho)\big) = 2\pi\tau\left(\frac{1}{2\rho}+\gamma\rho\right)+O(\gamma^{1.5}). \tag{6.13}$$

If we minimize $I\big(w(\cdot;\rho)+\phi(\cdot;\rho)\big)$ with respect to ρ, then a minimum is found at

$$\rho=\rho_\gamma=\frac{1}{\sqrt{2\gamma}}+o(\gamma^{-1/2}). \tag{6.14}$$

One can show that at this ρ_γ

$$S\big(w(\cdot;\rho_\gamma)+\phi(\cdot;\rho_\gamma)\big)=0, \tag{6.15}$$

i.e., (1.8) is exactly solved.

$n\geq 4$. The case $n\geq 4$ is very similar. We use the same approximate solution w. The main difference occurs in the last step:

$$I(w+\phi)=\omega_{n-1}\tau\left(\frac{(n-1)^2\rho^{n-3}}{2}+\gamma\rho^{n-1}\right)+o(\gamma^{(3-n)/2}) \tag{6.16}$$

where ω_{n-1} is the area of the $(n-1)$-dimensional unit sphere. If $\gamma<0$, the above quantity has a *maximum* at

$$\rho_\gamma=\sqrt{\frac{(n-1)(n-3)}{-2\gamma}}+o(\gamma^{-1/2}).$$

This gives a solution $w(\cdot;\rho_\gamma)+\phi(\cdot;\rho_\gamma)$ of (1.8).

$n=1$. The case $n=1$ is more complex. Let

$$\rho\in\left(\frac{1}{4a}\log\frac{1}{\gamma},\frac{1}{a}\log\frac{1}{\gamma}\right). \tag{6.17}$$

For each ρ satisfying (6.17) we define an approximate solution w. Compared to the $n=2$ and $n\geq 4$ cases, the construction of w is more involved. We let

$$\alpha(x;\rho)=c_{0,\rho}\mathrm{e}^{-ax},\quad \text{where } c_{0,\rho}=\frac{H'(-\rho)}{a},\ \text{so that } H'(-\rho)+\alpha'(0)=0. \tag{6.18}$$

Now we define a function $g(y;\rho)$ on $(-\infty,\infty)$ which is the solution of

$$g''-f'(H)g+\alpha(y+\rho)\big(f'(1)-f'(H)\big)=d_\rho H',\quad g(0)=-\alpha(\rho). \tag{6.19}$$

Note that g is motivated by u_1 in Section 4. In (6.19) the constant d_ρ is chosen so that

$$d_\rho\int_{\mathbb{R}}(H')^2\,\mathrm{d}y=\int_{\mathbb{R}}\alpha(y+\rho)\Big(f'(1)-f'\big(H(y)\big)\Big)H'(y)\,\mathrm{d}y. \tag{6.20}$$

This ensures that (6.19) is solvable. The condition $g(0)=-\alpha(\rho)$ gives a unique solution. We calculate the right side of (6.20):

$$\begin{aligned}\int_{\mathbb{R}}\alpha(y+\rho)\Big(f'(1)-f'\big(H(y)\big)\Big)H'(y)\,\mathrm{d}y &= \alpha(\rho)\int_{\mathbb{R}}\mathrm{e}^{-ay}(f'(1)H'-H''')\,\mathrm{d}y\\ &= \alpha(\rho)[-\mathrm{e}^{-ay}H''(y)-a\mathrm{e}^{-ay}H'(y)]\Big|_{y=-\infty}^{y=\infty}\\ &= \alpha(\rho)\lim_{y\to-\infty}[\mathrm{e}^{-ay}H''(y)+a\mathrm{e}^{-ay}H'(y)]\\ &= 2ak\alpha(\rho)=2k^2\mathrm{e}^{-2a\rho}+o(\mathrm{e}^{-2a\rho}).\end{aligned}$$

Therefore

$$d_\rho = \frac{2k^2 e^{-2a\rho}}{\tau} + o(e^{-2a\rho}). \tag{6.21}$$

We include g in the construction of w. One last term is β which is given as

$$\beta(x;\rho) = c_{1,\rho} e^{-ax} + c_{2,\rho} x e^{-ax} + \frac{f''(1)c_{0,\rho}^2}{6a^2} e^{-2ax}. \tag{6.22}$$

It is a solution of

$$\left(D^2 - f'(1)\right)^2 \beta = \left(D^2 - f'(1)\right) \frac{f''(1)\alpha^2}{2}. \tag{6.23}$$

The constants $c_{1,\rho}$ and $c_{2,\rho}$ are chosen so that

$$\beta'(0) = -H'(-\rho) - \alpha'(0) - g'(-\rho) = -g'(-\rho), \tag{6.24}$$

$$\beta'''(0) = -H'''(-\rho) - \alpha'''(0) - g'''(-\rho). \tag{6.25}$$

Here

$$c_{1,\rho} = O(e^{-2a\rho}), \quad c_{2,\rho} = O(e^{-2a\rho}). \tag{6.26}$$

Now we set

$$w(x;\rho) = H(x-\rho) + \alpha(x;\rho) + g(x-\rho;\rho) + \beta(x;\rho). \tag{6.27}$$

Our choice of β ensures that $w'(0) = w'''(0) = 0$. Note that this β is different from the one (6.3) used in the $n = 2$ case.

As in the proof of the $n = 2$ case one can find ϕ so that

$$\pi_\rho S(w+\phi) = 0. \tag{6.28}$$

The free energy of $w + \phi$ is

$$I(w+\phi) = I(w) + o(\gamma^2) = \frac{4k^4 e^{-4a\rho}}{\tau} + \gamma\left(2\tau - \frac{2k^2 e^{-2a\rho}}{a}\right) + o(\gamma^2). \tag{6.29}$$

We minimize $I\big(w(\cdot;\rho) + \phi(\cdot;\rho)\big)$ with respect to ρ and find that $I(w+\phi)$ is minimized at some

$$\rho_\gamma = -\frac{1}{2a}\log\frac{\gamma\tau}{4ak^2} + o\left(\log\frac{1}{\gamma}\right). \tag{6.30}$$

At this ρ_γ,

$$S\big(w(\cdot;\rho_\gamma) + \phi(\cdot;\rho_\gamma)\big) = 0. \tag{6.31}$$

$\boldsymbol{n = 3}$**.** Recall the function l in Section 5, We let

$$\rho \in \left(\frac{l(\gamma)}{2}, 2l(\gamma)\right). \tag{6.32}$$

We define a family of approximate solutions

$$w(r;\rho) = H(r-\rho) + \beta(r;\rho) \tag{6.33}$$

where β is the same as the one (6.3) used in the $n = 2$ case. Here β plays a more important role than in the $n = 2$ case. One can again find ϕ so that $\pi_\rho S(w+\phi) = 0$ and calculations show that

$$I(w+\phi) = 4\pi\left[2\tau - \frac{2k^2 e^{-2a\rho}}{a} + \gamma\tau\rho^2\right] + O(\gamma^{1+\delta}) \tag{6.34}$$

for some $\delta > 0$ independent of ρ and γ. We now maximize the above with respect to ρ to find a particular ρ_γ at which the equation (1.8) is solved.

References

1. S. E. Allen and J. W. Cahn, *A microscopic theroy for antiphase boundary motion and its application to antiphase domain coarsening*, Acta Metallurgica **27** (1979), no. 6, 1085–1095.
2. P. W. Bates, E. N. Dancer, and J. Shi, *Multi-spike stationary solutions of the Cahn–Hilliard equation in higher dimension and instability*, Adv. Differential Equations **4** (1999), no. 1, 1–69.
3. R. G. Casten and C. J. Holland, *Instability results for reaction diffusion equations with Neumann boundary conditions*, J. Differential Equations **27** (1978), no. 2, 266–273.
4. E. N. Dancer and S. Yan, *Multi-layer solutions for an elliptic problem*, J. Differential Equations **194** (2003), no. 2, 382–405.
5. ______, *A minimization problem associated with elliptic systems of FitzHugh–Nagumo type*, Ann. Inst. H. Poincaré Anal. Non Linéaire **21** (2004), no. 2, 237–257.
6. E. De Giorgi, *Sulla convergenza di alcune successioni d'integrali del tipo dell'area*, Rend. Mat. (6) **8** (1975), 277–294.
7. C. Gui and J. Wei, *Multiple interior peak solutions for some singular perturbed Neumann problems*, J. Differential Equations **158** (1999), no. 1, 1–27.
8. ______, *On multiple mixed interior and boundary peak solutions for some singularly perturbed Neumann problems*, Canad. J. Math. **52** (2000), no. 3, 522–538.
9. R. V. Kohn and P. Sternberg, *Local minimisers and singular perturbations*, Proc. Roy. Soc. Edinburgh Sect. A **111** (1989), no. 1-2, 69–84.
10. H. Matano, *Asymptotic behavior and stability of solutions of semilinear diffusion equations*, Publ. Res. Inst. Math. Sci. **15** (1979), no. 2, 401–454.
11. L. Modica, *The gradient theory of phase transitions and the minimal interface criterion*, Arch. Rational Mech. Anal. **98** (1987), no. 2, 123–142.
12. L. Modica and S. Mortola, *Un esempio di Γ^--convergenza*, Boll. Un. Mat. Ital. B (5) **14** (1977), no. 1, 285–299.
13. R. Moser, *A higher order asymptotic problem related to phase transitions*, SIAM J. Math. Anal. **37** (2005), no. 3, 712–736.
14. X. Ren and J. Wei, *Nucleation in the FitzHugh–Nagumo system: interface-spike solutions*, J. Differential Equations **209** (2005), no. 2, 266–301.
15. ______, *Droplet solutions in the diblock copolymer problem with skewed monomer composition*, Calc. Var. Partial Differential Equations **25** (2006), no. 3, 333–359.
16. ______, *On a phase field problem driven by interface tension and interface curvature*, preprint.
17. L. Simon, *Existence of surfaces minimizing the Willmore functional*, Comm. Anal. Geom. **1** (1993), no. 2, 281–326.
18. H. Tang and K. F. Freed, *Free energy functional expansion for inhomogeneous polymer blends*, J. Chem. Phys. **94** (1991), no. 2, 1572–1583.
19. T. J. Willmore, *Riemannian geometry*, Oxford Sci. Publ., Oxford Univ. Press, New York, 1993.

Department of Mathematics and Statistics, Utah State University, Logan, UT 84322-3900, USA

Current address: Department of Mathematics, The George Washington University, Washington, DC 20052, USA

E-mail address: ren@gwu.edu

Centre de Recherches Mathématiques
CRM Proceedings and Lecture Notes
Volume **44**, 2008

Rare Events, Action Minimization, and Sharp Interface Limits

Maria G. Westdickenberg

ABSTRACT. Estimating the probability of and mechanism for rare events is a physically relevant and mathematically interesting question. Focusing on issues related to "constrained switching events," we describe differences from the classical Arrhenius picture and how action minimization can be used as a tool. As a specific example, we turn to the stochastic Allen–Cahn equation and explain recent results for the scaling of the action problem and the sharp interface limit.

1. Motivation

Small random forces are unexpectedly powerful: In so-called rare events, a small perturbation drives a system to violate deterministic behavior in a "large" way. We set the stage with a few examples.

Three examples. As our canonical example, consider the deterministic ordinary differential equation (ODE)

$$\dot{x}(t) = x(t) - x(t)^3, \quad x(0) = x_0, \tag{1}$$

which is the gradient flow for the double-well potential

$$V(x) = \tfrac{1}{4}(1 - x^2)^2. \tag{2}$$

Notice that the energy has two absolute minimizers, $x_- = -1$ and $x_+ = +1$. Equation (1) is tame: Choosing $x_0 = x_-$ leads to $x(t) \equiv x_-$, and choosing x_0 near x_- leads to the relaxation $x(t) \to x_-$ as $t \to \infty$. Even an arbitrarily small stochastic perturbation spices things up, however, destroying the stationarity and stability of x_-. Consider the stochastic ODE

$$\mathrm{d}X(t) = (X(t) - X(t)^3)\,\mathrm{d}t + \sqrt{2\delta}\,\mathrm{d}B(t), \quad X(0) = x_0, \tag{3}$$

where $B(t)$ is a Brownian motion and $\delta > 0$. Because of the perturbation, the solution for $x_0 = x_-$ is not stationary, but instead makes small, random hops. The more dramatic change, however, is that with probability one, $X(t)$ will eventually

2000 *Mathematics Subject Classification.* Primary 35R60; Secondary 49J45, 60F10.

This work was partially supported by the National Science Foundation under Grant No. 0402762.

This is the final form of the paper.

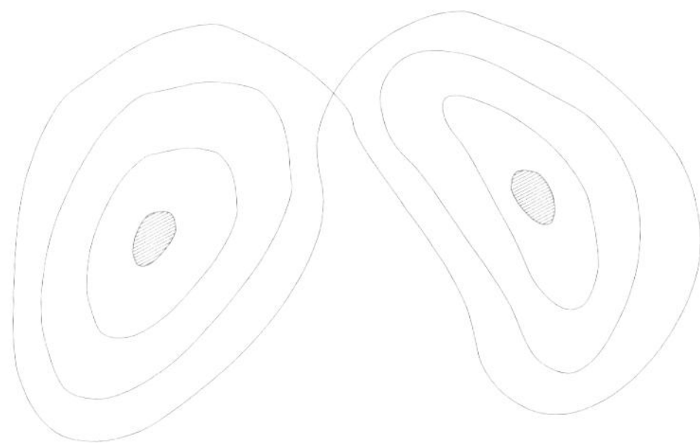

FIGURE 1. The level curves of a potential $V: \mathbb{R}^2 \to \mathbb{R}$. The shaded regions represent small neighborhoods of the absolute energy minimizers.

visit *a small neighborhood of* x_+. In other words, an order $\sqrt{\delta}$ perturbation will force an order 1 increase in energy, driving the process against the gradient flow and over the energy barrier. The expected value of the time for this switching event to occur, which we denote by τ_δ, is exponentially large in δ^{-1} (hence the descriptor "rare"). Even for much shorter timescales T, however, the probability of the switching event occurring for $t \leq T$ is positive. We will be interested in estimating switching probabilities.

New structure emerges in the two-dimensional analogue. Consider the energy with level curves as in Figure 1. We again denote the two absolute minimizers by x_- and x_+. The stochastically perturbed gradient flow is

$$\mathrm{d}X(t) = -\nabla V\big(X(t)\big)\,\mathrm{d}t + \sqrt{2\delta}\,\mathrm{d}B(t), \quad X(0) = x_0, \tag{4}$$

where $X(t), x_0 \in \mathbb{R}^2$ and $B(t)$ is a vector of two independent Brownian motions. As in the first example, we can ask about the probability to switch from x_- to a neighborhood of x_+ in time $t \leq T$. A new question that becomes relevant in this example is *how* the process switches. It turns out that there is a distinguished pathway, the so-called most likely switching pathway, such that when the process does transition from one minimizer to the other, it almost certainly stays close to this pathway. We will be interested in identifying switching pathways.

For the third example, we turn to the physical problem of developing reliable, nanoscale sized magnetic memory devices. In small devices, thermal noise is significant: Finite temperature effects can cause magnetic switching on physically relevant timescales. Micromagnetics is a continuum theory for magnetic behavior in terms of the magnetization vector $m\colon \Omega \to \mathbb{S}^2$, where Ω represents the magnetic sample and $\mathbb{S}^2$ represents the unit sphere. Associated to a given magnetization is the micromagnetic energy, which includes an exchange term that penalizes spatial variation, an anisotropy term that prefers certain orientations, and a nonlocal stray field term that favors divergence-free configurations. This rich energy functional has a host of local energy minimizers; an example of one minimizer in a thin-film sample is shown in Figure 2. The associated dynamic equations are the stochastically perturbed Landau – Lifshitz – Gilbert equations. We emphasize two features of the stochastic switching problem: (i) critical points of the energy have a nontrivial spatial structure, and (ii) the most likely switching pathway depends on a rich parameter regime including, for instance, the size of the sample, the time of the

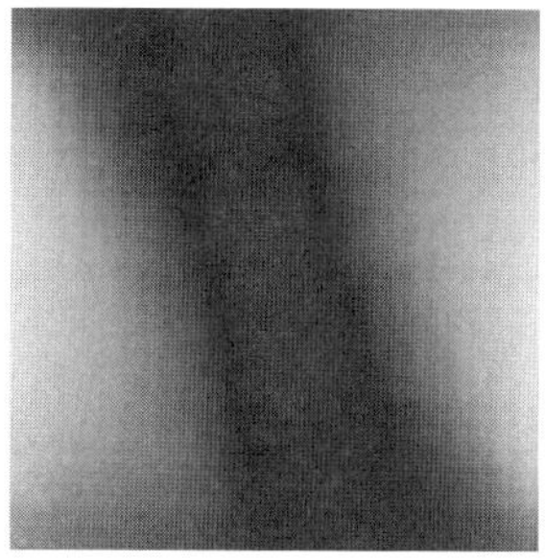

FIGURE 2. A local minimizer of the micromagnetic energy in a $200\,\text{nm} \times 200\,\text{nm} \times 10\,\text{nm}$ thin film sample. Here the magnetization vector in the dark shaded region is $(1,0,0)$, and the magnetization in the lighter region is $(0,1,0)$. This figure has been calculated numerically by Weiqing Ren, and is reproduced with his permission. Cf. `http://math.nyu.edu/~weiqing/MM/MM.html`

experiment, and the coefficient of the exchange term. We will be interested in how spatial structure and competition between system parameters affect the switching problem.

Timescales and switching probabilities. There is a classical analysis of the switching problem, well-known in the physics community and later analyzed in the mathematics literature in the 1980s, that we now describe. To fix ideas, consider (4). Let

$$\mathcal{A}^\epsilon = \{x : \mathbb{R}^+ \to \mathbb{R}^2 \mid x(0) = x_-,\ x(t) \in B_\epsilon(x_+) \text{ for some } t \le T\},$$

where $\epsilon > 0$ and $B_\epsilon(x_+)$ denotes a ball of radius ϵ around x_+. In this case we have

$$\mathbb{E}(\inf\{T \text{ such that } X \in \mathcal{A}^\epsilon\}) =: \tau_\delta \asymp \exp(\delta^{-1}\Delta E)$$

where the energy barrier ΔE is the difference in energy of x_- and the saddle point with smallest energy, and $f_\delta \asymp g_\delta$ denotes logarithmic equivalence, i.e., $\ln f_\delta \sim \ln g_\delta$ as $\delta \to 0$. The classical analysis says that the process is governed by the Arrhenius law

$$\text{(5)} \qquad \operatorname{Prob}(X \notin \mathcal{A}^\epsilon) \approx \exp(-T/\tau_\delta).$$

This statement was put on mathematically rigorous ground in [17]: Roughly speaking, (5) is proved *in the case* $T \asymp \tau_\delta$. However, and this will be important for us, the Arrhenius approximation (5) gives no information in the case $\ln T \ll \ln \tau_\delta$.

At first glance, considering switching on the timescale $T \asymp \tau_\delta$ may seem like the *only* reasonable question: Switching for $\ln T \ll \ln \tau_\delta$ is far less likely than switching for $T \asymp \tau_\delta$. Indeed, if one starts the dynamics and waits for a rare event, the timescale τ_δ is the relevant one. On the other hand, one can also fix $T > 0$ and decide a priori to wait *only* that long to see whether there is a rare event. Why do that? One motivation comes from the magnetic problem: Nanoscale magnetic devices are subjected to short, applied-field pulses that are intended to switch the magnetization in certain elements, but not others. The "nontargeted elements" must reliably withstand the pulses without switching. The pulse time T will be short compared to their expected switching time, so the probability of switching is exceedingly small, but there will be many of these pulses in their lifetime, so even a small probability matters. What one seeks in this case is a quantitative estimate of the very small switching probability. See [23] for further discussion.

The Freidlin – Wentzell theory of large deviations [15] analyzes switching probabilities for fixed time T in terms of the so-called large deviation action functional. For (4), the statement is

$$\text{Prob}(X \in \mathcal{A}^\epsilon) \asymp \exp\Big(-\delta^{-1} \inf_{\mathcal{A}^\epsilon} S(x)\Big), \quad \text{where } S(x) = \frac{1}{4}\int_0^T |\dot{x}(t) + \nabla V\big(x(t)\big)|^2 \, \mathrm{d}t. \tag{6}$$

Freidlin – Wentzell theory is not the only tool by which the switching problem for (4) can be analyzed; indeed, (6) only estimates the exponential factor in the switching probability, and more refined estimates are possible by other methods. See for instance [2]. What is remarkable, however, and the source of our interest in the tools of large deviation theory, is how naturally statements such as (6) generalize to the case of stochastic *partial* differential equations [6, 12, 14].

Timescales and switching pathways. The timescale of the experiment is also critical in determining the most likely switching pathway. It is almost folkloric that the optimal pathway for switching "should follow the time-reversed gradient flow connecting x_- and the minimal energy saddle point." For T sufficiently large, this fact can be deduced from large deviation theory, as we now explain. Large deviation theory says that the action minimizer is the most likely switching pathway. In other words, if $X \in \mathcal{A}^\epsilon$, then with high probability, X stays within a small neighborhood of the pathway in $\mathcal{A}^\epsilon$ that minimizes $S(\cdot)$; cf. [15, Chapter 3, Theorem 3.4]. On the other hand, for any $s \leq T$, we have

$$\begin{aligned} S(x) &= \frac{1}{4}\int_0^T \big|\dot{x}(t) + \nabla V\big(x(t)\big)\big|^2 \, \mathrm{d}t \\ &\geq \frac{1}{4}\int_0^s \big|\dot{x}(t) + \nabla V\big(x(t)\big)\big|^2 \, \mathrm{d}t \qquad (7) \\ &= \frac{1}{4}\int_0^s \big|\dot{x}(t) - \nabla V\big(x(t)\big)\big|^2 \, \mathrm{d}t + \int_0^s \dot{x}(t) \cdot \nabla V\big(x(t)\big) \, \mathrm{d}t \\ &\geq \int_0^s \dot{x}(t) \cdot \nabla V\big(x(t)\big) \, \mathrm{d}t = V\big(x(s)\big) - V\big(x(0)\big). \qquad (8) \end{aligned}$$

Hence, (8) implies that the action cost is at least as large as the difference in energy between x_- and the minimal energy saddle point. Moreover, this bound can be achieved if (i) the inequality in (8) can be made tight by satisfying the time-reversed gradient flow:

$$\dot{x}(t) = +\nabla V\big(x(t)\big) \quad \text{for } t \leq s \tag{9}$$

and (ii) the bound in (7) can be made tight by satisfying

$$\dot{x}(t) = -\nabla V\big(x(t)\big) \quad \text{for } s < t \leq T. \tag{10}$$

Using this construction actually requires $T = \infty$, but for T sufficiently large compared to the timescale of the deterministic gradient flow, an approximation of this construction (with patching near the minima and saddle point) gives a good bound.

The classical picture changes when one considers shorter timescales. Indeed, it is easy to see via a scaling argument that for $T \downarrow 0$,

$$S(x) \sim \frac{1}{T} \int_0^T |\dot{x}(t)|^2 \, \mathrm{d}t,$$

so that the action minimizer in $\mathcal{A}^\epsilon$ is a straight line connecting x_- and the closest point in $B_\epsilon(x_+)$. Similarly, in the case of magnetic switching, it has been observed that switching probabilities in short-time experiments and long-time experiments are different (see [23] and the references therein). In particular, the *pathway* in short-time experiments seems to be different. The optimal short-time switching pathway involves new spatial structure that seems to be related to the time of the experiment.

We will be interested in using action minimization to predict most likely switching pathways. We will be particularly interested in quantifying the way that the spatial structure of the switching pathway depends on the parameter regime.

2. Simplest interesting PDE example

Because we are interested in spatial structure, we want to consider a stochastic *partial* differential equation. Because the magnetic problem is complicated (vector-valued order parameter, nonconvex constraint $|m| = 1$, nonlocal stray field term), we want to study a simpler problem. We consider the simplest interesting PDE example, namely the stochastic Allen-Cahn equation.

2.1. Introduction. The deterministic Allen-Cahn equation

$$\dot{u} = \Delta u - V'(u) \tag{11}$$

is the L^2-gradient flow for the energy

$$E(u) = \int_\Omega \left(\frac{|\nabla u|^2}{2} + V(u) \right) \mathrm{d}x. \tag{12}$$

The energy consists of two terms, a gradient term that penalizes spatial variation and a potential term in which the minima of V represent the preferred states of the order parameter. For simplicity, let V be given by (2) and let $\Omega = [0, L]^d$. The stochastic PDE is

$$\dot{u} = \Delta u - V'(u) + \sqrt{2\delta}\eta. \tag{13}$$

Here δ is a measure of the temperature of the system and η is a space-time noise. Choosing a space-time *white* noise is natural in the sense that formally, the associated invariant measure is the Gibbs distribution with the energy (12) and temperature δ. However a solution to (13) with white noise exists only in the case $d = 1$.

Consider the functional

$$S(u) = \frac{1}{4} \int_0^T \int_{[0,L]^d} \left(\dot{u} - \Delta u + V'(u) \right)^2 \mathrm{d}x \, \mathrm{d}t. \tag{14}$$

Notice that (14) depends on two parameters, L and T. That (14) is the large deviation action functional for (13) with white space-time noise in dimension $d = 1$ is proved in [12]. In higher space dimensions, some regularization of the noise in space is necessary. We are interested in (14) in $d > 1$ because we expect it to capture the leading order behavior of (13) in an appropriate regime in the space of L, T, δ, and the regularization parameter of the noise.

With the stochastic problem as motivation, we study the following leading order problem. We replace $\mathcal{A}^\epsilon$ by

$$\mathcal{A} := \{u \in C^\infty([0,L]^d \times [0,T] \mid u(\cdot,0) \equiv -1, u(\cdot,T) \equiv +1\}. \tag{15}$$

To fix ideas, we equip $\mathcal{A}$ with Neumann boundary conditions; periodic or Dirichlet boundary conditions are also possible. We are interested in the value and minimizers of the action:

$$\mathcal{S} := \inf_{\mathcal{A}} S(u). \tag{16}$$

2.2. Scaling regimes and heuristics. Four scaling regimes for the problem (16) were analyzed in [21]. See Figure 3 for an overview. The method in each case is to develop an upper bound and a lower bound that agree on the level of scaling. The first ingredient is an explicit construction; the value of $S(\cdot)$ for this construction gives an upper bound for the infimum (16). One tries to produce a bound that is close to optimal by using heuristics to develop a good construction. The second ingredient is to develop a lower bound for the infimum that says "no function can do better than the construction." We summarize the results below; see [21] for details.

Short-time limit: $T \downarrow 0$. In this limit, the action scales as

$$\mathcal{S} \sim \frac{L^d}{T}.$$

The heuristic, as discussed above, is that for $T \downarrow 0$, the $(\dot{u})^2$ term dominates the action. The construction in this case is constant in space and linear in time. The lower bound uses the spatial mean and Jensen's inequality.

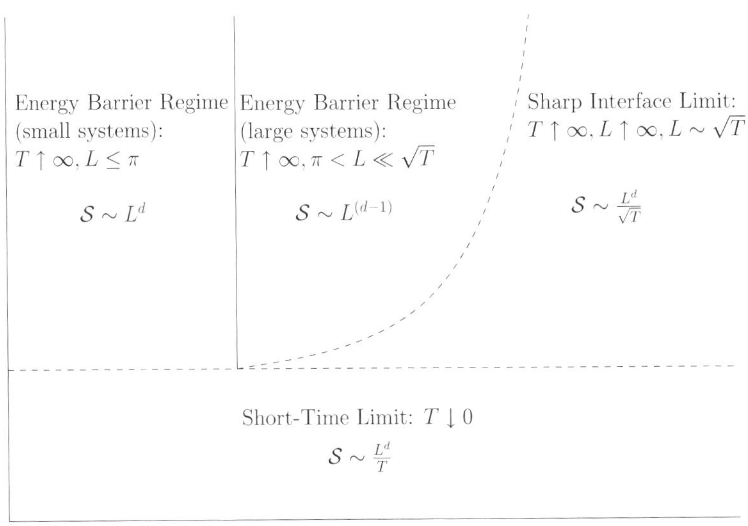

FIGURE 3. The scaling of the action in the four parameter regimes.

Energy barrier regime for small systems: $T \uparrow \infty$, $L \leq \pi$. In this limit, the action scales as

$$\mathcal{S} \sim L^d.$$

Because T is large, we expect to use the long-time most likely switching pathway, cf. Section 1. Because the system is small, the minimal energy saddle point is $u \equiv 0$. Since the path switches between spatially uniform critical points *via* a spatially uniform critical point, the heuristic is that developing spatial structure is unnecessary. Hence, the construction is constant in space, as in the short-time limit, but the time dynamic is completely different: $u(t)$ uses the gradient flow dynamics (1) in backward and then forward time, according to the heuristics discussed near (9) and (10) above.

Energy barrier regime for large systems: $T \uparrow \infty$, $\pi < L \leq T^{1/2}$. In this limit, the action scales as

$$\mathcal{S} \sim L^{d-1}.$$

For $L > \pi$, the minimal energy saddle point is spatially nonuniform and its energy scales like perimeter: L^{d-1}. This fact provides an easy lower bound in this region, as in (8). Because the saddle is nonuniform, any nearly optimal construction is also expected to be nonuniform. Extending the ideas used for small systems, one would like to use the time-reversed gradient flow connecting $u \equiv -1$ to the saddle. These ideas work fine when L is fixed or bounded. We have to be careful when considering limits in which $L \uparrow \infty$, however.

The "catch" for $L \uparrow \infty$ is that the deterministic timescale of this pathway is extremely slow; this is due to the slow motion of transition layers in the one-dimensional deterministic Allen Cahn equation [4, 5, 16]. (Even in $d > 1$, the minimal energy saddle point seems to have one-dimensional structure [21].) Hence, we would expect only to be able to use such a pathway for L small enough, which means $L \sim \log T$ in this case. Indeed, for $L \gg \log T$, we are forced to use constructions that artificially accelerate the motion of the deterministic gradient flow. When L grows sufficiently quickly with respect to T, accelerating these transition layers will lead to an additional cost in the action functional. However it turns out that as long as $L \ll \sqrt{T}$, this additional propagation cost is negligible; see [21] for details.

Sharp interface limit: $T \uparrow \infty$, $L \uparrow \infty$, $L \sim T^{1/2}$. In this limit, the action scales as

$$\mathcal{S} \sim \frac{L^d}{T^{1/2}}. \tag{17}$$

Heuristically, the sharp interface limit consists of a limit in which the system size grows so quickly with respect to T that the optimal action includes a propagation cost. Because the system is so large, the optimal path is expected to have the following character: A small region of $+1$ is formed somewhere in the system, connected to the surrounding -1 phase by transition layers. The transition layers then spread across the system, increasing the support of the $+1$ phase (cf. Figures 4 and 6 below). The propagation cost is the cost in the action functional that is associated to forcing the transition layers to cover the large system in the relatively short time T. The nucleation cost, on the other hand, is the action cost to form the transition layers. The optimal configuration balances the competing costs of nucleation and propagation, forming additional transition layers in larger systems so that each does not have to cover as much distance. For larger and larger systems there is a "cascade" of optimal configurations with more and more spatial structure. These ideas were first explored in [9], using a combination of numerical investigation and formal arguments. See also [13] for the one dimensional case.

In [21], the scaling (17) is proved by the method of upper and lower bounds. The construction used to develop the upper bound has an arbitrary number N of transition layers, each of which has the shape of an *energy minimizer* (i.e., a hyperbolic tangent), and the layers are propagated across the system with a constant velocity. Elementary calculations produce a lower bound that matches in terms of scaling. To develop a *sharp* lower bound, however, requires a more careful analysis. We describe this work below.

3. The sharp interface limit

The sharp interface limit has its name because on the scale $L \gg 1$, the order-one transition layers look like sharp interfaces connecting ± 1. It will be convenient to rescale so that the interface length goes to zero while the global length and time scales of the system remain finite; let

$$\varepsilon := L^{-1} \quad \text{and} \quad x \to \varepsilon x,\ t \to \varepsilon^2 t.$$

Under this rescaling, the action functional (14) becomes $\varepsilon^{-(d-1)} S_\varepsilon(u)$, where

$$S_\varepsilon(u) = \frac{1}{4}\int_0^T \int_{[0,L]^d} \left(\varepsilon^{1/2}\dot{u} - \varepsilon^{-1/2}\left(\varepsilon\Delta u - \varepsilon^{-1}V'(u)\right)\right)^2 \mathrm{d}x\,\mathrm{d}t. \tag{18}$$

According to (17), the action value

$$\mathcal{S} \sim \varepsilon^{-(d-1)},$$

so that in particular, the infimum

$$\inf_{u\in\mathcal{A}} S_\varepsilon(u) \tag{19}$$

is expected to have a well-defined limit as $\varepsilon \downarrow 0$. We are interested in the minimum value and the action minimizers. More broadly, we would like to identify the Γ-limit of $S_\varepsilon(\cdot)$.

3.1. Connections with other sharp interface problems.

This sharp interface limit of the action functional is related to a family of previously studied sharp interface problems, namely, the sharp interface limit of the energy functional

$$E_\varepsilon(u) = \int_{\Omega_L} \left(\frac{\varepsilon}{2}|\nabla u|^2 + \varepsilon^{-1} V(u)\right) \mathrm{d}x \tag{20}$$

and the sharp interface limit of the associated gradient flow

$$\dot{u} = \varepsilon\Delta u - \varepsilon^{-1}V'(u). \tag{21}$$

The Γ-convergence of (20) to the perimeter functional is studied in [26, 27]. Roughly speaking, Γ-convergence is a notion of convergence for functionals that implies in particular that minimizers of the finite ε problem converge to minimizers of the limit problem as $\varepsilon \downarrow 0$. (See [3] for a nice introduction.) In order for the second term in the energy to remain finite as $\varepsilon \downarrow 0$, u must converge almost everywhere to ± 1. The Γ-convergence of $E_\varepsilon(\cdot)$ to the perimeter functional is a way of making precise the idea that the leading order contribution to the energy in the limit $\varepsilon \downarrow 0$ is the length of the interface separating the two phases.

Given the convergence of $E_\varepsilon(\cdot)$, a natural question is whether the gradient flow (21) converges to the gradient flow of the perimeter functional, i.e., flow by mean curvature: interfaces with normal velocity v_n and mean curvature κ evolving according to

$$v_n = -\kappa. \tag{22}$$

The answer is yes, although only after rescaling time by ε: The dynamics (21) leads to the algebraically slow mean curvature flow $v_n = -\varepsilon\kappa$. See [8, 11, 20, 34].

The case of one space dimension is special, since the "perimeter of the interface" reduces to the number of jump discontinuities of u. The limiting energy is degenerate in the sense that the leading order energy of two N-layer configurations is the same, regardless of the location of the layers. Because of this degeneracy, the associated dynamics (21) in $d = 1$ features transition layers that move *exponentially* slowly with respect to ε^{-1} and the distance between layers [4, 5, 16, 31].

With this background in mind, we would like to ask whether the sharp interface limit of (18) is, in some sense, the action functional corresponding to the sharp interface dynamics (22). More generally, we would like to understand connections among the sharp interface limits of (20), (21), and (18). One expects the situation in $d = 1$ to be special because of the degenerate energy and exponentially slow dynamics.

3.2. Rigorous results in $d = 1$. The sharp interface limit of the upper bound constructions from [21] is sketched in Figure 4. In [22], a sharp lower bound on the action value is proved. Assuming Neumann boundary conditions, it is shown that for any sequence of smooth functions u_ε, one has

$$\liminf_{\varepsilon\to 0} S_\varepsilon(u_\varepsilon) \geq c_0 \min_{N\in\mathbb{Z}^+}\left(N + \frac{L^2}{4NT}\right),$$

where

$$c_0 = \int_{-1}^{+1} \sqrt{2\,V(u)}\,du = \frac{2\sqrt{2}}{3}. \tag{23}$$

Since this lower bound matches the upper bound from [21], this proves that the constructions with sharp interface limit sketched in Figure 4 do achieve the optimal value.

Are the constructions used for the upper bound in fact *the only* configurations that achieve the optimal action? The answer is yes, which follows from the Γ-convergence of $S_\varepsilon(\cdot)$ proved in [32]. Letting $S_0(\cdot)$ and $\mathcal{A}_0$ denote an appropriately defined limit functional and space, respectively, the upper and lower bounds of Γ-convergence are proved, namely:

Theorem 1 (Upper bound). *For every $u_0 \in \mathcal{A}_0$, there exists a sequence $u_\varepsilon \in \mathcal{A}$ with $u_\varepsilon \to u_0$ in $L^\infty(L^2)$ such that*

$$\limsup_{\varepsilon\to 0} S_\varepsilon(u_\varepsilon) \leq S_0(u_0).$$

Theorem 2 (Lower bound). *Given any sequence $u_\varepsilon \in \mathcal{A}$ with*

$$\limsup_{\varepsilon\to 0} S_\varepsilon(u_\varepsilon) < \infty \quad \text{and} \quad u_\varepsilon \to u_0 \text{ in } L^\infty(L^2),$$

we have that $u_0 \in \mathcal{A}_0$ and moreover,

$$\liminf_{\varepsilon\to 0} S_\varepsilon(u_\varepsilon) \geq S_0(u_0). \tag{24}$$

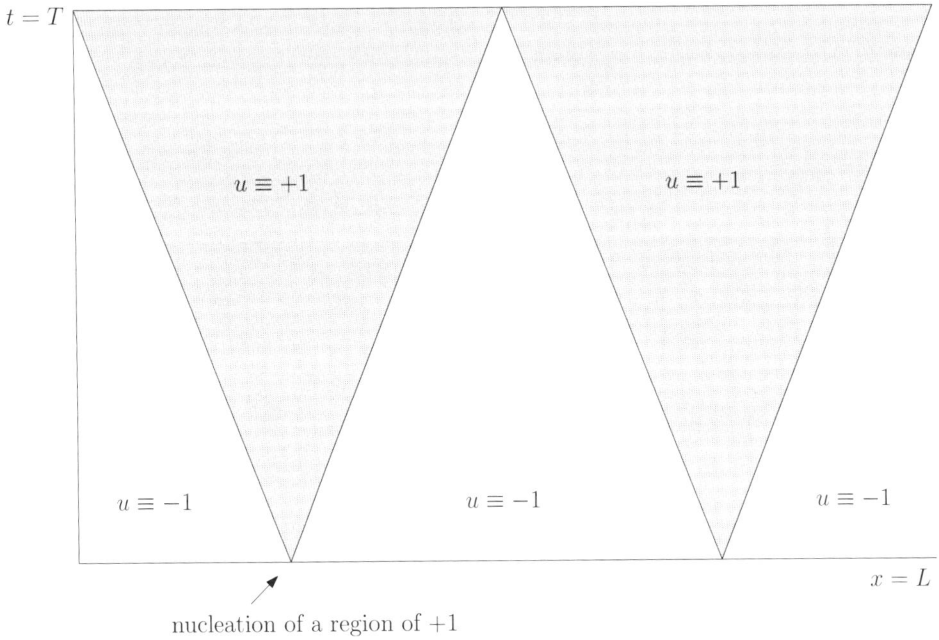

FIGURE 4. The sharp interface limit of the upper bound constructions. The N interfaces are formed at $t = 0$ and move linearly across the system in time T.

In order to explain S_0 and $\mathcal{A}_0$, it is convenient to begin by explaining a bit about the method of proof in [22] and [32]; see Subsection 3.3 below.

3.3. Details and method. It is an elementary observation that for any sequence of functions u_ε with uniformly bounded action, the energy $E_\varepsilon\big(u_\varepsilon(\cdot,t)\big)$ is uniformly bounded for every $t \in [0,T]$. Therefore it makes sense to consider the energy measures

$$\mathrm{d}\mu_\varepsilon^t := \left(\frac{\varepsilon}{2}(u_x)^2 + \varepsilon^{-1}V(u)\right)\mathrm{d}x$$

together with the space-time energy measure

$$\mathrm{d}\mu_\varepsilon := \left(\frac{\varepsilon}{2}(u_x)^2 + \varepsilon^{-1}V(u)\right)\mathrm{d}x\,\mathrm{d}t$$

and action measure

$$\mathrm{d}\nu_\varepsilon := \frac{1}{4}\Big(\varepsilon^{1/2}\dot{u} - \varepsilon^{-1/2}\big(\varepsilon\Delta u - \varepsilon^{-1}V'(u)\big)\Big)^2\,\mathrm{d}x\,\mathrm{d}t.$$

In [22], the behavior of the measures μ_ε^t and μ_ε are studied using tools from [19,35]. It is proved that the measures do what one would expect: μ_ε^t converges to a finite sum of delta masses with coefficient $c_0\,J$ where J is a positive integer and c_0 is given by (23). A delta mass with $J = 1$ at location x occurs when the limit function u_0 has a jump discontinuity at x and u_ε has a simple transition layer structure near x as $\varepsilon \downarrow 0$. Higher multiplicity (i.e., $J > 1$) of a delta mass is also possible if u_ε has multiple transition layers converging to the same point x in the limit; cf. Figure 5.

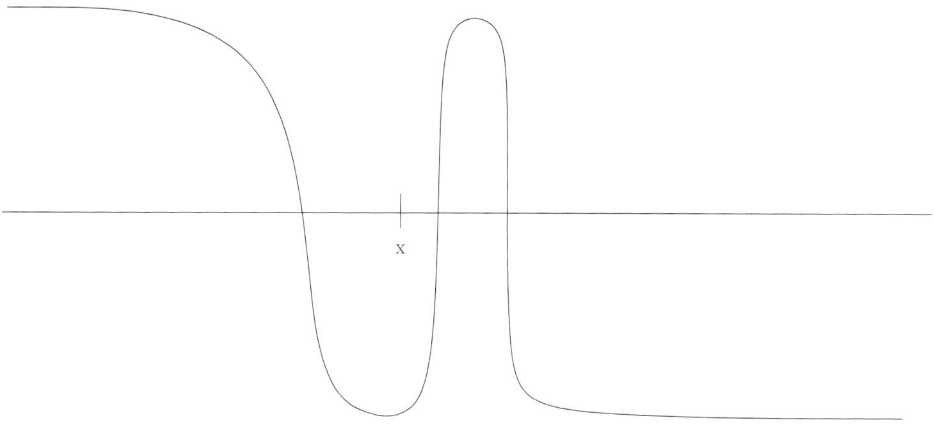

FIGURE 5. Three interfaces accumulating at the same point x in the limit. The energy will have a delta mass at x with weight $3c_0$.

For J odd, the limit function u_0 has a jump discontinuity at x, while for J even there is no jump in u_0 at x (a "hidden" interface).

A lower bound on the action functional can be derived using the convergence of the measures μ_ε^t, μ_ε, ν_ε. In [22], the lower bound is suboptimal for higher multiplicity interfaces; hence, an assumption of single multiplicity interfaces is made. In [32], the single multiplicity assumption is removed, using an improved technique for the lower bound and properties of continuous functions in $d = 1$. The improved (optimal) bound can be described roughly in the following way. Let the sequence u_ε converge in $L^\infty(L^2)$ to u_0 with $u_0(\cdot, t) = \pm 1$ almost everywhere in space for every time. Suppose for simplicity that the associated measures μ_ε^t converge to

$$c_0 \sum_{j=1}^{N} \delta_{g_j(t)}$$

where N is constant for $t \in [0, T]$ and $g_1(t) \leq g_2(t) \leq \cdots \leq g_N(t)$. (Higher multiplicity means that two or more interfaces are equal.) Then the action satisfies the lower bound:

$$S_\varepsilon(u_\varepsilon) \geq c_0 N + \frac{c_0}{4} \int_0^T \sum_{j=1}^{N} \big(\dot{g}_j(t)\big)^2 \, \mathrm{d}t.$$

To make things precise in Theorems 1 and 2, we introduce the notion of admissible functions and compatible measures. The limiting functional on the space of compatible measures is defined as

$$S^{\mathcal{M}}(\mu) = \frac{1}{2} \sum_{k=1}^{M} |\mu^{T_k^+} - \mu^{T_k^-}|_{\mathrm{TV}} + \frac{c_0}{4} \sum_{k=1}^{M-1} \int_{(T_k, T_{k+1})} \sum_{j=1}^{N(k)} (\dot{g}_j)^2 \, \mathrm{d}t,$$

where $|\cdot|_{\mathrm{TV}}$ denotes the total variation norm and

$$\mu^{T_k^+} = \lim_{t \downarrow T_k} \mu^t, \quad \mu^{T_k^-} = \lim_{t \uparrow T_k} \mu^t.$$

The limiting functional on the space of *functions* is then defined by taking the infimum over compatible measures:

$$S_0(u) := \inf_{\mu_u} S^{\mathcal{M}}(\mu_u).$$

We refer the reader to [32] for full details.

3.4. Progress in $d > 1$. The results from [9] and the upper bound constructions from [21] suggest that the Γ-limit of $S_\varepsilon(\cdot)$ in $d > 1$ should have the form

$$S_{\text{nuc}} + \frac{c_0}{4} \int_0^T \int_{\Gamma(t)} (v_n + \kappa)^2 \, d\sigma \, dt, \tag{25}$$

where S_{nuc} represents a nucleation cost for introducing a $(d-1)$–dimensional interface; $\Gamma(t)$ represents the interface at time t; and v_n, κ denote the normal velocity and mean curvature of the interface, respectively. Such a limit would answer in the affirmative the question at the end of Section 3.1: The sharp interface limit of the action functional is, formally, the action functional of a stochastically perturbed version of mean curvature flow. Moreover, (25) suggests that action minimizers should use reverse and forward curvature flow; see (9), (10) above and the accompanying discussion. A sketch is given in Figure 6. This is consistent with numerical observations [9]. For additional details, see [21].

Convergence of (18) to (25) is related to a conjecture of De Giorgi that says, roughly speaking,

$$\varepsilon^{-1} \int_\Omega \big(\varepsilon \Delta u - \varepsilon^{-1} V'(u)\big)^2 \, dx \xrightarrow{\Gamma} c_0 \int_\Gamma \kappa^2 \, d\sigma.$$

This conjecture has recently been proved in two and three space dimensions in [30, 33]; see also [1, 28] for earlier work.

4. Outlook

As explained above, the Allen – Cahn problem has been studied here in part as a "warm up" problem with the full Landau – Lifshitz – Gilbert (LLG) equations as one long range goal or inspiration. Interesting numerical investigations of the magnetic problem include [10, 25]. To what extent can the analysis capture the phenomena observed in the numerics? Already in the Allen Cahn problem, we find that the spatial structure of action minimizers depends critically on parameter space. A similar phenomenon is analyzed in [7] for a nonlocal evolution equation in one space dimension. A natural next step is to analyze the action functional for other equations that are even more closely related to the LLG equations, for instance reduced models such as [24]. See also [29] for a generalization of the Allen – Cahn result to the case in which V is a double-well potential with unequal wells.

Approaches based on large deviation theory can be useful for building efficient numerical schemes to quantify the effect of small noise [9, 18]. Although we have focused on gradient systems, it is important to emphasize that gradient structure is not required by large deviation theory. See [18] for examples. Numerical investigation of action minimization for physically relevant, nongradient systems such as spin-induced magnetic switching may produce interesting results.

There are many stochastic questions here that have been side-stepped. The theory of large deviations operates in the zero noise limit. We have considered *subsequently* sending the interface parameter to zero. Can one reverse the limit?

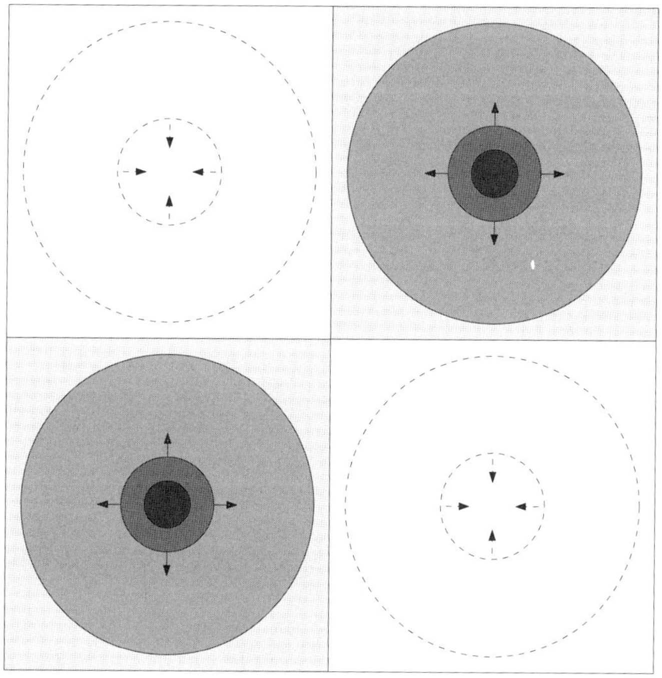

FIGURE 6. The sharp interface limit of a construction in $d = 2$. In the gray squares, interfaces grow by time-reversed curvature flow until a "checkerboard" pattern is formed. Then in the white squares, interfaces shrink by forward curvature flow.

Is there a stochastically perturbed sharp interface process that has S_0 as its action functional? An extra layer of questions arises in $d > 1$ about the role of the correlation length or regularization parameter of the noise.

Acknowledgments. We have described joint work with Bob Kohn, Felix Otto, Yoshihiro Tonegawa, and Eric Vanden-Eijnden, whose collaboration we gratefully acknowledge. We would also like to thank Weiqing Ren together with Weinan E and Eric Vanden-Eijnden for the numerical study that served as a basis for the work described in Section 3. Finally, we thank Geoff Grinstein and Roger Koch for inspiring discussions about magnetic switching phenomena.

References

1. G. Bellettini and L. Mugnai, *On the approximation of the elastica functional in radial symmetry*, Calc. Var. Partial Differential Equations **24** (2005), no. 1, 1–20.
2. A. Bovier, M. Eckhoff, V. Gayrard, and M. Klein, *Metastability in reversible diffusion processes.* I: *Sharp asymptotics for capacities and exit times*, J. Eur. Math. Soc. (JEMS) **6** (2004), no. 4, 399–424.
3. A. Braides, *Γ-convergence for beginners*, Oxford Lecture Ser. Math. Appl., vol. 22, Oxford Univ. Press, Oxford, 2002.
4. L. Bronsard and R. V. Kohn, *On the slowness of phase boundary motion in one space dimension*, Comm. Pure Appl. Math. **43** (1990), no. 8, 983–997.
5. J. Carr and R. L. Pego, *Metastable patterns in solutions of* $u_t = \epsilon^2 u_{xx} - f(u)$, Comm. Pure Appl. Math. **42** (1989), no. 5, 523–576.

6. G. Da Prato and J. Zabczyk, *Stochastic equations in infinite dimensions*, Encyclopedia Math. Appl., vol. 44, Cambridge Univ. Press, Cambridge, 1992.
7. A. De Masi, N. Dirr, and E. Presutti, *Interface instability under forced displacements*, Ann. Henri Poincaré **7** (2006), no. 3, 471–511.
8. P. de Mottoni and M. Schatzman, *Évolution géométrique d'interfaces*, C. R. Acad. Sci. Paris Sér. I Math. **309** (1989), no. 7, 453–458.
9. W. E, W. Ren, and E. Vanden-Eijnden, *Minimum action method for the study of rare events*, Comm. Pure Appl. Math. **57** (2004), no. 5, 637–656.
10. ———, *Energy landscape and thermally activated switching of submicron-size ferromagnetic elements*, Journal of Applied Physics **93** (2003), no. 4, 2275–2282.
11. L. C. Evans, H. M. Soner, and P. E. Souganides, *Phase transitions and generalized motion by mean curvature*, Comm. Pure Appl. Math. **45** (1992), no. 2, 1097–1123.
12. W. G. Faris and G. Jona-Lasinio, *Large fluctuations for a nonlinear heat equation with noise*, J. Phys. A **15** (1982), no. 10, 3025–3055.
13. H. C. Fogedby, J. Hertz, and A. Svane, *Domain wall propagation and nucleation in a metastable two-level system*, Phys. Rev. E (3) **70** (2004), no. 3, 031105.
14. M. I. Freidlin, *Random perturbations of reaction-diffusion equations: the quasideterministic approximation*, Trans. Amer. Math. Soc. **305** (1988), no. 2, 665–697.
15. M. I. Freidlin and A. D. Wentzell, *Random perturbations of dynamical systems*, 2nd ed., Grundlehren Math. Wiss., vol. 260, Springer, New York, 1998.
16. G. Fusco and J. K. Hale, *Slow-motion manifolds, dormant instability, and singular perturbations*, J. Dynam. Differential Equations **1** (1989), no. 1, 75–94.
17. A. Galves, E. Olivieri, and M. E. Vares, *Metastability for a class of dynamical systems subject to small random perturbations*, Ann. Probab. **15** (1987), no. 4, 1288–1305.
18. M. Heymann and E. Vanden-Eijnden, *The minimum action method: a least action principle on the space of curves*, Comm. Pure Appl. Math., to appear.
19. J. Hutchinson and Y. Tonegawa, *Convergence of phase interfaces in the van der Waals–Cahn–Hilliard theory*, Calc. Var. Partial Differential Equations **10** (2000), no. 1, 49–84.
20. T. Ilmanen, *Convergence of the Allen–Cahn equation to Brakke's motion by mean curvature*, J. Differential Geom. **38** (1993), no. 2, 417–461.
21. R. V. Kohn, F. Otto, M. G. Reznikoff, and E. Vanden-Eijnden, *Action minimization and sharp-interface limits for the stochastic Allen–Cahn equation*, Comm. Pure Appl. Math. **60** (2007), no. 3, 393–438.
22. R. V. Kohn, M. G. Reznikoff, and Y. Tonegawa, *Sharp-interface limit of the action functional for Allen–Cahn in one space dimension*, Calc. Var. Partial Differential Equations **25** (2006), no. 4, 503–534.
23. R. V. Kohn, M. G. Reznikoff, and E. Vanden-Eijnden, *Magnetic elements at finite temperature and large deviation theory*, J. Nonlinear Sci. **15** (2005), no. 4, 223–253.
24. M. Kurzke, *Boundary vortices in thin magnetic films*, Calc. Var. Partial Differential Equations **26** (2006), no. 1, 1–28.
25. D. Liu and C. Garcia-Cervera, *Magnetic switching of ferromagnetic thin films under thermal perturbation*, Journal of Applied Physics **98** (2005), no. 2, 023903.
26. L. Modica, *The gradient theory of phase transitions and the minimal interface criterion*, Arch. Rational Mech. Anal. **98** (1987), no. 2, 123–142.
27. L. Modica and S. Mortola, *Il limite nella* Γ*-convergenza di una famiglia di funzionali ellittici*, Boll. Un. Mat. Ital. A (5) **14** (1977), no. 3, 526–529.
28. R. Moser, *A higher order asymptotic problem related to phase transitions*, SIAM J. Math. Anal. **37** (2005), no. 3, 712–736.
29. Y. Nagase, *The action minimization problem for an Allen–Cahn equation with an unequal double-well potential*, in preparation.
30. Y. Nagase and Y. Tonegawa, *A singular perturbation problem with integral curvature bound*, Hiroshima Math. J., to appear.
31. F. Otto and M. G. Reznikoff, *Slow motion of gradient flows*, J. Differential Equations **237** (2007), no. 2, 372–420.

32. M. G. Reznikoff and Y. Tonegawa, *Higher multiplicity in the one-dimensional Allen Cahn action functional*, Indiana Univ. Math. J. **56** (2007), no. 6.
33. M. Röger and R. Schätzle, *On a modified conjecture of De Giorgi*, Math. Z. **254** (2006), no. 4, 675–714.
34. J. Rubinstein, P. Sternberg, and J. B. Keller, *Fast reaction, slow diffusion, and curve shortening*, SIAM J. Appl. Math. **49** (1989), no. 1, 116–133.
35. Y. Tonegawa, *Phase field model with a variable chemical potential*, Proc. Roy. Soc. London Ser. A **132** (2002), no. 4, 993–1019.

School of Mathematics, Georgia Institute of Technology, 686 Cherry Street, Atlanta, GA 30332-0160, USA
E-mail address: maria@math.gatech.edu

Centre de Recherches Mathématiques
CRM Proceedings and Lecture Notes
Volume **44**, 2008

The Gauss – Green Theorem for Weakly Differentiable Vector Fields

William P. Ziemer

ABSTRACT. This paper responds to a question posed by Gui Qiang Chen and Monica Torres concerning the validity of the Gauss – Green formula for weakly differentiable vector fields. They posed this question because of their need for such a formula in their research on nonlinear conservation laws, cf. [22]. The now classical Gauss – Green theorem (De Giorgi – Federer) states that

$$\int_E \operatorname{div} \boldsymbol{F} \, \mathrm{d}x = \int_{\partial^* E} \boldsymbol{F}(y) \cdot \nu(y) \, \mathrm{d}\mathcal{H}^{N-1}(y)$$

where E is a set of finite perimeter and $\boldsymbol{F}\colon \mathbb{R}^N \to \mathbb{R}^N$ is Lipschitz. Although a general Gauss – Green formula will have many applications, the interested reader will find a direct application to the work of Chen and Torres in [24], which completely subsumes this paper. In this paper we investigate a class of vector fields $\boldsymbol{F}\colon \mathbb{R}^N \to \mathbb{R}^N$ with the property that $\boldsymbol{F} \in L^\infty$ and $\operatorname{div} \boldsymbol{F}$ is a totally finite signed measure, μ, in the sense of distributions. The primary focus of our investigation is to introduce a suitable notion of normal trace to obtain a general version of the Gauss – Green theorem for $\boldsymbol{F}$ which holds for any set of finite perimeter. Thus, we show that there is a bounded field $\boldsymbol{F}^*\colon \partial^* E \to \mathbb{R}^N$ such that

$$\int_E \operatorname{div} \boldsymbol{F} := \mu(E) = \int_{\partial^* E} \boldsymbol{F}^*(y) \cdot \nu(y) \, \mathrm{d}\mathcal{H}^{N-1}(y).$$

The function $\boldsymbol{F}^*(y) \cdot \nu(y)$ can be interpreted as the normal trace. In the event that $\boldsymbol{F}$ is also continuous, we show that $\boldsymbol{F}^* = \boldsymbol{F}$.

1. Introduction

In this paper we investigate a class of vector fields $\boldsymbol{F}\colon \mathbb{R}^N \to \mathbb{R}^N$ with the property that $\boldsymbol{F} \in L^\infty$ and $\operatorname{div} \boldsymbol{F}$ is a Radon measure μ (i.e., a totally finite signed measure). These fields, denoted by $\mathcal{DM}^\infty$, are called *bounded divergence measure fields*. The primary focus of our investigation is to introduce a suitable notion of normal trace to obtain a general version of the Gauss – Green theorem for $\boldsymbol{F}$ which holds for any set of finite perimeter. Clearly, this investigation will be

2000 *Mathematics Subject Classification.* Primary 28C05, 35L65, 35L50; Secondary 26B12, 28A75.

Key words and phrases. Measure-theoretical analysis, geometric measures, conservation laws, Cauchy fluxes, balance laws, entropy solutions, divergence-measure fields, sets of finite perimeter, Gauss – Green formula, normal traces, oriented surfaces.

This is the final form of the paper.

closely related to the theory of BV functions in $\mathbb{R}^N$; in fact, it would be completely subsumed by the BV theory if the fields were of the form $\boldsymbol{F} = (F_1, F_2, \ldots, F_N)$ with each $\boldsymbol{F}_k \in \mathrm{BV}(\mathbb{R}^N)$, since

$$\operatorname{div} \boldsymbol{F} = \sum_{k=1}^{N} \frac{\partial F_k}{\partial x_k}$$

is then a Radon measure μ. However, in general, the condition $\operatorname{div} \boldsymbol{F} = \mu$ allows for cancellation, thus making the problem more difficult and accordingly more important for applications (see [24, Sections 10 – 12]). For the Gauss – Green theorem in the BV setting, we refer to Burago – Maz′ja [15], Vol′pert [86], and the references therein. The Gauss – Green theorem for Lipschitz vector fields over sets of finite perimeter was first obtained by De Giorgi [29, 30] and Federer [36, 40–42].

Among the many who have have contributed to this and related problems are, Anzellotti [7], Chen – Frid [18, 19] , Chen – Torres [22], Pfeffer [32, 76, 77, 79, 80], Jurkat – Nonnemacher [58–62].

Some earlier efforts were made on generalizing the Gauss – Green theorem for some special situations of divergence-measure fields, where relevant results can be found in Anzellotti [7] for an abstract formulation for $\boldsymbol{F} \in L^\infty$ over a set with C^1 boundary and Ziemer [92] for a related problem for $\operatorname{div} \boldsymbol{F} \in L^1$. In Chen – Frid [18, 19], an explicit way to calculate the suitable normal traces over Lipschitz deformable surfaces was first observed for $\boldsymbol{F} \in D^\infty$, under which the Gauss – Green theorem was shown to hold for any set with Lipschitz deformable boundary. In particular, it was shown in [18, 19] that the normal trace over a Lipschitz deformable surface, oriented by the unit normal vector ν, is determined completely by the information near the positive side of the surface oriented by ν and is independent of the information from the opposite side. This is one of the primary motivations for our further investigation on divergence-measure fields. In Chen – Torres [22]), the normal trace for bounded divergence-measure fields over sets of finite perimeter and the corresponding Gauss – Green theorem was first obtained. One of the main objectives of this paper is to provide a completely different approach to the Chen – Torres result and answer whether the normal trace of a set of finite perimeter can also be determined completely by the information near the boundary, but also essentially from the interior of the set, so that the Gauss – Green theorem holds for any set of finite perimeter.

We recall a very general approach was initiated by Fuglede [48] in which the following result was established: Suppose $\boldsymbol{F} \in L^p(\mathbb{R}^N; \mathbb{R}^N)$, $1 \le p \le \infty$, is a vector field with $\operatorname{div} \boldsymbol{F} = \mu$, where μ is a signed measure. Then

$$\int_E \operatorname{div} \boldsymbol{F} := \mu(E) = -\int_{\partial^* E} \boldsymbol{F}(y) \cdot \nu(y) \, \mathrm{d}\mathcal{H}^{n-1}(y) \tag{1.1}$$

for "almost all" sets of finite perimeter E. The term "almost all" is expressed in terms of a condition that resembles "extremal length," a concept used in complex variable and potential theory. Perhaps one way of summarizing our work is to say that we wish to extend Fuglede's result so that set E is of finite perimeter. Of course, this requires a suitable notion of the normal trace $\boldsymbol{F}$ on $\partial^* E$. This is really the crux of the problem as $\boldsymbol{F}$, being only measurable, cannot be redefined on an arbitrary set of dimension $N - 1$. To achieve our goal we first establish an approximation theorem for any set of finite perimeter by a family of sets with smooth boundary essentially from the interior of the set with respect to the Radon measure $\mu = \operatorname{div} \boldsymbol{F}$; we then employ this approximation theorem to derive the normal trace of $\boldsymbol{F}$ on the boundary of any set of finite perimeter to thus establish the Gauss–Green theorem for any set of finite perimeter.

In a subsequent paper, [24], we apply our results on divergence-measure fields to analyze Cauchy fluxes that are bounded by a Radon measure μ over an oriented surface (i.e., an $(N-1)$-dimensional surface that is a part of the boundary of a set of finite perimeter) and thereby develop a general mathematical formulation of the balance law. In the classical setting of the balance law, Cauchy first discovered that the flux density is necessarily a linear function of the interior normal (equivalently, the exterior normal) under the assumption that the flux density through a surface depends on the surface solely through the normal at that point. Noll showed [69] (see also [70–73]) that Cauchy's assumption follows directly from the balance law. Ziemer [92] provided a first formulation of the balance law for the flux function $\boldsymbol{F} \in L^1$ in the general context of sets with finite perimeter (see also [25, 26, 52, 55]).

One of the new features in our formulation is to allow the presence of exceptional surfaces, "shock waves", on which the Cauchy flux has a discontinuity. When σ reduces to the N-dimensional Lebesgue measure $\mathcal{L}^N$, this formulation reduces to Ziemer's formulation, which shows its consistency with the classical setting. We show that for a Cauchy flux $\mathcal{F}$ bounded by a measure σ, there exists a bounded divergence-measure field $\boldsymbol{F}\colon \mathbb{R}^N \to \mathbb{R}^N$, defined $\mathcal{L}^N$-a.e., such that

$$\mathcal{F}(S) = -\int_S \boldsymbol{F}(y) \cdot \nu(y)\, \mathrm{d}\mathcal{H}^{N-1}(y)$$

for almost any oriented surface S, oriented by the interior unit normal ν. Then we employ our results on divergence-measure fields to recover the values of the Cauchy flux on the exceptional surfaces directly via the vector field $\boldsymbol{F}$. The value of $\boldsymbol{F}$ on the exceptional surface is the unique limit of the normal traces of $\boldsymbol{F}$ on the nonexceptional surfaces essentially from the positive side of the exceptional surface oriented by ν. Finally, we apply this general framework to the derivation of systems of balance laws with measure-valued source terms from the formulation of the balance law. We also apply our results to the recovery of Cauchy entropy fluxes through the entropy inequality for entropy solutions of hyperbolic conservation laws by capturing entropy dissipation.

The organization of this paper is the following. In Section 2, we introduce some basic properties of sets of finite perimeter, oriented surfaces, and divergence-measure fields. In Section 3, we prove that the normal trace of $\boldsymbol{F}$ on a smooth boundary exists, which then leads to the generalized Gauss–Green theorem for $\boldsymbol{F}$. In Section 4, we establish a fundamental approximation theorem which implies that, given a Radon measure $\mu = \operatorname{div} \boldsymbol{F}$ for any $\boldsymbol{F} \in \mathcal{DM}^\infty$, any set of finite perimeter can be approximated by a sequence of sets with smooth boundary essentially from the measure theoretic interior of the set.

In Section 5, we introduce the normal trace of a divergence-measure field $\boldsymbol{F}$ on the boundary ∂E of any set of finite perimeter as the limit of the normal traces of $\boldsymbol{F}$ on the smooth surfaces that were constructed in Section 4. This yields the corresponding main result, the Gauss–Green theorem for any set of finite perimeter. Section 6 is devoted to a further analysis of properties of divergence-measure fields, especially showing, in particular, the consistency of our normal trace with the classical trace when the vector field is continuous.

In Section 7, we first introduce a class of Cauchy fluxes and then prove that such a Cauchy flux induces a divergence-measure field $\boldsymbol{F}$ so that the Cauchy flux over almost every oriented surface by ν can be recovered through $\boldsymbol{F} \cdot \nu$. We then recover the values of the Cauchy flux on the exceptional surfaces, "shock waves," through the corresponding divergence-measure field via the normal traces over all oriented surfaces.

2. Preliminaries: BV functions and sets of finite perimeter

In this section we recall the fundamental properties of BV functions and sets of finite perimeter and direct the reader to some standard sources for this material [4, 33, 42, 49, 93]. We then introduce the notion of oriented surfaces and develop some of the basic properties of divergence-measure fields.

We start with some basic notations and definitions. First, $\mathcal{H}^M$ with $M \leq N$ denotes the M-dimensional Hausdorff measure in $\mathbb{R}^N$, and $\mathcal{L}^N$ denotes the Lebesgue measure in $\mathbb{R}^N$. We recall that $\mathcal{L}^N = \mathcal{H}^N$. For any $\mathcal{L}^N$-measurable set $E \subset \mathbb{R}^N$, we will often denote $|E|$ as the $\mathcal{L}^N$-Lebesgue measure of the set E. For such E, we denote its topological boundary as ∂E. Also, we denote $B(x, r)$ as the open ball of radius r and center at x. The symmetric difference of sets is denoted by $A \Delta B$. When $\Omega \subset \mathbb{R}^N$, we let $E \Subset \Omega$ denote that the closure of E is compact and is contained in Ω. The L^p norm of a function u defined on a set E with integration taken with respect to a measure μ is denoted by $\|u\|_{p;E,\mu}$. In the event μ is Lebesgue measure, we will simply write $\|u\|_{p;E}$. If X is a Hausdorff space, we let $C_c(X)$ denote the space of compactly supported continuous function on X with $\|\varphi\|_0 := \sup\{|\varphi(x)| : x \in X\}$.

A Radon measure, μ, is a Borel measure which has the property that $\mu(K) < \infty$ for each compact set $K \subset \mathbb{R}^N$. The space of Radon measures supported on an open set Ω is denoted by

$$\mathscr{M}(\Omega). \tag{2.1}$$

Any Radon measure μ can be decomposed in the difference of two positive Radon measures

$$\mu = \mu^+ - \mu^-,$$

for which we denote by $\|\mu\| = \mu^+ + \mu^-$, the *total variation* of μ. Equivalently, If μ is a signed Radon measure on $\mathbb{R}^N$, the total variation of μ can be shown to be

$$\begin{aligned} \|\mu\|(\mathbb{R}^N) &= \sup\left\{\int_{\mathbb{R}^N} \varphi \, d\mu : \varphi \text{ continuous with } \|\varphi\|_0 \leq 1\right\} \\ &= \sup\{\mu(B) : B \subset \mathbb{R}^N,\ B \text{ a Borel set}\}. \end{aligned} \tag{2.2}$$

Since the space of Radon measures can be identified with the dual of $C_c(\mathbb{R}^N)$, we may consider a measure μ as a linear functional on $C_c(\mathbb{R}^N)$; thus we will write

$$\mu(\varphi) := \int_{\mathbb{R}^N} \varphi \, d\mu. \tag{2.3}$$

We recall the familiar weak* topology on $\mathscr{M}(\Omega)$, which when restricted to sequences, yields

$$\mu_k(\varphi) \rightharpoonup \mu(\varphi) \quad \text{for each } \varphi \in C_c(\Omega);$$

that is, μ_k converges to μ in the weak* topology if and only if

$$\int_\Omega \varphi \, \mathrm{d}\mu_k \to \int_\Omega \varphi \, \mathrm{d}\mu \quad \text{or equivalently } \mu_k(\varphi) \to \mu(\varphi) \quad \text{for each } \varphi \in C_c(\Omega). \tag{2.4}$$

The space $L^p(X, \mu)$, $1 \le p \le \infty$, denotes all functions f with the property that $|f|^p$ is μ-integrable. The conjugate of p is $p' := p/(p-1)$.

Theorem 2.1 (Uniform boundedness principle). *Let $\mathcal{F}$ be a family of continuous linear mappings from a Banach space X into a normed linear space Y such that*

$$\sup_{T \in \mathcal{F}} \|T(x)\| < \infty$$

for each $x \in X$. Then

$$\sup_{T \in \mathcal{F}} \|T\| < \infty.$$

This result implies the following which will be of great importance to us.

Theorem 2.2. *If B is a Banach space and if T_k is a sequence of liner functionals that converge weak* to T, then*

$$\limsup_{k \to \infty} \|T_k\| < \infty.$$

Corollary 2.3. *Let μ_k be a sequence of Radon measures that converge to μ in the weak* topology, then*

$$\limsup_{k \to \infty} \|\mu_k\| < \infty.$$

Next, we quote a familiar result concerning weak* convergence.

Lemma 2.4. *Let μ be a positive Radon measure on $\mathbb{R}^N$ and suppose μ_k is a sequence of Radon measures converging weak* to μ. Let f be a bounded (by $M > 0$), μ-measurable function with compact support whose set of discontinuities are assumed to lie in a μ-null set. Then*

$$\lim_{k \to \infty} \int_{\mathbb{R}^N} f \, \mathrm{d}\mu_k = \int_{\mathbb{R}^N} f \, \mathrm{d}\mu.$$

In particular this is true whenever f is the characteristic function of a μ-measurable set, A, with $\mu(\partial A) = 0$.

We will have use for the following basic fact which is a direct consequence of Fubini's theorem.

Lemma 2.5. *If $u \ge 0$ is measurable, $p > 0$, and $A_t = \{x : u(x) > t\}$, then*

$$\int_\Omega u(x)^p \, \mathrm{d}x = \int_0^\infty |A_t| \, \mathrm{d}t^p = p \int_0^\infty t^{p-1} |A_t| \, \mathrm{d}t.$$

The next result, Federer's coarea formula, is of critical importance to us.

Theorem 2.6 ([42, Theorem 3.1])**.** *If $f \colon \mathbb{R}^N \to \mathbb{R}^k$, $N \ge k$, is a Lipschitzian map, then*

$$\int_{\mathbb{R}^N} g(x) Jf(x) \, \mathrm{d}H^N(x) = \int_{\mathbb{R}^k} \int_{f^{-1}(y)} g(x) \, \mathrm{d}H^{N-k}(x) \, \mathrm{d}H^k(y) \tag{2.5}$$

whenever $g \colon \mathbb{R}^N \to \mathbb{R}^1$ is H^N-integrable. Here, $Jf(x)$ denotes the k-dimensional Jacobian of f at x, namely, the square root of the sum of the squares of the determinants of the $k \times k$ minors of the differential of f at x.

Definition 2.7. For every $\alpha \in [0,1]$ and every $\mathcal{L}^N$-measurable set $E \subset \mathbb{R}^N$, define

$$E^\alpha := \{y \in \mathbb{R}^N : D(E,y) = \alpha\}, \tag{2.6}$$

where

$$D(E,y) = \lim_{r\to 0} \frac{|E \cap B(y,r)|}{|B(y,r)|}. \tag{2.7}$$

E^α is the set of all *points with density* α. We define the *measure-theoretic boundary* of E, $\partial^m E$, as

$$\partial^m E = \mathbb{R}^N \setminus (E^0 \cup E^1). \tag{2.8}$$

Definition 2.8. A function $f\colon \mathbb{R}^N \to \mathbb{R}^1$ is called a function of bounded variation if each partial derivative of f is a totally finite signed measure. Notationally, we write $f \in \mathrm{BV}(\mathbb{R}^N)$. Let E be an $\mathcal{L}^N$-measurable subset of $\mathbb{R}^N$. We say that E is a *set of finite perimeter* if $\mathcal{H}^{N-1}(\partial^m E) < \infty$. Equivalently, E is of finite perimeter if $\chi_E \in BV(\mathbb{R}^N)$. Consequently, if E is a set of finite perimeter, then $\nabla\chi_E$ is a Radon (vector-valued) measure and we denote its total variation by $\|\nabla\chi_E\|$.

Definition 2.9. Let E be a set of finite perimeter. The **reduced boundary** of E, denoted as $\partial^* E$, is the set of all points $x \in \mathbb{R}^N$ such that

(i) $\|\nabla\chi_E\|\big(B(x,r)\big) > 0$ for all $r > 0$,
(ii) the limit exists $\nu_E(x) := \lim_{r\to 0} \nabla\chi_E\big(B(x,r)\big)/\big\|\nabla\chi_E\big(B(x,r)\big)\big\|$,
(iii) For $\mathcal{H}^{N-1}$-a.e. $y \in \partial^* E$,

$$\lim_{r\to 0} \frac{\|\nabla\chi_E\|\big(B(y,r)\big)}{\omega(N-1)r^{N-1}} = 1.$$

The generalized gradient of χ_E enjoys the following basic relationship with H^{N-1}:

$$\|\nabla\chi_E\| = H^{N-1} \llcorner \partial^* E \tag{2.9}$$

The unit vector, $\nu_E(x)$, is called the *measure-theoretic interior normal to E at x*.

Also, we recall that the reduced boundary, $\partial^* E$, is an $(n-1)$-rectifiable set which implies, in particular, that there exists a countable family of C^1-manifolds M_k of dimension $(N-1)$ and a $\mathcal{H}^{N-1}$-nullset, $\mathcal{N}$, such that

$$\partial^* E \subset \bigcup_{k=1}^{\infty} M_k \cup \mathcal{N} \tag{2.10}$$

In view of the following we see that ν_E is aptly named because ν (we sometimes write ν instead of ν_E for notational simplicity) is the interior normal to E provided that E (in the limit and in measure) lies in the appropriate half-space determined by the hyperplane orthogonal to ν; that is, ν is the interior normal to E at x provided

$$\lim_{r\to 0} r^{-n} |B(x,r) \cap \{y : (y-x)\cdot\nu > 0, y \notin E\}| = 0$$

and

$$\lim_{r\to 0} r^{-n} |B(x,r) \cap \{y : (y-x)\cdot\nu < 0, y \in E\}| = 0,$$

The sets E^0 and E^1 may be considered as the measure-theoretic exterior and interior of E, which motivate the definition of measure-theoretic boundary. Indeed, for any set $E \subset \mathbb{R}^N$, the definitions imply that $\mathbb{R}^N = E^1 \cup \partial^m E \cup E^0 \cup \mathcal{N}$ where $\mathcal{H}^{N-1}(\mathcal{N}) = 0$. If we define a set E to be "open" if E is both measurable and

$D(E, x) = 1$ for all $x \in E$, then this concept of openness defines a topology, called the *density topology*. It is an interesting exercise to prove that the open sets are closed under arbitrary unions. The crux of the problem is to prove that the arbitrary union is, in fact, measurable. This topology has interest because it is smallest topology (the one with the smallest number of open sets) for which the approximately continuous functions are continuous cf. [50, 51]. The following result, which is easily verified (although tedious), will be needed in the sequel.

Lemma 2.10. *If $A, B \subset \mathbb{R}^N$ are sets of finite perimeter, then*

$$\partial^m(A \cap B) + \partial^m(A \cup B) \subset \partial^m A + \partial^m B.$$

Remark 2.11. If E is a set of finite perimeter, then clearly

$$\partial^* E \subset E^{1/2} \subset \partial^m E, \tag{2.11}$$

and

$$\mathcal{H}^{N-1}\{(\mathbb{R}^N \setminus (E^0 \cup E^1)) \setminus \partial^* E\} = 0. \tag{2.12}$$

Lemma 2.12. *Let $u \colon \mathbb{R}^N \to \mathbb{R}^1$ be Lipschitz function and $A \subset \mathbb{R}^N$ a set of measure 0. Then*

$$\mathcal{H}^{N-1}(\partial A_{k;s}) \cap A) = \mathcal{H}^{N-1}(u_k^{-1}(s) \cap A) = 0 \quad \textit{for almost all } s.$$

while

$$\mathcal{H}^{N-1}(\partial A_{k;s}) \cap A) = \mathcal{H}^{N-1}(u_k^{-1}(s) \cap A) \quad \textit{for any measurable set } A,$$

because the set of critical values of u are of measure 0.

Proof. From the coarea formula,

$$0 = \int_A |\nabla u(x)| \, dx = \int_{\mathbb{R}^1} \mathcal{H}^{N-1}(A \cap u^{-1}(s)) \, ds. \qquad \square$$

If $E \subset \mathbb{R}^N$ is a set of finite perimeter, than we know $\mathcal{H}^{N-1}(\partial^* E) < \infty$ and therefore, the following is immediate from the previous result.

Corollary 2.13. *Let $u \colon \mathbb{R}^N \to \mathbb{R}^1$ be Lipschitz function and $E \subset \mathbb{R}^N$ a set of finite perimeter. Then*

$$\mathcal{H}^{N-1}(u_k^{-1}(s) \cap \partial^* E) = 0 \quad \textit{for almost all } s.$$

Definition 2.14. Let $\rho \in C_c^\infty(\mathbb{R}^N)$ be a standard symmetric *mollifying kernel*; that is, ρ is a nonnegative function with support in the unit ball which satisfies $\|\rho\|_{1;\mathbb{R}^N} = 1$. With $u \in L^1(\mathbb{R}^N)$, we let $u_\varepsilon := u * \rho_\varepsilon$ where $\rho_\varepsilon(x) := (1/\varepsilon^N)\rho(x/\varepsilon)$.

Recall the following properties of mollification (cf. [93]):

Theorem 2.15. (i) *If $u \in L^1_{\text{loc}}(\mathbb{R}^N)$, then for every $\varepsilon > 0$, $u_\varepsilon \in C^\infty(\mathbb{R}^N)$ and $D^\alpha(\varphi_\varepsilon * u) = (D^\alpha \varphi_\varepsilon) * u$ for each multi-index α,*

(ii) *With $k := 1/\varepsilon$ and $\varepsilon \to 0$, $u_k(x) \to u(x)$ whenever x is a Lebesgue point for u. In case u is continuous then u_ε converges uniformly to u on compact subsets of $\mathbb{R}^N$,*

(iii) *If $u \in L^p(R^N)$ and $1 \le p < \infty$, then $u_\varepsilon \in L^p(R^N)$, $\|u_\varepsilon\|_p \le \|u\|_p$, and $\lim_{\varepsilon \to 0} u_\varepsilon - u$.*

When u is taken as χ_E, E a set of finite perimeter, this result can be considerably strengthened.

Lemma 2.16. *If u_k is a sequence of mollifiers of χ_E, where E is of finite perimeter, then*

(i) $u_k \in C^\infty(\mathbb{R}^N)$,

(ii) *there is a set $\mathcal{N}$ with $\mathcal{H}^{N-1}(\mathcal{N}) = 0$ and a function $u_E \in \mathrm{BV}$ such that for all $x \notin \mathcal{N}$, $u_k(x) \to u_E(x)$ where*

$$u_E(x) = \begin{cases} 1 & x \in E^1 \\ \frac{1}{2} & x \in \partial^* E \\ 0 & x \in E^0, \end{cases}$$

(iii) $\nabla u_k \rightharpoonup \nabla u_E$ *in the weak* topology,*

(iv) *there exists a number $M > 0$ such that*

$$\limsup_{k\to\infty} \|\nabla u_k\|(\mathbb{R}^N) \leq M,$$

(v) $\|\nabla u_k\|(U) \to \|\nabla u_E\|(U)$, *whenever U is an open set with $\|\nabla u_E\|(\partial U) = 0$,*

(vi) *Note, since $u_E = \chi_E$ almost everywhere (in fact, everywhere except for an $\mathcal{H}^{N-1}$-null set), that $\nabla \chi_E = \nabla u_E$,*

(vii) *Finally, note that $\{x : u_E(x) > \frac{1}{2}\} \subset E^1$ while $\{x : 0 < u_E(x) < \frac{1}{2}\} \subset E$.*

PROOF. Only (iii) and (iv) require a proof since (i), (ii) and (v) are results from standard BV theory while (vi) and (vii) are immediate from definitions.

(iii) Since $u_k \to u_E$ in $L^1(\mathbb{R}^N)$, then $u_k \rightharpoonup u_E$ when considered as distributions; that is,

$$\int_{\mathbb{R}^N} \varphi \, \mathrm{d}\mu_k \to \int_{\mathbb{R}^N} \varphi \, \mathrm{d}\mu$$

for each test function $\varphi \in C_c^\infty$. Therefore, $\nabla u_k \rightharpoonup \nabla u_E$ as distributions and consequently as measures since ∇u_k and ∇u_E are already in the space of measures.

(iv) This follows from Corollary 2.3. This result is valid in a more general context, namely, that whenever $u_k \to u$ in L^1, where $u_k, u \in BV$, then $\|\nabla u_k\| \leq M$. However, when $u = u_E$, then M can be taken as $M = \|\nabla u_E\| = \|\nabla \chi_E\|$, see Lemma 2.18 below. □

Remark 2.17. Using different techniques, the previous result was proved in [22]. In fact, it was used to prove their main result (Theorem 1) that $2\overline{\chi_E F \cdot \nabla \chi_E}$, which is absolutely continuous with respect to $\mathcal{H}^{N-1} \llcorner \partial^* E$, is the normal trace. Thus

$$\mu(E^1) := \int_{E^1} \operatorname{div} \boldsymbol{F} = \int_{\partial^m E} 2\overline{\chi_E F \cdot \nabla \chi_E} \, \mathrm{d}\mathcal{H}^{N-1}. \tag{2.13}$$

Later, we will establish this result (Theorem 5.2) as part of our development.

The next result affirms the notion that mollification is, in general, a norm reducing operation.

Lemma 2.18. *Let E be a set of finite perimeter and let u_k denote a sequence of mollifiers of χ_E. Then*

$$\|\nabla u_k\|_1 \leq \|\nabla \chi_E\|.$$

PROOF. We know that

$$f_\varepsilon(x) = \int_{\mathbb{R}^N} \varphi_\varepsilon(x-y) f(y)\, \mathrm{d}y$$

and this implies, in the sense of measures

$$Df_\varepsilon = \varphi_\varepsilon * Df.$$

Note that $f_\varepsilon \in C^\infty$. so that

$$Df_\varepsilon(x) = \int_{\mathbb{R}^N} \varphi_\varepsilon(x-y)\, \mathrm{d}m(y) \quad \text{where } m \text{ is the measure } m := Df$$

and therefore

$$|Df_\varepsilon(x)| \leq \int_{\mathbb{R}^N} \varphi_\varepsilon(x-y)\, \mathrm{d}\|m\| = \|m\|_\varepsilon(\mathbb{R}^N) \leq \|m\|(\mathbb{R}^N).$$

Thus, the derivative of f_ε is bounded, the bound being independent of ε.

Now replace f by χ_E and f_ε by u_k to obtain

$$|Du_k(x)| \leq \int_{\mathbb{R}^N} \varphi_\varepsilon(x-y)\, \mathrm{d}\|m\| = \|m\|_\varepsilon(\mathbb{R}^N) \leq \|m\|(\mathbb{R}^N), \quad \text{where } m := Du_E,$$

for all $x \in \mathbb{R}^N$. That is

$$\int_{\mathbb{R}^N} |\nabla u_k|\, \mathrm{d}x \leq \int_{\mathbb{R}^N} \varphi_\varepsilon(x-y)\, \mathrm{d}\|m\| = \|m\|_\varepsilon(\mathbb{R}^N) \leq \|m\|(\mathbb{R}^N). \qquad \square$$

One of the fundamental results of Geometric Measure theory is that sets of finite perimeter possess a measure-theoretic interior normal which is suitably general to ensure the validity of the Gauss – Green theorem.

Theorem 2.19 (De Giorgi – Federer (1945 – 58)). *If E has finite perimeter, then with ν denoting the interior normal*

$$\int_E \operatorname{div} \boldsymbol{F}\, \mathrm{d}x = -\int_{\partial^* E} \boldsymbol{F}(y)\cdot \nu(y)\, \mathrm{d}H^{n-1}(y)$$

whenever $\boldsymbol{F}\colon \mathbb{R}^N \to \mathbb{R}^N$ is Lipschitz.

Remark 2.20. The De Giorgi – Federer result shows that integration by parts hold on a very large and rich family of sets, but only for fields, $\boldsymbol{F}$, that are "regular," (Lipschitz). We contrast their work with that of the following result of Fuglede [48].

Theorem 2.21 (Fuglede (1960)). *Suppose $\boldsymbol{F} \in L^p(\mathbb{R}^n;\mathbb{R}^n)$, $1 \leq p \leq \infty$, is a vector field with $\operatorname{div} \boldsymbol{F} = \mu$, where μ is a signed measure. Then there exists a function $g\colon \mathbb{R}^n \to \mathbb{R}$ with $g \in L^{p'}$ such that*

$$\int_E \operatorname{div} \boldsymbol{F} := \mu(E) = -\int_{\partial^* E} \boldsymbol{F}(y)\cdot \nu(y)\, \mathrm{d}H^{n-1}(y) \tag{2.14}$$

for all sets of finite perimeter E except possibly those for which

$$\int_{\partial^* E} g(y)\, \mathrm{d}H^{n-1}(y) = \infty.$$

Remark 2.22. following, which is a direct consequence of Fuglede's result, will be of great use to us. Suppose $g\colon \mathbb{R}^N \to \mathbb{R}^1$ is Lipschitz. For $s < t$, consider the "annulus" determined by g, $\mathcal{A}_{s,t} := \{x : s < g(s) \leq t\}$. Then, by appealing to the coarea formula, we see that $\mathcal{A}_{s,t}$ is a set finite perimeter for almost all $s < t$. Moreover, again appealing to coarea, we see for almost all $s < t$ that

$$\mu(A_{s,t}) := \int_{\mathcal{A}_{s,t}} \operatorname{div} \boldsymbol{F} = \int_{\partial^* \mathcal{A}_{s,t}} \boldsymbol{F}(y) \cdot \nu(y) \, \mathrm{d}H^{n-1}(y). \tag{2.15}$$

Remark 2.23. One of the main objectives of this paper is to show that when $\boldsymbol{F} \in L^\infty(\mathbb{R}^N; \mathbb{R}^N)$, we can extend Fuglede's result by showing (2.14) and (2.15) hold for all sets of finite perimeter and not merely for "almost all" sets in the sense of Fuglede. Although we do not employ Fuglede's theorem directly, his result provided the motivation and insight for the development of our method.

We recall that the BV space, the space of functions of bounded variation, in fact represents equivalence classes of functions so that when a function in a class is changed on a set of $\mathcal{L}^N$-measure zero, it remains in this class. The same is true for sets of finite perimeter because a set of finite perimeter, E, is by definition a function whose characteristic function, χ_E, is a function of bounded variation. Thus it follows that E may be altered by a set of $\mathcal{L}^N$-measure zero and still determine the same essential boundary $\partial^m E$.

Throughout, we will choose a preferred representative for E and thereby adopt the following convention.

Definition 2.24. $E := \{y : D(E, y) = 1\} \cup \partial^m E\}$.

Definition 2.25. Let $\boldsymbol{F} \in L^p(\Omega; \mathbb{R}^N)$, $1 \leq p \leq \infty$ and assume that $\operatorname{div} \boldsymbol{F} = \mu$ where μ is a signed Radon measure with finite variation. For $\Omega' \Subset \Omega$, the trace of the normal component of $\boldsymbol{F}$ on $\partial\Omega'$ is a functional defined by

$$T\boldsymbol{F}_{\partial\Omega'}(\varphi) = \int_{\Omega'} \nabla\varphi \cdot \boldsymbol{F} + \int_{\Omega'} \varphi \, \mathrm{d}\mu \tag{2.16}$$

for all test functions $\varphi \in C_c^\infty(\Omega)$. Clearly, $T\boldsymbol{F}_{\partial\Omega'}$ is a distribution defined on Ω. One of the main objectives of his paper is to show that when $p = \infty$ and Ω' has finite perimeter, then $T\boldsymbol{F}_{\partial\Omega'}$ is a bounded field on $\partial\Omega'$.

Lemma 2.26. $\operatorname{spt} T\boldsymbol{F}_{\partial\Omega'} \subset \partial\Omega'$. *That is, if ψ and φ are test functions in $\mathcal{D}(\Omega)$ with $\psi = \varphi$ on $\partial\Omega'$, then $T\boldsymbol{F}_{\partial\Omega'}(\psi) = T\boldsymbol{F}_{\partial\Omega'}(\varphi)$.*

PROOF. If the support were not contained in $\partial\Omega'$, there would be a point $x_0 \notin \partial\Omega'$ with $x_0 \in \operatorname{spt} T\boldsymbol{F}_{\partial\Omega'}$. This implies that for each open set U containing x_0, there exists a test function $\varphi \in C_c^\infty(U \cap \Omega)$ such that $T\boldsymbol{F}_{\partial\Omega'}(\varphi) \neq 0$. Choose U so that $U \subset \mathbb{R}^N \setminus \partial\Omega'$. Let $\boldsymbol{F}_k$ denote a sequence of mollifiers of $\boldsymbol{F}$, (see Theorem 2.15). Then

$$\begin{aligned}
0 &= \int_{\partial\Omega'} (\boldsymbol{F}^\varepsilon \varphi) \cdot \nu = \int_{\Omega'} \operatorname{div}(\boldsymbol{F}^\varepsilon \varphi) = T\boldsymbol{F}^\varepsilon_{\partial\Omega'}(\varphi) = \int_{\Omega'} \boldsymbol{F}^\varepsilon \cdot \nabla\varphi + \int_{\Omega'} \varphi \operatorname{div} \boldsymbol{F}^\varepsilon \\
&\to \int_{\Omega'} \boldsymbol{F} \cdot \nabla\varphi + \int_{\Omega'} \varphi \, \mathrm{d}\mu, \quad \text{because } \operatorname{spt}\varphi \cap \partial\Omega' = \varnothing. \\
&= T\boldsymbol{F}_{\partial\Omega'}(\varphi) \neq 0.
\end{aligned}$$

□

Definition 2.27. For $1 \leq p \leq N$, the *p-capacity* of an arbitrary set $A \subset \mathbb{R}^N$ is defined as

$$\gamma_p(A) := \inf\left\{\int_{\mathbb{R}^N} |\nabla\varphi|^p\right\}$$

where the infimum is taken over all test functions $\varphi \in C_c^\infty(\mathbb{R}^N)$ that are identically one in a neighborhood of A. It is well known, cf. [92], that $\gamma_p(A) = 0$ for $1 < p < \infty$ implies that $\mathcal{H}^{N-p+\varepsilon}(A) = 0$, for each $\varepsilon > 0$ and that, conversely, if $\mathcal{H}^{N-p}(A) < \infty$, then $\gamma_p(A) = 0$. Furthermore, $\gamma_1(A) = 0$ if and only if $\mathcal{H}^{N-1}(A) = 0$.

Lemma 2.28. *If $\boldsymbol{F} \in L^p(\mathbb{R}^N;\mathbb{R}^N)$, $1 \le p \le \infty$, and $\operatorname{div}\boldsymbol{F} = \mu$, then $\mu(A) = 0$ whenever $\gamma_{p'}(A) = 0$, $(p' := p/(p-1))$. In particular, when $p = \infty$, then $\mu(A) = 0$ whenever $\mathcal{H}^{N-1}(A) = 0$.*

PROOF. If $\gamma_{p'}(A) = 0$, then, for each $\varepsilon > 0$, there exists $\varphi = \varphi_\varepsilon \in C_c^\infty(\mathbb{R}^N)$ such that $\varphi = 1$ on an open set $U \supset A$ and

$$\int_{\mathbb{R}^N} |\nabla\varphi|^{p'}\, dx < \varepsilon.$$

Thus,

$$|\mu(A)| = \left|\int_A \operatorname{div}\boldsymbol{F}\right| \le \int_{\mathbb{R}^N} |\boldsymbol{F}\cdot\nabla\varphi|\, dx \le \|\boldsymbol{F}\|_p \|\nabla\varphi\|_{p'} \le C\varepsilon,$$

which implies our desired result, $\mu(A) = 0$. □

Corollary 2.29. *If $1 < p' \le N$ and $\mathcal{H}^{N-p'}(A) < \infty$, then $\gamma_{p'}(A) = 0$, and therefore that $\mu(A) = 0$.*

Definition 2.30. A vector field $\boldsymbol{F} \in L^p(\mathbb{R}^N;\mathbb{R}^N)$, $1 \le p \le \infty$, is called a *divergence-measure field*, written as $\boldsymbol{F} \in \mathcal{DM}^p(\mathbb{R}^N)$, if $\operatorname{div}\boldsymbol{F}$ is a Radon measure on $\mathbb{R}^N$, in the sense of distributions.

We will let $\{u_k\}$ denote a sequence of mollifiers for χ_E. The following result was established [19]. For sake of completeness, we will include its proof. The one given here is slightly different from that given in [19]. For more properties of these spaces $\mathcal{DM}^p$, see Chen – Frid [19].

Lemma 2.31 (Chen – Frid). *Let $\boldsymbol{F} \in \mathcal{DM}^\infty(\mathbb{R}^N)$. If $E \subset \mathbb{R}^N$ is a set of finite perimeter, then*

$$\operatorname{div}(\chi_E\boldsymbol{F}) = u_E \operatorname{div}\boldsymbol{F} + \overline{\boldsymbol{F}\cdot\nabla u_E} \tag{2.17}$$

where $\overline{\boldsymbol{F}\cdot\nabla u_E}$ denotes the weak limit of the measures $\boldsymbol{F}\cdot\nabla u_k$. Furthermore, the measure $\overline{\boldsymbol{F}\cdot\nabla u_E}$ is absolutely continuous with respect to the measure $\|\nabla\chi_E\|$, which, in turn, is absolutely continuous with respect to $\mathcal{H}^{N-1}$-measure.*

PROOF. Let $\boldsymbol{F}_\varepsilon$ be a mollifier of $\boldsymbol{F}$. Since the mollifiers are smooth, the classical product rule yields

$$\operatorname{div}(\boldsymbol{F}_\varepsilon\cdot u_k) = u_k \operatorname{div}\boldsymbol{F}_\varepsilon + \boldsymbol{F}_\varepsilon\cdot\nabla u_k = \mu_\varepsilon(u_k) + \boldsymbol{F}_\varepsilon\cdot\nabla u_k \underset{\varepsilon}{\to} \mu(u_k) + \boldsymbol{F}_\varepsilon\cdot\nabla u_k. \tag{2.18}$$

Here we have used the fact that $\operatorname{div}\boldsymbol{F}_\varepsilon = (\operatorname{div}\boldsymbol{F})_\varepsilon = \mu_\varepsilon \underset{\varepsilon}{\to} \mu$. We also have

$$u_k \operatorname{div}\boldsymbol{F}_\varepsilon + \boldsymbol{F}_\varepsilon\cdot\nabla u_k \underset{k}{\to} u_E \operatorname{div}\boldsymbol{F}_\varepsilon + \boldsymbol{F}_\varepsilon\cdot\nabla u_E. \tag{2.19}$$

Since $\boldsymbol{F}_\varepsilon\cdot u_k \underset{k}{\to} \boldsymbol{F}_\varepsilon\cdot u_E$ in $L^1(\mathbb{R}^N)$, with the help of (2.18) we have (in the sense of distributions)

$$\operatorname{div}(\boldsymbol{F}_\varepsilon\cdot u_k) \underset{k}{\to} \operatorname{div}(\boldsymbol{F}_\varepsilon\cdot u_E) = \mu(u_E) + \boldsymbol{F}_\varepsilon\cdot\nabla u_E \underset{k}{\leftarrow} \mu(u_k) + \boldsymbol{F}_\varepsilon\cdot\nabla u_k$$

This and (2.19) therefore yield

$$\operatorname{div}(\boldsymbol{F}_\varepsilon \cdot u_E) = u_E \operatorname{div} \boldsymbol{F} + \boldsymbol{F}_\varepsilon \cdot \nabla u_E.$$

That is, $u_E \operatorname{div} \boldsymbol{F} = \operatorname{div}(\boldsymbol{F}_\varepsilon \cdot u_E) - \boldsymbol{F}_\varepsilon \cdot \nabla u_E$; furthermore

$$u_E \operatorname{div} \boldsymbol{F}_\varepsilon = \int u_E \, \mathrm{d}\mu \leftarrow \int u_k \, \mathrm{d}\mu = u_k \operatorname{div} \boldsymbol{F} = \operatorname{div}(u_k \boldsymbol{F}) - u_k \operatorname{div} \boldsymbol{F}, \tag{2.20}$$

thus yielding (2.17)

Owing to the fact that $\boldsymbol{F} \in L^\infty$, we see that $\boldsymbol{F} \cdot \nabla u_k$ is a bounded sequence in $L^1(\mathbb{R}^N)$ and therefore, there is a subsequence such that $\boldsymbol{F} \cdot \nabla u_k$ converges weak* to some measure, denoted by $\overline{\boldsymbol{F} \cdot \nabla u_E}$. It is easy to see that $\overline{\boldsymbol{F} \cdot \nabla u_E}$ is well-defined. Also

$$\boldsymbol{F} \cdot u_k \rightharpoonup \overline{\boldsymbol{F} \cdot \chi_E} \quad \text{weak*}$$

Recall that $\|\nabla \chi_E\| = H^{N-1} \llcorner \partial^* E$, (2.9). Since $u_E \boldsymbol{F}$ is a divergence measure field, we see that $\operatorname{div}(u_E \boldsymbol{F}) << H^{N-1}$; and so is $u_E \operatorname{div} \boldsymbol{F}$. Therefore, so is $\overline{\boldsymbol{F} \cdot \nabla \chi_E}$. □

3. The generalized Gauss – Green for smoothly bounded sets

As the first step in establishing our main result, Theorem 5.2 below, we extend Fuglede's result (Theorem 2.21) by showing that it is valid for all smooth sets. This is done with the help of a result by Whitney, [90], which states a compact C^1-manifold possesses a C^1 vector field with the property it is uniformly close to the (continuous) normal field of the C^1-manifold and that the elements of the field do not intersect.

Theorem 3.1. *Let $U \subset \mathbb{R}^N$ be a bounded open set whose boundary is a compact, C^1, $(N-1)$-dimensional manifold, M, with finite $\mathcal{H}^{N-1}$-measure. Then, if $\boldsymbol{F}\colon \mathbb{R}^N \to \mathbb{R}^N$ is a bounded, measurable vector field whose distributional divergence is a measure μ, then there is a signed measure σ supported on ∂U such that sigma $\ll \mathcal{H}^{N-1} \llcorner \partial U$ and*

$$\mu(U) = -\sigma(\partial U) = -\int_{\partial U} \mathscr{F}_{\mathrm{i}}(y) \cdot \nu(y) \, \mathrm{d}\mathcal{H}^{N-1}(y), \tag{3.1}$$

for some field $\mathscr{F}_{\mathrm{i}} \in L^\infty(\partial U; \mathbb{R}^N)$ that satisfies

$$\|\mathscr{F}_{\mathrm{i}}\|_{\infty;\mathbb{R}^N} \leq \|\boldsymbol{F}\|_\infty.$$

Remark 3.2. We may regard $\mathscr{F}_{\mathrm{i}}$ as the *inner normal trace* of $\boldsymbol{F}$ on ∂U and thus write

$$\mathscr{F}_{\mathrm{i}}(y) \cdot \nu(y) = \boldsymbol{F}(y) \cdot \nu(y),$$

where $\nu(y)$ is the unit inner normal to M at y. Thus, with this convention, we may write (3.1) as

$$\int_U \operatorname{div} \boldsymbol{F} = -\int_{\partial U} \boldsymbol{F}(y) \cdot \nu(y) \, d\mathcal{H}^{N-1}(y).$$

while bearing in mind, that since $\boldsymbol{F}$ is merely a measurable field and thus defined only up to a Lebesgue null set, it may not even be defined on ∂U. We use the term "inner normal trace" to suggest that the field $\mathscr{F}_{\mathrm{i}}$ is determined by the behavior of $\boldsymbol{F}$ on the interior of the manifold M. The proof will reveal that In a similar way, it is possible to define the concept of "outer normal trace". This will discussed more fully below in Theorem 5.2.

PROOF. According to [90], there is a C^1 unit vector field V defined on M which is close to the inner normal to M. That is, $V: M \to U$ is of class C^1 and there is a continuous function δ defined on M with the property that with

$$\lambda(p) := \{q : q = tV(p) + (1-t)p,\ 0 \le t \le 1\}$$

and

$$Q_p^* := \lambda(p) \cap B\big(p, \delta(p)\big),$$

then the Q_p^* fill out a neighborhood U^* of M in a one-to-one way. We will take only half of this neighborhood by defining $U_{\mathrm{i}}^* := M \cap U^*$[1] which is the half corresponding to the interior of M. The mapping $\pi: U_{\mathrm{i}}^* \to M$ defined by

$$\pi(q) = p \quad \text{if } q \in Q_p^*$$

is also of class C^1 and thus, so is the mapping $\psi: U_{\mathrm{i}}^* \to \mathbb{R}^1$ defined by

$$\psi(q) := |\pi(q) - p|, \text{ where } q \in Q_p^*.$$

Consequently, Sard's theorem implies that the critical values, $\mathbb{V} \subset \mathbb{R}^1$, of ψ are of measure zero. Thus, the Implicit function theorem implies that $M_t := \psi^{-1}(t)$ is a C^1 manifold for almost all $t \in (0, \delta) \setminus \mathbb{V}$ where $\delta := \min\{\delta(p) : p \in M\}$. The manifold M_t can be considered as a deformation of M along the line segments $\lambda(p)$. Thus, the sets $U_t := \{\psi > t\}$ are open subsets of U with smooth boundaries, $\partial U_t = M_t = \psi^{-1}(t)$. Observe that $\psi^{-1}(t) = \{q \in U : q = t[V(p) - p], p \in M\}$. Since $V \in C^1$ and M is compact, it follows that V is Lipschitz with Lipschitz constant, say L, and therefore that

$$\mathcal{H}^{N-1}\big(\psi^{-1}(t)\big) \le L\mathcal{H}^{N-1}(M). \tag{3.2}$$

To see this, consider a local coordinate system on M expressed as the inverse of its projection onto the tangent plane at a point $p \in M$. Consider $T_p(M)$ as a subspace in $\mathbb{R}^N$ with the notation $p = (p', 0) \in \mathbb{R}^N$ as identified with $p' \in \mathbb{R}^{N-1}$. Thus, $g: T_p(M) \to M$ will render M as the graph of g; thus $g(p') = \big(p', y(p')\big)$, where the "height" function $y \in C^1$. Now define $h: M \to M_t$ by $h\big(p', y(p')\big) = tV\big(p', y(p')\big) + (1-t)\big(p', y(p')\big)$. Clearly, all derivatives of $h \circ g$ are uniformly bounded on compact subsets of $T_p(M)$ and therefore (3.2) is evident. An application of Lemma 2.21 shows that

$$\mu(U_t) = -\int_{\partial U_t} \boldsymbol{F}(y) \cdot \nu(y)\, \mathrm{d}\mathcal{H}^{N-1}(y), \quad \text{for a.e. } t > 0.$$

Thus, the function $\boldsymbol{F} \cdot \nu \llcorner \partial U_t$ induces a measure on $\mathbb{R}^N$, namely, for any Borel set $E \subset \mathbb{R}^N$, define

$$\sigma_t(E) := \int_{E \cap \partial U_t} \boldsymbol{F}(y) \cdot \nu(y)\, \mathrm{d}\mathcal{H}^{N-1}(y).$$

Since $\boldsymbol{F} \in L^\infty$ and $\mathcal{H}^{N-1}(\partial U) < \infty$, we see, (3.2), that $\mathcal{H}^{N-1}(\partial U_t) < M$ for some $M > 0$ and therefore that the measures σ_t, $t > 0$, form a bounded set of measures

[1]Here, we could have defined $U_{\mathrm{e}}^* := (\mathbb{R}^N \setminus M) \cap U^*$ to develop the notion of the outer normal trace.

on $\mathbb{R}^N$. Hence there is a sequence $\{t_k\} \to 0$ and a Radon measures σ, σ^+ and σ^- such that

$$\begin{aligned} &\{\sigma^+_{t_k}\} \to \sigma^+ \\ &\{\sigma^-_{t_k}\} \to \sigma^- \\ &\{\sigma_{t_k}\} \to \sigma \ \text{ weak*, where } \sigma = \sigma^+ - \sigma^- \end{aligned} \tag{3.3}$$

To see that σ^+ is supported on ∂U, suppose the contrary: Let $x \in \operatorname{spt} \sigma^+ \setminus \partial U$ and choose $B(x,r)$ such that $B(x,r) \cap \partial U = \varnothing$. Since $x \in \operatorname{spt} \sigma^+$, there exists $\varphi \in C_c\big(B(x,r)\big)$ such that $\int \varphi \, d\sigma^+ := \sigma^+(\varphi) \neq 0$. Then, since φ is continuous, we have $\sigma^+_t(\varphi) \to \sigma^+(\varphi) \neq 0$. This implies that $\sigma^+_t(\varphi) \neq 0$ for all small $t > 0$ which leads to a contradiction since $\partial U_t \cap B(x,r) = \varnothing$ and $\operatorname{spt} \sigma^+_t \subset \partial U_t$. Clearly,

$$\sigma^+_{t_k}(\partial U_{t_k}) \to \sigma^+(\partial U), \tag{3.4}$$

because

$$\liminf_{k\to\infty} \sigma^+_{t_k}(\partial U_{t_k}) = \liminf_{k\to\infty} \sigma^+_{t_k}(\mathbb{R}^N) \geq \sigma^+(\mathbb{R}^N) = \sigma^+(\partial U)$$

whereas

$$\limsup_{k\to\infty} \sigma^+_{t_k}(\partial U_{t_k}) = \limsup_{k\to\infty} \sigma^+_{t_k}(K) \leq \sigma^+(K) = \sigma^+(\partial U)$$

for any closed set K that contains U, in particular, when $K = \mathbb{R}^N$. In the same way we prove that

$$\sigma^-_{t_k}(\partial U_{t_k}) \to \sigma^-(\partial U), \tag{3.5}$$

and therefore

$$\sigma_{t_k}(\partial U_{t_k}) \to \sigma(\partial U). \tag{3.6}$$

Also, note that the measure σ is well-defined; for let $U_{t'_k}$ be another sequence of open sets with smooth boundaries defined by $U_{t'_k} := \{\psi > t'_k\}$ for which Theorem 2.21 applies. Then assuming that $t_k > t'_k$ for all k, we have

$$\begin{aligned} \mu(U_{t'_k} \setminus U_{t_k}) &= \int_{U_{t'_k}} \operatorname{div} \boldsymbol{F} - \int_{U_{t_k}} \operatorname{div} \boldsymbol{F} \\ &= -\int_{\partial(U_{t'_k})} \boldsymbol{F}(y)\cdot\nu(y) d\mathcal{H}^{N-1}(y) + \int_{\partial(U_{t_k})} \boldsymbol{F}(y)\cdot\nu(y) d\mathcal{H}^{N-1}(y) \\ &= -\sigma_{t'_k}(\partial U_{t'_k}) + \sigma_{t_k}(\partial U_{t_k}). \end{aligned}$$

Since $U_{t'_k} \setminus U_{t_k}$ is a monotone sequence of sets with $U_{t'_k} \setminus U_{t_k} \to \varnothing$, it follows that $\mu(U_{t'_k} \setminus U_{t_k}) \to 0$ and therefore that $\sigma_{t'_k}(\partial U_{t'_k}) - \sigma_{t_k}(\partial U_{t_k}) \to 0$, which shows that σ is well-defined.

To show that $\sigma \ll \mathcal{H}^{N-1} \llcorner \partial U$, let $A \subset \partial U$ with $\mathcal{H}^{N-1}(A) = 0$. Then, there exists an open set $G \subset \mathbb{R}^N$, $G \supset A$ with $\mathcal{H}^{N-1}(G) < \varepsilon$. We may take G as $G := \bigcup_{k=1}^N B_k(r_k)$ where $\sum_k r_k^{N-1} < \varepsilon$. Therefore

$$\|\sigma_t\|(G) \leq \int_{G\cap M_t} |\boldsymbol{F}(y)\cdot\nu(y)| \, d\mathcal{H}^{N-1}(y) \leq \|\boldsymbol{F}\|_\infty \mathcal{H}^{N-1}(G \cap M_t) < \varepsilon \|\boldsymbol{F}\|_\infty,$$

(ii) which implies $\|\sigma_t\|(G) < \varepsilon \|\boldsymbol{F}\|_\infty$ which leads to the conclusion $\|\sigma\|(A) = 0$ as desired. We note know that since the sets U_{t_k} are increasing we have

$$\mu(U_{t_k}) \to \mu(U) \tag{3.7}$$

Thus, using (3.6) and (3.7) send $k \to \infty$ in

$$\mu(U_{t_k}) = -\sigma_{t_k}(\partial U_{t_k}) \tag{3.8}$$

to obtain

$$\mu(U) = -\sigma(\partial U). \tag{3.9}$$

□

Lemma 3.3. *Let $\boldsymbol{F}\colon \mathbb{R}^N \to \mathbb{R}^N$ be a bounded, measurable vector field whose distributional divergence is a measure μ, and let $\boldsymbol{F}_\varepsilon$ be a mollifier of $\boldsymbol{F}$. If*

(a) $\boldsymbol{F}_\varepsilon(x) \to \boldsymbol{F}(x)$ *for $\mathcal{H}^{N-1}$-a.e. $x \in \partial^* E$ and*
(b) $\mu(\partial E) = 0$,

then

$$\mu(E) = -\int_{\partial^* E} \boldsymbol{F}(y) \cdot \nu(y)\, \mathrm{d}\mathcal{H}^{N-1}(y),$$

where $\nu(y)$ denotes the interior measure-theoretic interior unit normal to E at y.

Proof. For any set of finite perimeter E, the classical De Giorgi – Federer divergence theorem holds:

$$\int_E \operatorname{div} \boldsymbol{F}_\varepsilon = -\int_{\partial^* E} \boldsymbol{F}_\varepsilon(y) \cdot \nu(y)\, \mathrm{d}\mathcal{H}^{N-1}(y).$$

Since $\mu_\varepsilon := \operatorname{div} \boldsymbol{F}_\varepsilon = \operatorname{div}(\boldsymbol{F}_\varepsilon) \to \operatorname{div} \boldsymbol{F} = \mu$ as measures with respect to weak* convergence, we refer to Lemma 2.4 to conclude that $\mu_k(\partial E) \to \mu(\partial E)$. Then since $\boldsymbol{F}$ is bounded, by the dominated convergence theorem, we obtain our desired result,

$$\mu_\varepsilon(E) = \int_{\partial^* E} \boldsymbol{F}_\varepsilon(y) \cdot \nu(y)\, \mathrm{d}\mathcal{H}^{N-1}(y) \to -\int_{\partial^* E} \boldsymbol{F}(y) \cdot \nu(y)\, \mathrm{d}\mathcal{H}^{N-1}(y).$$

□

Remark 3.4. We recall that $\boldsymbol{F}_\epsilon(x) \to \boldsymbol{F}(x)$ for $\mathcal{H}^N$-a.e. $x \in \mathbb{R}^N$. Let $\mathcal{N}_0$ be the set of $\mathcal{L}^N$-measure zero where the pointwise convergence fails and let g be any Lipschitz function. Then, using the coarea formula, we have

$$0 = \int_{\mathcal{N}_0} |\nabla g|\, \mathrm{d}x = \int_{-\infty}^{\infty} \mathcal{H}^{N-1}(g^{-1}(t) \cap \mathcal{N}_0)\, \mathrm{d}t,$$

where $g^{-1}(t) = \{x \in \mathbb{R}^N : g(x) = t\}$. This implies that, for $\mathcal{H}^1$-a.e. t, $\mathcal{H}^{N-1}(g^{-1}(t) \cap \mathcal{N}_0) = 0$. Thus, Theorem 2.21 holds for *almost every set $g^{-1}(t)$ of finite perimeter.* One of our main goals of this paper is to introduce a notion of normal traces and establish the Gauss – Green formula for *every set of finite perimeter* by approximation from "good" sets with smooth boundaries.

4. Smooth approximation of sets of finite perimeter

In this section we show that the superlevel sets of the mollifiers of sets of finite perimeter provide good approximations with respect to $\|\mu\|$ measure; this will allow us to employ Theorem 3.1 which, after passage to a limit, will lead to our main result, Theorem 5.2.

Lemma 4.1. *Let μ be a (signed) Radon measure on $\mathbb{R}^N$ such that $\mu \ll \mathcal{H}^{N-1}$. Let E be a set of finite perimeter, and let u_k be a sequence of mollifiers for χ_E. Then, for any $\varepsilon > 0$, $t > 0$ and with $A_{k;t} := \{x : u_k(x) > t\}$, there exists $k^* = k^*(\varepsilon)$ such that*

(i) $\|\mu\|(A_{k;t} \setminus E := E^1 \cup E^{1/2}) < \varepsilon$ *for all* $1 > t > \frac{1}{2}$,
(ii) $\|\mu\|(A_{k;t} \setminus E^1) < \varepsilon$ *for all* $t > \frac{1}{2}$,

(iii) $\|\mu\|(E^1 \setminus A_{k;t}) < \varepsilon$ *for all* $0 < t < 1 - \varepsilon$ *and all* $k \geq k^*$,

(iv) $\|\mu\|(E \setminus A_{k;t}) < \varepsilon$ *for all* $0 < t < \frac{1}{2}$ *and all* $k \geq k^*$.

PROOF. The proof of (i) is very similar to that of (ii).

(ii) With $t > \frac{1}{2}$, choose $0 < \varepsilon < t - \frac{1}{2}$. Let u_k denote (as usual) a sequence of mollifiers for χ_E. We know that $u_k(x) \to u_E(x)$ for $\mathcal{H}^{N-1}$-a.e. x and therefore the same is true for $\|\mu\|$ as well. By Egorov's theorem, there is an open set U_ε such that $\|\mu\|(U_\varepsilon) < \varepsilon$ and that $|u_k(x) - u_E(x)| < \varepsilon$ for all $x \notin U_\varepsilon$ and for all $k \geq k^*(\varepsilon) = k^*(\varepsilon, t)$. On $A_{k;t} \setminus (U_\varepsilon \cup E^1)$, we have

$$t < u_k(x);$$

and so, since $u_k(x) < u_E(x) + \varepsilon$ on $\mathbb{R}^N \setminus U_\varepsilon$, we have

$$\tfrac{1}{2} < t - \varepsilon < u_E(x) \implies u_E(x) = 1 \implies x \in E^1 \quad \text{if } t - \varepsilon > \tfrac{1}{2}.$$

This yields

$$A_{k;t} \setminus (U_\varepsilon \cup E^1) \subset E^1 \implies A_{k;t} \setminus E^1 \subset U_\varepsilon \cup E^1 \implies A_{k;t} \setminus E^1 \subset U_\varepsilon$$

Since $\|\mu\|(U_\varepsilon) < \varepsilon$, our desired result, (ii), follows.

The proof of (iii) is similar to the following.

(iv) For all large k, we have $|u_k(x) - u_E(x)| < \varepsilon$ for all $x \notin U_\varepsilon$. Thus,

$$\begin{aligned} \tfrac{1}{2} - u_k(x) \leq u_E(x) - u_k(x) \leq \varepsilon \text{ on } E \setminus U_\varepsilon \\ &\implies u_k(x) \geq \tfrac{1}{2} - \varepsilon \text{ for all } x \in E \setminus U_\varepsilon \\ &\implies E \setminus U_\varepsilon \subset A_{k;1/2-\varepsilon} \subset A_t \text{ whenever } 0 < t < \tfrac{1}{2} - \varepsilon. \quad \square \end{aligned}$$

Corollary 4.2. *For each* $\varepsilon > 0$ *and* $t > 0$, *there exists* $k^* = k^*(\varepsilon) > 0$ *such that*

(i) $\|\mu\|(A_{k;,t} \Delta E) < \varepsilon$ *whenever* $0 < t < \frac{1}{2}$ *and* $k \geq k^*$,

(ii) $\|\mu\|(A_{k;,t} \Delta E^1) < \varepsilon$ *whenever* $\frac{1}{2} < t < 1$ *and* $k \geq k^*$.

Remark 4.3. In the previous result, we used open super-levelsets $A_{k;t} := \{x : u_k(x) > t\}$. However, we could have used closed super-levelsets $\bar{A}_{k;t} := \{x : u_k(x) \geq t\}$, as well, because for an arbitrary Radon measure, ω, we have

$$\omega(\bar{A}_{k;t}) - \omega(A_{k;t}) = \omega(\partial A_{k;t}) = 0 \tag{4.1}$$

for all but countably many t, because the family of sets $\{\partial A_{k;t} : t \in \mathbb{R}^1\}$ is pairwise disjoint, and any Radon measure, ω, can assign positive values to only a countable number of such a family.

Corollary 4.4. *For each* $\varepsilon > 0$ *and* $t > 0$, *there exists* $k^* = k^*(\varepsilon) > 0$ *such that*

(i) $\|\mu\|(A_{k;,t} \Delta E) = \|\mu\|(\bar{A}_{k;,t} \Delta E) < \varepsilon$ *for all but countably many* $t \in (\frac{1}{2}, 1)$ *and for* $k \geq k^*$,

(ii) $\|\mu\|(A_{k;,t} \Delta E^1) = \|\mu\|(\bar{A}_{k;,t} \Delta E^1) < \varepsilon$ *for all but countably many* $t \in (0, \frac{1}{2})$ *and for* $k \geq k^*$,

(iii) $\|\mu\|(\partial A_{k;,t} \Delta E) = \|\mu\|(u_k^{-1}(t) \Delta E) < \varepsilon$ *for almost all* $t \in (\frac{1}{2}, 1)$ *and for* $k \geq k^*$,

(iv) $\|\mu\|(\partial A_{k;,t} \Delta E^1) = \|\mu\|(u_k^{-1}(t) \Delta E^1) < \varepsilon$ *for almost all* $t \in (0, \frac{1}{2})$ *and for* $k \geq k^*$.

In the following, with $A_{k;s}$ a set with smooth boundary and $\boldsymbol{F} \in L^\infty(\partial A_{k;s}; \mathbb{R}^N)$, we define a measure for all Borel sets $B \subset \mathbb{R}^N$ by

$$\sigma_{k,s}(B) := \int_{B \cap \partial A_{k,s}} \boldsymbol{F}(y) \cdot \nu(y) \, \mathrm{d}\mathcal{H}^{N-1}(y), \tag{4.2}$$

where $\nu(y)$ denotes the classical unit interior normal to $A_{k,s}$.

Lemma 4.5. *With $\sigma_{k;t}$ defined by (4.2), there exists a number $0 \leq M < \infty$ such that for all positive integers k and almost all $t \in \mathbb{R}^1$*

$$\mathcal{H}^{N-1}(u_k^{-1}(t)) + \|\sigma_{k;t}\|(\mathbb{R}^N) \leq M, \tag{4.3}$$

PROOF. Clearly, we may restrict our attention to $t \in (0,1)$. From Corollary 4.2 it follows for almost all $t \in (0,1)$ that there exists a sequence of smoothly bounded sets $A_{k;t}$ such that either $\chi_{A_{k;t}} \to \chi_{E^1}$ $\|\mu\|$-a.e. (if $1 > t > \frac{1}{2}$) or that $\chi_{A_{k;t}} \to \chi_E$ $\|\mu\|$-a.e. (if $0 < t < \frac{1}{2}$). Since $\|\mu\| \ll \mathcal{H}^{N-1}$ it follows that for a.e. $t \in (0,1)$, that $\chi_{A_{k;t}} \to \chi_E$ everywhere except for a set of Lebesgue measure zero (in fact, for a set of $\mathcal{H}^{N-1}$-measure zero). Since these functions are integrable, we may consider these functions as distributions and thus by appealing to Lebesgue's dominated convergence theorem, we may conclude

$$\chi_{A_{k;t}} \to \chi_{E^1} \quad \text{convergence in the sense of distributions}$$

and therefore

$$\nabla\chi_{A_{k;t}} \to \nabla\chi_{E^1} \quad \text{convergence also in the sense of distributions.}$$

Since all functions are in BV, we find that

$$\nabla\chi_{A_{k;t}} \rightharpoonup \nabla\chi_{E^1} \quad \text{weak* convergence in the sense of measures.}$$

Therefore, the uniform boundedness theorem (Corollary 2.3) implies that the norm of these measures are uniformly bounded; that is, for almost all $t \in (0,1)$ there exists $0 < M < \infty$, independent of t, such that

$$\sup_k \|\nabla\chi_{A_{k;t}}\|(\mathbb{R}^N) \leq M.$$

In view of the facts that

$$\|\nabla\chi_{A_{k;t}}\| = \mathcal{H}^{N-1} \llcorner u_k^{-1}(t)$$

and $\boldsymbol{F} \in L^\infty$, our result follows. □

Remark 4.6. We offer another proof of this result for the purpose of broadening the context of our development. By using the theory of integral currents [44] the result is immediate. Rather than actually introducing integral currents, we will introduce a small structure that will reflect the argument from integral currents. For this, let $\mathcal{V}$ denote the Banach space of C^∞-vector fields ψ on $\mathbb{R}^N$ endowed with norm

$$\|\psi\| := \sup_{x \in K} \left\{ |\psi(x)| + \sum_{i=1}^N |\nabla\psi_i(x)| \right\},$$

where K is a compact set containing the bounded set E.

Let

$$T_E(\varphi) := \int_E \varphi(x)\,\mathrm{d}x$$

and for regular $t \in (0,1)$ let

$$T_{k;t}(\varphi) := \int_{\{x:u_k(x)>t\}} \varphi = \int_{A_{k;t}} \varphi \quad \text{for } \varphi \in \mathcal{V}.$$

Then, for $\psi \in \mathcal{V}$, define

$$\partial T_k(\psi) := T_k(\operatorname{div}\psi) = \int_{\partial^* E} \psi \cdot \nu = \int_E \operatorname{div}\psi$$

Because each $u_k^{-1}(t)$ is a C^∞-manifold, note that

$$\|\partial T_{k;t}\| := \sup_{\|\psi\|\leq 1} \partial T_{k;t}(\psi) = \mathcal{H}^{N-1}\big(u_k^{-1}(t)\big).$$

Indeed, with $\psi := \nu_k/|\nu_k|$ defined on the manifold, $u_k^{-1}(t)$, the norm-preserving extension of ψ to all of $\mathbb{R}^N$ by Whitney's extension theorem yields the inequality

$$\|\partial T_{k;t}\| := \sup_{\|\psi\|\leq 1} \partial T_{k;t}(\psi) \geq \mathcal{H}^{N-1}\big(u_k^{-1}(t)\big).$$

The opposite inequality is obvious.

Refer to Theorem 5.2 to conclude that

$$\lim_{k\to\infty} T_{k;t}(\varphi) \to T_E(\varphi), \quad \varphi \in \mathcal{V}$$

and therefore, in particular,

$$T_{k;t}(\operatorname{div}\psi) \to T_E(\operatorname{div}\psi) < \infty.$$

That is, $\sup_k\{\partial T_{k;t}(\psi)\} < \infty$. By the uniform boundedness principle, we see that since $\partial T_{k;t}$ is a linear functional on $\mathcal{V}$, we have

$$\sup_k \|\partial T_{k;t}\| = \sup_k \mathcal{H}^{N-1}\big(u_k^{-1}(t)\big) < \infty,$$

which gives our desired result.

The above argument simply rephrases the following basic fact from the theory of currents. We know that since E has finite perimeter, T_E is an integral current. Moreover, the currents T_k converge to T_E weakly and therefore so do their boundaries, $\partial T_k \to \partial T_E$; that is

$$\int_{u_k^{-1}(t)} \sigma\,\mathrm{d}\mathcal{H}^{N-1} = \int_{\partial A_{k;t}} \sigma\,\mathrm{d}\mathcal{H}^{N-1} \to \int_{\partial^* E} \sigma\,\mathrm{d}\mathcal{H}^{N-1}$$

for each smooth differential $(N-1)$-form σ. Appealing to the uniform boundedness theorem yields our result.

5. The main result

In this section we establish our main result, Theorem 5.2. We begin with a lemma that will lead to several of the assertions in this result.

Lemma 5.1. *If $E \subset \mathbb{R}^N$ is an arbitrary set of finite perimeter, then with our usual notation of u_k denoting the mollifiers of u_E, we have for any $t \in (\frac{1}{2}, 1)$,*

$$(5.1)\quad 2\overline{\chi_E F \cdot \nabla u_E}(\mathbb{R}^N) := \lim_{k\to\infty} \int_E F \cdot \nabla u_k \, \mathrm{d}\mathcal{H}^{N-1}, = \lim_{k\to\infty} \int_{A_{k,t}} F \cdot \nabla u_k \, \mathrm{d}\mathcal{H}^{N-1},$$
$$= \lim_{k\to\infty} \int_0^1 \int_{A_{k,t}} F \cdot \nabla \nu_k \, \mathrm{d}\mathcal{H}^{N-1} \, \mathrm{d}t, = \lim_{k\to\infty} \sigma_{k,t}(A_{k,t}).$$

Moreover, if we define the $A_{k,t}$ as closed sets: that is, $A_{k,t} := \{x : u_k(x) \geq t\}$, then

$$\lim_{t\to 1/2^+} \lim_{k\to\infty} \int_t^1 \int_{A_{k,t}} F \cdot \nu_k \mathcal{H}^{N-1} \, \mathrm{d}t = \lim_{k\to\infty} \int_{1/2}^1 \int_{A_{k,1/2}} F \cdot \nabla \nu_k \, \mathrm{d}\mathcal{H}^{N-1} \, \mathrm{d}t,$$
$$= \frac{1}{2} \int_E F \cdot \nabla u_k \, \mathrm{d}\mathcal{H}^{N-1} \, \mathrm{d}t.$$

PROOF. Recall our convention, Definition 2.24. Then, for any arbitrary set of finite perimeter, A, we have

$$(5.2)\quad \int_A F \cdot \nabla u_k \, \mathrm{d}x = \left(\int_{A \setminus \mathcal{N}} F \cdot \nabla u_k \, \mathrm{d}x + \int_{\mathcal{N}} F \cdot \nabla u_k \right),$$
$$= \left(\int_{A \setminus \mathcal{N}} |\nabla u_k| \frac{F \cdot \nabla u_k}{|\nabla u_k|} \, \mathrm{d}x + \int_{\mathcal{N}} F \cdot \nabla u_k \right),$$
$$= \int_{A \setminus \mathcal{N}} |\nabla u_k| \frac{F \cdot \nabla u_k}{|\nabla u_k|} \, \mathrm{d}x + 0,$$
$$= \int_A |\nabla u_k| g \, \mathrm{d}x, \quad \text{where } g = \left(\frac{\chi_{A \setminus \mathcal{N}} F \cdot \nabla u_k}{|\nabla u_k|} \right),$$
$$= \int_0^1 \int_{u_k^{-1}(t) \cap A} g(y) \, \mathrm{d}\mathcal{H}^{N-1}(y) \, \mathrm{d}t \quad \text{by the coarea formula},$$
$$= \int_0^1 \int_{u_k^{-1}(t) \cap A} \boldsymbol{F}(y) \cdot \nu_k(y) \, \mathrm{d}\mathcal{H}^{N-1}(y) \, \mathrm{d}t,$$
$$\text{because } \nu_k(y) = \frac{\nabla u_k(y)}{|\nabla u_k(y)|} \text{ for } y \in u_k^{-1}(t) \cap (A \setminus \mathcal{N}),$$
$$= \int_0^1 \int_{u_k^{-1}(t) \cap A} \boldsymbol{F}(y) \cdot \nu_k(y) \, \mathrm{d}\mathcal{H}^{N-1}(y) \, \mathrm{d}t = \int_0^1 \sigma_{k,t}(A) \, \mathrm{d}t.$$

Now refer to Corollary 4.2 with the signed measure μ_k (temporarily) defined as

$$\mu_k(B) := \int_B \boldsymbol{F} \cdot \nabla u_k \, \mathrm{d}\mathcal{H}^{N-1} \quad \text{for each Borel set } B \subset \mathbb{R}^N,$$

to conclude since $\mu_k \ll \mathcal{H}^{N-1}$, $\|\nabla u_k\|_1 \leq M$ (in fact $M = \|\nabla \chi_E\|$, by Lemma 2.18) and $\boldsymbol{F}$ is bounded, that for any $t \in (\frac{1}{2}, 1)$,

$$(5.3)\quad 2\overline{\chi_E F \cdot \nabla u_E}(\mathbb{R}^N) := \lim_{k\to\infty} \int_E \boldsymbol{F} \cdot \nabla u_k \, \mathrm{d}\mathcal{H}^{N-1} = \lim_{k\to\infty} \int_{A_{k,t}} \boldsymbol{F} \cdot \nabla u_k \, \mathrm{d}\mathcal{H}^{N-1}.$$

Since $A_{k,t}$ has finite perimeter for almost all $t \in (\frac{1}{2})$, we appeal to (5.1) to conclude that

$$
\begin{aligned}
(5.4)\qquad \int_{A_{k,1/2}} \boldsymbol{F} \cdot \nabla u_k \, \mathrm{d}\mathcal{H}^{N-1} &\leftarrow \int_{A_{k,t}} \boldsymbol{F} \cdot \nabla u_k \, \mathrm{d}\mathcal{H}^{N-1} \\
&= \int_{\{u_k > t\}} \boldsymbol{F} \cdot \nabla u_k \, \mathrm{d}\mathcal{H}^{N-1} \\
&= \int_t^1 \int_{u_k^{-1}(t) \cap A_{k,t}} \boldsymbol{F}(y) \cdot \nu_k(y) \, \mathrm{d}\mathcal{H}^{N-1}(y) \, \mathrm{d}t \\
&= \int_t^1 \int_{u_k^{-1}(t)} \boldsymbol{F}(y) \cdot \nu_k(y) \, \mathrm{d}\mathcal{H}^{N-1}(y) \, \mathrm{d}t \\
&:= \int_t^1 \sigma_{k;t}(\mathbb{R}^N) \\
&= (1-t)\sigma_{k,t_k}(\mathbb{R}^N), \quad \text{for some } t_k \in (\tfrac{1}{2}, 1), \\
&\to \tfrac{1}{2}\sigma_{\mathrm{i}}(\mathbb{R}^N).
\end{aligned}
$$

In order to justify both limits above, for $t \in (0,1)$ consider

$$
\begin{aligned}
g_{k;t} &:= \int_{u_k^{-1}(t)} \boldsymbol{F}(y) \cdot \nu_k(y) \, \mathrm{d}\mathcal{H}^{N-1}(y), \\
\int_0^1 g_{k;t} \, \mathrm{d}t &:= \int_0^1 \int_{u_k^{-1}(t)} \boldsymbol{F}(y) \cdot \nu_k(y) \, \mathrm{d}\mathcal{H}^{N-1}(y) \, \mathrm{d}t \\
&= \int_0^1 \boldsymbol{F}(y) \cdot \nabla u_k(y) \, \mathrm{d}\mathcal{H}^{N-1}(y) \, \mathrm{d}t.
\end{aligned}
$$

The third equation follows from the second by Lemma 5.2. Thus, it follows

$$
\begin{aligned}
\int_{1/2}^1 g_k(t) \, \mathrm{d}t - \int_t^1 g_k(t) \, \mathrm{d}t = \int_{1/2}^t g_k(t) \, \mathrm{d}t &\leq (t - \tfrac{1}{2}) \|\boldsymbol{F}\|_\infty \|\nabla u_k\|_1 \\
&\to (t - \tfrac{1}{2}) \|\boldsymbol{F}\|_\infty \|\nabla \chi_E\|_1 \to 0
\end{aligned}
$$

as $t \to \frac{1}{2}$.

Similarly, note that

$$\bigcap_{t > 1/2} A_{k,t} = A_{k,1/2}$$

provided we define the $A_{k,t}$ as closed sets; namely as $A_{k,t} := \{x : u_k(x) \geq t\}$, and therefore the limit on the left is clear. However

$$A_{k,1/2} = E \cup \mathcal{N} \quad \text{where } \mathcal{H}^{N-1}(\mathcal{N}) = 0$$

and thus we conclude that

$$2 \int_E \boldsymbol{F} \cdot \nabla u_k \, \mathrm{d}\mathcal{H}^{N-1} = \sigma_{k,t_k}(\mathbb{R}^N).$$

Since the left side is independent of t, the same is true of the right side, a fact that can be easily verified by employing integration by parts using Theorem 3.1.

(b) Let $A \subset \mathbb{R}^N \setminus \partial^* E$ be an arbitrary Borel set. Referring to (2.9), we see that

$$\|\nabla \chi_E\|(A) = 0.$$

From general measure theory considerations we know that

$$
\begin{aligned}
(5.5) \qquad 0 &= \|\nabla\chi_E\|(A), \\
&= \inf\{\|\nabla\chi_E\|(U) : A \subset U,\ U \text{ open}\}, \\
&= \inf\{\|\nabla\chi_E\|(U) : A \subset U,\ U \text{ open},\ \|\nabla\chi_E\|(\partial U) = 0\}.
\end{aligned}
$$

In order to prove that $\|\sigma\|(A) = 0$ we proceed by selecting an open set $U \supset A$ such that $\|\nabla\chi_E\|(\partial U) = 0$ and

$$
(5.6) \qquad \|\nabla\chi_E\|(U) < \varepsilon.
$$

Now, (5.2) implies that

$$
(5.7) \qquad \int_U \boldsymbol{F} \cdot \nabla u_k \,\mathrm{d}x = \int_0^1 \int_{u_k^{-1}(t)\cap U} \boldsymbol{F}(y) \cdot \nu_k(y) \,\mathrm{d}\mathcal{H}^{N-1}(y) \,\mathrm{d}t
$$

and therefore, by virtue of the conditions satisfied by U, we then have a

$$
\begin{aligned}
\|\sigma\|(A) &= \int_0^1 \|\sigma\|(A) \,\mathrm{d}t \\
&\le \int_0^1 \|\sigma\|(U) \,\mathrm{d}t \\
&\le \int_0^1 \liminf_{k\to\infty} \|\sigma_{k;t}\|(U) \,\mathrm{d}t && \text{because } \sigma_{k,t} \rightharpoonup \sigma \text{ weak*}, \\
&\le \liminf_{k\to\infty} \int_0^1 \|\sigma_{k;t}\|(U) \,\mathrm{d}t && \text{by Fatou's lemma}, \\
&\le \liminf_{k\to\infty} \int_0^1 \int_{u_k^{-1}(t)\cap U} |\boldsymbol{F}(y) \cdot \nu(y)| \,\mathrm{d}\mathcal{H}^{N-1}(y) \,\mathrm{d}t && \text{by (5.2)}, \\
&= \liminf_{k\to\infty} \int_{u_k^{-1}(t)\cap U} |\boldsymbol{F}(y) \cdot \nabla u_k| \,\mathrm{d}\mathcal{H}^{N-1}(y) \,\mathrm{d}t && \text{from definition}, \\
&\le \liminf_{k\to\infty} \|\boldsymbol{F}\|_\infty \|\nabla u_k\|(U), \\
&\le \|\boldsymbol{F}\|_\infty \|\nabla\chi_E\|(U), && \text{by Lemma 2.18}, \\
&< \|\boldsymbol{F}\|_\infty \varepsilon
\end{aligned}
$$

which leads to our result. □

With the help of Theorem 3.1, Corollary 4.2, and Lemma 4.5 we now are in a position to establish our main result, Theorem 5.2.

Theorem 5.2. *Let $\Omega \subset \mathbb{R}^N$ be an open set. Suppose $\boldsymbol{F} \in L^\infty(\Omega; \mathbb{R}^N)$ is a vector field with* $\operatorname{div} \boldsymbol{F} = \mu$ *where μ is a signed measure with finite variation and let $E \subset\subset \Omega$ be a bounded set of finite perimeter. Then,*

(1) *for almost every $s \in (\frac{1}{2}, 1)$, there exists a signed measure σ_{i} and a family of sets $A_{k;s}$ with smooth boundaries such that*
   (a) $\|\mu\|(A_{k;s}\Delta E^1) \to 0$
   (b) *σ_{i} is carried by $\partial^* E$ in the sense that* $\|\sigma_{\mathrm{i}}\|(\mathbb{R}^N \setminus \partial^* E) = 0$
   (c) $\|\sigma_{\mathrm{i}}\| \ll H^{N-1}$
   (d) $\limsup_{k\to\infty} \mathcal{H}^{N-1}(\partial A_{k;s} \cap E^0) = \limsup_{k\to\infty} \mathcal{H}^{N-1}(u_k^{-1}(s) \cap E^0) = 0$
   (e) $\limsup_{k\to\infty} \sigma_{k;s}(\mathbb{R}^N \setminus E) = \limsup_{k\to\infty} \sigma_{k;s}(E^0) = 0$

(f) *The density of* σ_{i}, $\mathscr{F}_{\mathrm{i}}$, *is called the* inner normal trace relative to E *of* $\boldsymbol{F}$ *on* $\partial^* E$, *and satisfies*

$$\int_{E^1} \operatorname{div} \boldsymbol{F} =: \mu(E^1) = -\sigma_{\mathrm{i}}(\partial^* E) = -\int_{\partial^* E} \mathscr{F}_{\mathrm{i}}(y) \cdot \nu(y)\, d\mathcal{H}^{N-1}(y). \tag{5.8}$$

(g) *If we consider* $\boldsymbol{F}\cdot\nabla u_k \chi_E$ *as a sequence of measures (as we may), then these measures converge weak* to the measure* $\mathscr{F}_{\mathrm{i}} \cdot \nu_{\partial^* E} \mathcal{H}^{N-1} \llcorner \partial^* E$,; *i.e.,*

$$(\boldsymbol{F} \cdot \nabla u_k)\chi_E \rightharpoonup (\mathscr{F}_{\mathrm{i}} \cdot \nu_{\partial^* E})\mathcal{H}^{N-1} \llcorner \partial^* E,$$

(h) $\|\sigma_{\mathrm{i}}\| = \|\mathscr{F}_{\mathrm{i}}\|_{\partial^* E;\infty} \leq \|\boldsymbol{F}\|_{\mathbb{R}^N;\infty}$,

(i) *The measure* σ_{i} *is the weak* limit of* $\sigma_{k,t}$ *for any* $t \in (\frac{1}{2}, 1)$, *thus showing that* σ_{i} *is independent of such* t.

(2) *for almost every* $s \in (0, \frac{1}{2})$, *there exists a signed measure* σ_{e} *and a family of sets* $A_{k;s}$ *with smooth boundaries such that*

(a) $\|\mu\|(A_{k;s} \Delta E) \to 0$

(b) σ_{e} *is carried by* $\partial^* E$ *in the sense that* $\|\sigma_{\mathrm{e}}\|(\mathbb{R}^N \setminus \partial^* E) = 0$

(c) $\sigma_{\mathrm{e}} << H^{N-1}$

(d) $\limsup_{k\to\infty} \mathcal{H}^{N-1}(\partial A_{k;s} \cap E) = \limsup_{k\to\infty} \mathcal{H}^{N-1}(u_k^{-1}(s) \cap E) = 0$

(e) $\limsup_{k\to\infty} \sigma_{k;s}(\mathbb{R}^N \setminus E^0) = \limsup_{k\to\infty} \sigma_{k;s}(E) = 0$

(f) *The density of* σ_{e}, $\mathscr{F}_{\mathrm{e}}$, *is called the* outer normal trace relative to E *of* $\boldsymbol{F}$ *on* $\partial^* E$, *and satisfies*

$$\int_{E} \operatorname{div} \boldsymbol{F} =: \mu(E) = -\sigma_{\mathrm{e}}(\partial^* E) = -\int_{\partial^* E} \mathscr{F}_{\mathrm{e}}(y) \cdot \nu(y)\, d\mathcal{H}^{N-1}(y). \tag{5.9}$$

(g) *If we consider* $\boldsymbol{F} \cdot \nabla u_k \chi_{E^0}$ *as a sequence of measures (as we may), then these measures converge weak* to the measure* $\mathscr{F}_{\mathrm{e}} \cdot \nu_{\partial^* E} \mathcal{H}^{N-1} \llcorner \partial^* E$; *i.e.,*

$$(\boldsymbol{F} \cdot \nabla u_k)\chi_E \rightharpoonup (\mathscr{F}_{\mathrm{e}} \cdot \nu_{\partial^* E})\mathcal{H}^{N-1} \llcorner \partial^* E,$$

(h) $\|\sigma_{\mathrm{e}}\| = \|\mathscr{F}_{\mathrm{e}}\|_{\partial^* E;\infty} \leq \|\boldsymbol{F}\|_{\mathbb{R}^N;\infty}$,

(i) *The measure* σ_{e} *is the weak* limit of* $\sigma_{k,t}$ *for any* $t \in (0, \frac{1}{2})$, *thus showing that* σ_{e} *is independent of such* t.

PROOF. We will prove only part (1), the proof of part (2) being virtually identical.

For notational simplicity, we will use the notation

$$\sigma \text{ rather than } \sigma_{\mathrm{i}} \tag{5.10}$$

in the proof of part (1). Throughout the proof we will consider only those values of $s \in (\frac{1}{2}, 1)$ for which Theorem 2.21 are valid for all mollifiers, u_k, of χ_E thus omitting at most a set of measure zero. Without increasing the measure of this exceptional set, call it $\mathcal{S}$, we will also include those values of s for which $\|\mu\|(u_k^{-1}(s) \neq 0$ for all k. Utilizing (4.3), Corollary 4.2 (ii) and then proceeding as in (3.3), we obtain a signed measure σ_s such that $\sigma_{k;s}(\mathbb{R}^N) \to \sigma_s(\mathbb{R}^N)$ and $\mu(A_{k;s}) \to \mu(E^1)$. This also shows that σ_s is independent of s.

We begin with the proof of part (a).

(a) This follows directly from Corollary 4.2.

(b) Let $A \subset \mathbb{R}^N \setminus \partial^* E$ be an arbitrary Borel set. Referring to (2.9), we see that

$$\|\nabla\chi_E\|(A) = 0.$$

From general measure theory considerations we know that

$$\begin{aligned} 0 &= \|\nabla\chi_E\|(A) \\ &= \inf\{\|\nabla\chi_E\|(U) : A \subset U,\, U \text{ open}\} \\ &= \inf\{\|\nabla\chi_E\|(U) : A \subset U,\, U \text{ open}, \|\nabla\chi_E\|(\partial U) = 0\}. \end{aligned} \tag{5.11}$$

In order to prove that $\|\sigma\|(A) = 0$ we proceed by selecting an open set $U \supset A$ such that both $\|\nabla\chi_E\|(\partial U) = 0$ and $\|\sigma\|(\partial U) = 0$

$$\|\nabla\chi_E\|(U) < \varepsilon/2. \tag{5.12}$$

Now, (5.2) implies that

$$\int_U F \cdot \nabla u_k \,\mathrm{d}x = \int_0^1 \int_{u_k^{-1}(t)\cap U} \boldsymbol{F}(y)\cdot \nu_k(y)\,\mathrm{d}\mathcal{H}^{N-1}(y)\,\mathrm{d}t \tag{5.13}$$

and therefore, by virtue of the conditions satisfied by U, we then have

$$\begin{aligned} \|\sigma\|(A) &= \int_0^1 \|\sigma\|(A)\,\mathrm{d}t \\ &\le 2\int_{1/2}^1 \|\sigma\|(U)\,\mathrm{d}t \\ &\le 2\int_{1/2}^1 \liminf_{k\to\infty} \|\sigma_{k;t}\|(U)\,\mathrm{d}t && \text{because } \sigma_{k,t} \rightharpoonup \sigma \text{ weak* and } t > \tfrac{1}{2}, \\ &\le \liminf_{k\to\infty} 2\int_{1/2}^1 \|\sigma_{k;t}\|(U)\,\mathrm{d}t && \text{by Fatou's lemma,} \\ &\le \liminf_{k\to\infty} 2\int_{1/2}^1 \int_{u_k^{-1}(t)\cap U} |\boldsymbol{F}(y)\cdot\nu(y)|\,\mathrm{d}\mathcal{H}^{N-1}(y)\,\mathrm{d}t \\ &= \liminf_{k\to\infty} 2\int_{u_k^{-1}(t)\cap U} |\boldsymbol{F}(y)\cdot\nabla u_k|\,\mathrm{d}\mathcal{H}^{N-1}(y)\,\mathrm{d}t \\ &\le 2\liminf_{k\to\infty} \|\boldsymbol{F}\|_\infty \|\nabla u_k\|(U) \\ &\le 2\|\boldsymbol{F}\|_\infty \|\nabla\chi_E\|(U) && \text{by Lemma 2.18,} \\ &< 2\|\boldsymbol{F}\|_\infty \varepsilon \end{aligned}$$

which leads to our result.

(c) Let $A \subset \mathbb{R}^N$ be a Borel set with $\mathcal{H}^{N-1}(A) = 0$. Then, appealing to (2.9), we find that $\|\nabla\chi_E\|(\mathbb{R}^N) = 0$. From this, the proof proceed precisely as in (b) to yield our desired conclusion.

(d) Using that Lebesgue measure is absolutely continuous with respect to $\mathcal{H}^{N-1}$ along with Corollary 4.2(ii) with $t > \frac{1}{2}$ leads to

$$|A_{k,t}\Delta E^1| \to 0 \quad \text{as } k \to \infty.$$

Therefore, since

$$R_{k,t} := A_{k,t} \setminus E^1 = A_{k,t} \cap (E^0 \cup \partial^* E^0)$$

it follows, in particular, that

(5.14) $$|R_{k,t}| \to 0, \quad \text{provided that } t > \tfrac{1}{2}.$$

Recall (Remark 4.3) that we have the option of defining $A_{k,t} := \{x : u_k(x) \geq t\}$ without altering the development. With this option in force, we have $u_k^{-1}(t) \subset A_{k,t}$ and consequently, by Lemma 5.2,

$$\begin{aligned}\int_{R_{k,t}} |\nabla u_k| &= \int_0^1 \mathcal{H}^{N-1}(u_k^{-1}(t) \cap R_{k,t})\, \mathrm{d}t,\\ &= \int_0^1 \mathcal{H}^{N-1}(u_k^{-1}(t) \cap (E^0 \cup \partial^* E^0)\, \mathrm{d}t.\end{aligned}$$

Since $\nabla u_k \rightharpoonup \nabla \chi_E$, we have $\|\nabla u_k\|_1 \leq \|\nabla \chi_E\|$ (Lemma 2.18) and therefore it follows from Vitali's convergence theorem for $t > \frac{1}{2}$, that

$$\int_{R_{k,t}} |\nabla u_k| \to 0.$$

Thus, for a subsequence, we can conclude that

$$\mathcal{H}^{N-1}(u_k^{-1}(t) \cap (E^0 \cup \partial^* E^0) \to 0 \quad \text{as } k \to \infty.$$

The dependence on the subsequence is illusory since if there were a subsequence such that

$$\mathcal{H}^{N-1}(u_k^{-1}(t) \cap (E^0 \cup \partial^* E^0) \to L \neq 0 \quad \text{as } k \to \infty,$$

then one could appeal to our previous argument to conclude for some further subsequence, that

$$\mathcal{H}^{N-1}(u_k^{-1}(t) \cap (E^0 \cup \partial^* E^0) \to 0 \quad \text{as } k \to \infty,$$

contrary to our assertion that $L \neq 0$.

(e) In view of the definition

$$\sigma_{k;s}(B) := \int_{\partial A_{k;s} \cap B} \boldsymbol{F}(y) \cdot \nu_k(y)\, \mathrm{d}\mathcal{H}^{N-1}(y) \quad \text{for every Borel set } B,$$

and the fact that $\boldsymbol{F}$ is bounded, the result follows immediately from (d).

(f) From (a) we have the existence of smoothly bounded sets such that

(5.15) $$\|\mu\|(A_{k;t} \Delta E^1) \to 0,$$

and from Lemma 3.1 we know that our desired result holds for all smoothly bounded sets, in particular for $A_{k,t}$ for a.e. t:

(5.16) $$\mu(A_{k;t}) := \int_{A_{k;t}} \operatorname{div} \boldsymbol{F} = -\int_{\partial^* A_{k;t}} \boldsymbol{F}(y) \cdot \nu(y)\, \mathrm{d}\mathcal{H}^{N-1}(y).$$

We note that with our notation in force, we may write (5.16) as

(5.17) $$\mu(A_{k;s}) = -\sigma_{k;s}(\mathbb{R}^N) = -\sigma_{k;s}(\partial A_{k;s}).$$

From (5.17) and (5.16), it follows that

$$\mu(A_{k;s}) := \int_{A_{k;s}} \operatorname{div} \boldsymbol{F} = -\int_{\partial A_{k;s}} \boldsymbol{F}(y) \cdot \nu(y)\, \mathrm{d}\mathcal{H}^{N-1}(y) = -\sigma_{k;s}(\mathbb{R}^N)$$

for almost each value of $s \in (\frac{1}{2}, 1)$. Since

(5.18) $$\mu(A_{k;s}) \to \mu(E^1) \quad \text{and} \quad \sigma_{k;s}(\mathbb{R}^N) \to \sigma(\mathbb{R}^N)$$

with the help of (5.15), we obtain

$$\mu(E^1) = -\sigma(\partial^* E) = -\sigma_{\mathrm{i}}(\partial^* E) \quad \text{see (5.10)}.$$

Because $\sigma \ll \mathcal{H}^{N-1}$, we know there exists $\mathscr{G} \in L^\infty(\partial^* E; \mathbb{R}^1)$ such that

$$\sigma(B) = \int_{B\cap\partial^* E} \mathscr{G}(y)\, \mathrm{d}\mathcal{H}^{N-1}(y).$$

Now $\mathscr{G}(y)$ is of the form

$$\mathscr{G}(y) := \mathscr{G}_{\mathrm{i}}(y) \cdot \nu(y),$$

for some $\mathscr{G} \in L^\infty(\partial^* E; \mathbb{R}^N)$. Thus, taking $\mathscr{F}_{\mathrm{i}} = \mathscr{G}_{\mathrm{i}}$ leads to our desired result, (5.8).

(g) In order to establish that $\boldsymbol{F} \cdot \nabla u_k \chi_E \rightharpoonup \mathscr{F}_{\mathrm{i}} \cdot \nu_{\partial^* E} \mathcal{H}^{N-1} \llcorner \partial^* E$ weak* in $\mathscr{M}(\mathbb{R}^N)$, we will show that (see (2.4))

$$\boldsymbol{F} \cdot \nabla u_k \chi_E(\varphi) \to (\mathscr{F}_{\mathrm{i}} \cdot \nu_{\partial^* E}) \mathcal{H}^{N-1} \llcorner \partial^* E(\varphi)$$

or equivalently,

$$\lim_{k\to\infty} \int_E \varphi(x) \boldsymbol{F}(x) \cdot \nabla u_k(x)\, \mathrm{d}x = \int_{\mathbb{R}^N} \varphi(x) \mathscr{F}_{\mathrm{i}}(x) \cdot \nu_{\partial^* E}\, \mathrm{d}\mathcal{H}^{N-1}(x)\, \mathrm{d}x$$

for each test function $\varphi \in C_{\mathrm{c}}(\mathbb{R}^N)$.

Since $\overline{\boldsymbol{F} \cdot \nabla u_k}$ is absolutely continuous with respect to $\|\nabla \chi_E\| = \mathcal{H}^{N-1} \llcorner \partial^* E$ (see (2.17)), and since, its density relative to $\partial^* E$ is equal to $\mathscr{F}_{\mathrm{i}} \cdot \nu_{\partial^* E}$, (see (5.8)) we have

$$\int_{\mathbb{R}^N} \varphi(x)\, \mathrm{d}\overline{\boldsymbol{F} \cdot \nabla \chi_E} = \int_{\mathbb{R}^N} \varphi(x) \mathscr{F}_{\mathrm{i}}(x) \cdot \nu_{\partial^* E}(x)\, \mathrm{d}\mathcal{H}^{N-1}(x).$$

Therefore, our result, Theorem 5.8(i), follows from the definition,

$$\lim_{k\to\infty} \int_E \varphi(x) \boldsymbol{F}(x) \cdot \nabla u_k(x)\, \mathrm{d}x = \lim_{k\to\infty} \overline{\boldsymbol{F} \cdot \nabla u_k}(\varphi) = \int_{\mathbb{R}^N} \varphi(x)\, \mathrm{d}\overline{\boldsymbol{F} \cdot \nabla \chi_E}. \tag{5.19}$$

(h) From Theorem 5.8, we know for $\mathcal{H}^{N-1}$-a.e. $x \in \partial^* E$, that for $y \in H^-(x) \cap E$ we have the following representation for the inner trace in terms of the approximate limit of $\boldsymbol{F} \llcorner E \cap H^-$, where denotes the half space orthogonal to $\nu_{u_E}(x)$

$$\mathscr{F}_{\mathrm{i}}(x) = \operatorname{ap} \lim_{y\to x} \boldsymbol{F}(y).$$

From this our result will easily follow.

(i) This follows from the foregoing discussion, especially (5.18). □

Subtracting (5.9) from (5.8), we obtain

Corollary 5.3. $\mu(\partial^* E) = \int_{\partial^* E} \big(\mathscr{F}_{\mathrm{e}}(y) - \mathscr{F}_{\mathrm{i}}(y)\big) \cdot \nu(y)\, \mathrm{d}\mathcal{H}^{N-1}(y)$.

If we replicate the proof of Theorem 5.2 with E replaced by $\widetilde{E} := E^0 \cup \partial^* E$ then

$$u_k(t) := \varphi_{1/k} * \chi_E = \fint_{B(1/k)} \chi_{A_{k;t}},$$

and so we obtain

$$\fint_{B(1/k)} \chi_E + \fint_{B(1/k)} \chi_{\widetilde{E}} = 1.$$

Hence, if we define

$$v_k(t) := \varphi_{1/k} * \chi_{\widetilde{E}}$$

we have

$$u_k(x) + v_k(x) = 1,$$

which implies $u_k^{-1}(t) = v_k^{-1}(1-t)$. Since $t \in (\frac{1}{2}, 1)$ if and only if $1 - t \in (0, \frac{1}{2})$, the following observation becomes evident.

Corollary 5.4. *The inner trace of* $\boldsymbol{F}$ *relative to* E *on* $\partial^* E$ *is the same as the outer trace of* $\boldsymbol{F}$ *relative to* $\widetilde{E}$ *on* $\partial^* E$.

We offer the following simple example to illustrate our result. This example also dramatically demonstrates the difference between the classical derivative and the weak (distributional) derivative.

Example 5.5. Consider the most elementary situation of $N = 1$ where $\Omega := [0, 2]$, $E := [0, 1]$ and f is a nondecreasing function defined on $[0, 2]$ which is continuous everywhere except at $x = 1$, at which point we assume right continuity of f.

(1) The case $\frac{1}{2} < s < 1$. Since f is BV, we know that $f' = \mu$, for some measure μ. Then, according to Theorem 5.2,

$$\mu(E^1) = \mu\big((0,1)\big) := \int_0^1 f' \, \mathrm{d}x = \mathscr{F}_{\mathrm{i}}(1) - \mathscr{F}_{\mathrm{i}}(0),$$

where $\mathscr{F}_{\mathrm{i}}(1) = \lim_{x \to 1^-} f(x)$ and $\mathscr{F}_{\mathrm{i}}(0) = \lim_{x \to 0^+} f(x)$. Indeed, the sets $A_{k;s}$, with fixed $\frac{1}{2} < s < 1$, form a nested family of open intervals contained in $[0, 2]$. The measures σ_k correspond to f evaluated on point masses located at x_k; thus, as in (4.2), $f(x_k)$ converges to a limit, $\mathscr{F}_{\mathrm{i}}(1)$.

(2) The case $0 < s < \frac{1}{2}$. Here the sets $A_{k;s}$, $0 < s < \frac{1}{2}$, form a nested family of open intervals containing $[0, 1]$. Similar to above, we have

$$\mu(E) = \mu([0,1]) := \int_0^1 f' \, \mathrm{d}x = \mathscr{F}_{\mathrm{e}}(1) - \mathscr{F}_{\mathrm{e}}(0),$$

the measures σ_k correspond to f evaluated on point masses located at x_k and thus $\mathscr{F}_{\mathrm{e}}(1) = \lim_{x \to 1^-} f(x)$ and $\mathscr{F}_{\mathrm{e}}(0) = \lim_{x \to 0^+} f(x)$.

Now we proceed to further analyze the inner trace measure σ_{i}.

Theorem 5.6. *If* $\boldsymbol{F}$ *is a bounded divergence measure field and* E *a set of finite perimeter, then the inner trace of* $\boldsymbol{F}$ *on* $\partial^* E$, σ_{i}, *can be expressed as*

$$\lim_{k \to \infty} \int_E \boldsymbol{F} \cdot \nabla u_k = \frac{1}{2} \sigma_{\mathrm{i}}. \tag{5.20}$$

For fixed k, we define

$$\mathscr{N} := \{x : |\nabla u_k(x)| = 0.\}$$

We thus conclude the result of Chen – Torres, [22].

Theorem 5.7 ([22, Theorem 1])**.** *If* $\boldsymbol{F}$ *is a divergence measure field and* E *a set of finite perimeter, then*

$$\sigma_{\mathrm{i}} = 2\overline{\chi_E \boldsymbol{F} \cdot \nabla u_E} := \lim_{k \to \infty} \int_E \boldsymbol{F} \cdot \nabla u_k \, \mathrm{d}x.$$

Since we have, by definition,

$$\boldsymbol{F}\cdot\nabla u_k \to \overline{\boldsymbol{F}\cdot\nabla\chi_E}\,,$$

and since $\overline{\boldsymbol{F}\cdot\nabla\chi_E}$ is absolutely continuous with respect to $\mathcal{H}^{N-1}\llcorner\partial^*E$ (Lemma 2.17) this all suggests the following result.

Theorem 5.8. *Let $\boldsymbol{F}\in\mathcal{DM}^\infty(\mathbb{R}^N)$. Let u_k be a sequence of mollifiers of χ_E. Then, as $\boldsymbol{F}$ is bounded, we may consider $\boldsymbol{F}\cdot\nabla u_k$ as a Radon measure as well as $(\boldsymbol{F}\cdot\nabla u_k)\chi_E$, and thus conclude*

(i) $(\boldsymbol{F}\cdot\nabla u_k)\chi_E \rightharpoonup (\mathscr{F}_{\mathrm{i}}\cdot\nu_{\partial^*E})\mathcal{H}^{N-1}\llcorner\partial^*E$,

(ii) $(\boldsymbol{F}\cdot\nabla u_k)\chi_{E^0} \rightharpoonup (\mathscr{F}_{\mathrm{e}}\cdot\nu_{\partial^*E})\mathcal{H}^{N-1}\llcorner\partial^*E$,

as $k\to\infty$.

PROOF. We will prove only the first assertion; the other proof is very similar.

In order to establish that $\boldsymbol{F}\cdot\nabla u_k\chi_E \rightharpoonup \mathscr{F}_{\mathrm{i}}\cdot\nu_{\partial^*E}\mathcal{H}^{N-1}\llcorner\partial^*E$ in $\mathscr{M}(\mathbb{R}^N)$, we will show that (see (2.4))

$$\boldsymbol{F}\cdot\nabla u_k\chi_E(\varphi)\rightharpoonup(\mathscr{F}_{\mathrm{i}}\cdot\nu_{\partial^*E})\mathcal{H}^{N-1}\llcorner\partial^*E(\varphi)$$

or equivalently,

$$\lim_{k\to\infty}\int_E\varphi(x)\boldsymbol{F}(x)\cdot\nabla u_k(x)\,\mathrm{d}x=\int_{\mathbb{R}^N}\varphi(x)\mathscr{F}_{\mathrm{i}}(x)\cdot\nu_{\partial^*E}\,\mathrm{d}\mathcal{H}^{N-1}(x)\,\mathrm{d}x$$

for each test function $\varphi\in C_{\mathrm{c}}(\mathbb{R}^N)$.

Since $\overline{F\cdot\nabla u_k}$ is absolutely continuous with respect to $\|\nabla\chi_E\|=\mathcal{H}^{N-1}\llcorner\partial^*E$ (see (2.17)), and since, its density relative to ∂^*E is equal to $\mathscr{F}_{\mathrm{i}}\cdot\nu_{\partial^*E}$, (see (5.8)) we have

$$\int_{\mathbb{R}^N}\varphi(x)\,\mathrm{d}\overline{F\cdot\nabla\chi_E}=\int_{\mathbb{R}^N}\varphi(x)\mathscr{F}_{\mathrm{i}}(x)\cdot\nu_{\partial^*E}(x)\,\mathrm{d}\mathcal{H}^{N-1}(x).$$

Therefore, our result, Theorem 5.8(i), follows from the definition,

$$\lim_{k\to\infty}\int_E\varphi(x)\boldsymbol{F}(x)\cdot\nabla u_k(x)\,\mathrm{d}x=\lim_{k\to\infty}\overline{\boldsymbol{F}\cdot\nabla u_k}(\varphi)=\int_{\mathbb{R}^N}\varphi(x)\,\mathrm{d}\overline{\boldsymbol{F}\cdot\nabla\chi_E}. \tag{5.21}$$

□

Since $\nabla u_k\rightharpoonup\nabla\chi_E$ and if $\boldsymbol{F}$ were continuous, it would then follow that

$$\lim_{k\to\infty}\int_E\boldsymbol{F}(x)\cdot\nabla u_k(x)\,\mathrm{d}x=\nabla\chi_E(\boldsymbol{F})=\int_{\mathbb{R}^N}\mathscr{F}_{\mathrm{i}}(x)\cdot\nu_{\partial^*E}\,\mathrm{d}\mathcal{H}^{N-1}(x)\,\mathrm{d}x.$$

By Lusin's theorem we know for each $\varepsilon>0$, that there is a closed set $K_\varepsilon\subset\partial^*E$ such that $\mathscr{F}_{\mathrm{i}}\llcorner K_\varepsilon$ is continuous and that $\mathcal{H}^{N-1}(\partial^*E\setminus K_\varepsilon)<\varepsilon$. Tietze's extension theorem implies the existence of a continuous extension of $\mathscr{F}_{\mathrm{i}}\llcorner(\partial^*E\setminus K_\varepsilon)$, $\boldsymbol{F}_\varepsilon$, to all of $\mathbb{R}^N$ with the property that its norm is not increased, that is, $\|\boldsymbol{F}_\varepsilon\|_0\le\|\mathscr{F}_{\mathrm{i}}\|_0$

$$\lim_{k\to\infty}\int_E\boldsymbol{F}_\varepsilon(x)\cdot\nabla u_k(x)\,\mathrm{d}x=\int_{\partial^*E}\boldsymbol{F}_\varepsilon(y)\cdot\nu_{\partial^*E}(y)\,\mathrm{d}\mathcal{H}^{N-1}(y).$$

Since $\boldsymbol{F}_\varepsilon$ is uniformly bounded and since $\boldsymbol{F}_\varepsilon\to\mathscr{F}_{\mathrm{i}}$ $\mathcal{H}^{N-1}$-a.e., an application of Lebesgue's dominated convergence theorem yields

$$\lim_{k\to\infty}\int_E\boldsymbol{F}_\varepsilon(x)\cdot\nabla u_k(x)\,\mathrm{d}x=\int_{\partial^*E}\mathscr{F}_{\mathrm{i}}\cdot\nu_{\partial^*E}(y)\,\mathrm{d}\mathcal{H}^{N-1}(y),$$

and therefore, from (5.21) with $\varphi = 1$ in a neighborhood of E, that

$$\lim_{k\to\infty} \int_E \boldsymbol{F}(x)\cdot\nabla u_k(x)\,\mathrm{d}x = \int_{\partial^* E} \mathscr{F}_{\mathrm{i}}\cdot\nu_{\partial^* E}(y)\,\mathrm{d}\mathcal{H}^{N-1}(y).$$

Lemma 5.9. *Let* $\mu = \operatorname{div}\boldsymbol{F}$ *where* $\boldsymbol{F}\in\mathcal{DM}^\infty(\mathbb{R}^N)\cap C(\mathbb{R}^N;\mathbb{R}^N)$*. Then*

$$\|\mu\|(G) = 0,$$

for any set G *that can be written as the graph of a Lipschitz function* f*.*

Proof. We have that $G := \{(x', f(x')), x'\in\Omega\subset\mathbb{R}^{N-1}\}$. We let P and P^c be a Hahn decomposition of $\mathbb{R}^N$ for μ. Therefore, we have that μ^+ and μ^-, the positive and negative parts of μ, satisfy

$$\mu^+(B) = \mu(B\cap P) \quad\text{and}\quad \mu^-(B) = -\mu(B\cap P^c),$$

for any Borel set B.

$$A^+ = \{z = (x', f(x'))\in G : \boldsymbol{F}(z)\cdot\nu(z) > 0\}, \tag{5.22}$$

where $\nu(z)$ is the unit normal to the graph.

We define

$$A^+ = G\cap P, \quad A^- = G\cap P^c.$$

We also define

$$T_t = \{(x', f(x') + t)\}, \quad B_t = \{(x', f(x') - t)\}.$$

Let $K\subset A^+$ be any compact set. Let $U_k\subset\mathbb{R}^N$ be a sequence of open sets such that

$$\mu(U_k\cap G)\to\mu(K). \tag{5.23}$$

We fix a set U_k and note that, by the Besicovitch – Morse covering theorem [12, 68], U_k can be written, up to a set of $\|\mu\|$-measure zero, as a countable union of disjoints open parallelepipeds I_i^k. That is,

$$\bigcup_{i=1}^\infty I_i^k\subset U_k \quad\text{and}\quad \|\mu\|(U_k\setminus\bigcup_{i=1}^\infty I_i^k) = 0. \tag{5.24}$$

We denote U_k simply as U and I_i^k as I_i. We fix an i for which $I_i\cap G\neq\varnothing$ and note that, for t small enough, the graphs T_t and B_t are contained in I_i. We let R_t be the region inside I_i bounded above and below by T_t and B_t, respectively. We define

$$\alpha_t = \int_{\partial R_t\setminus(T_t\cup B_t)} \boldsymbol{F}(z)\cdot\nu(z)\,\mathrm{d}\mathcal{H}^{N-1}(z),$$

where $\nu(z)$ is the unit outward normal to R_t at the sides $\partial R_t\setminus(T_t\cup B_t)$. Since Theorem 2.21 applies to R_t for a.e. t, we have

$$\begin{aligned}
\mu(R_t) &= \int_{R_t}\operatorname{div}\boldsymbol{F}\\
&= \int_{B_t}\boldsymbol{F}(z)\cdot\nu(z)\,\mathrm{d}\mathcal{H}^{N-1}(z) - \int_{T_t}\boldsymbol{F}(z)\cdot\nu(z)d\mathcal{H}^{N-1}(z) + \alpha_t\\
&= \int_{T_t}\boldsymbol{F}(z_1, z')\cdot\nu(z)\,\mathrm{d}\mathcal{H}^{N-1}(z) - \int_{T_t}\boldsymbol{F}(z_1, z'-2t)\cdot\nu(z)\,\mathrm{d}\mathcal{H}^{N-1}(z) + \alpha_t\\
&= \int_{T_t}\big(\boldsymbol{F}(z_1, z') - \boldsymbol{F}(z_1, z'-2t)\big)\cdot\nu(z)\,\mathrm{d}\mathcal{H}^{N-1}(z) + \alpha_t.
\end{aligned}$$

Since $\boldsymbol{F}$ is continuous and $\alpha_t \to 0$ as $t \to 0$ we have, for any $\varepsilon > 0$, that

$$\mu(R_t) \leq \varepsilon,$$

for all $t \leq t_0(\varepsilon, F, G)$.

Thus, from (5.22) we have, for $t \leq t_0$,

$$\mu^+(R_t) = \mu(R_t) + \mu^-(R_t) = \mu(R_t) + \mu^-(U_\delta) \leq \varepsilon + C(\delta).$$

We have

$$\mu(I_i \cap G) = \lim_{t \to 0} \mu(R_t) \leq \varepsilon$$

and since ε is arbitrary we obtain that $\mu(I_i \cap G) = 0$. Therefore, using (5.24) we obtain

$$\mu(K) = \lim_{k \to \infty} \mu(U_k \cap G) = \lim_{k \to \infty} \sum \mu(I_i^k \cap G) = 0,$$

which implies that $\mu^+(K) = \mu(K \cap P) = \mu(K) = 0$, $K \subset A^+ \subset P$. Since

$$\mu^+(A^+) = \sup_{K \subset A^+} \mu^+(K)$$

we obtain that $\mu^+(A^+) = 0$.

Proceeding in the same way we prove that $\mu^-(K) = 0$ for any compact set $K \subset A^-$. This implies that $\mu(A^+) = \mu(A^-) = 0$. Thus, we conclude

$$\begin{aligned} \|\mu\|(G) &= \mu^+(G) + \mu^-(G) = 0 \\ &= \mu(G \cap P) - \mu(G \cap P^c) = \mu(A^+) - \mu(A^-) = 0. \end{aligned} \tag{5.25}$$

□

6. One-sided approximation of sets of finite perimeter

Given an arbitrary set of finite perimeter, E, then we know that E can be approximated by sets with smooth boundaries. It is well-known that, in general, that such an approximation cannot be one-sided; that is, E cannot be approximated by smooth sets that lie in the interior of E.

The next result shows that if E is sufficiently regular, then there exists a one-sided approximation. The condition of regularity we impose is similar to Lewis's "uniformly fat" condition in potential theory [64]. Note that Theorem 5.2(d) states that for an arbitrary set of finite perimeter, E, that a one-sided approximation can "almost" be achieved.

Proposition 6.1. *Suppose that E is a bounded set of finite perimeter with the property that for all $x \in \partial E$, there are positive numbers c_0 and r_0 such that*

$$\frac{|E^0 \cap B(x,r)|}{B(x,r)} \geq c_0, \quad \textit{for all } r \leq r_0.$$

Then, there exists $t \in (0,1)$ such that

$$A_{k;t} \Subset E \quad \textit{for large } k.$$

Proof. Choose a mollifier ρ such that $\rho = 1$ on $B(0, \frac{1}{2})$. If $x \in \partial E$ we have

$$\begin{aligned} v_k(x) := \chi_{\mathbb{R}^N \setminus E} * \rho_{\varepsilon_k}(x) &= \frac{1}{\varepsilon_k^N} \int_{B(x,\varepsilon_k)} \chi_{\mathbb{R}^N \setminus E}(y) \rho\Big(\frac{y-x}{\varepsilon_k}\Big)\, dy \\ &\geq \frac{1}{\varepsilon_k^N} \int_{B(x,\varepsilon_k/2)} \chi_{\mathbb{R}^N \setminus E}(y)\, dy = \frac{|(\mathbb{R}^N \setminus E) \cap B(x, \varepsilon_k/2)|}{\varepsilon_k^N} \end{aligned}$$

$$= \frac{|E^0 \cap B(x, \varepsilon_k/2)|}{\varepsilon_k^N} = c_0/2^N := \tilde{c}_0$$

where $0 < \tilde{c}_0 < 1$ depends only on the dimension N and not on the point x. Note that $u_k(x) + v_k(x) = 1$ for all $x \in \mathbb{R}^N$ since

$$\frac{1}{\varepsilon_k^N} \int_{B(x,\varepsilon_k)} \chi_E(y) \rho\Big(\frac{y-x}{\varepsilon_k}\Big)\, dy + \frac{1}{\varepsilon_k^N} \int_{B(x,\varepsilon_k)} \chi_{\mathbb{R}^N \setminus E}(y) \rho\Big(\frac{y-x}{\varepsilon_k}\Big)\, dy = 1.$$

Therefore, for all $x \in \partial E$,

$$u_k(x) = 1 - v_k(x) \leq 1 - \tilde{c}_0,$$

and thus, taking $1 > t > 1 - \tilde{c}_0$, we see that $A_{k;t} \cap \partial E = \varnothing$. Consequently, each connected component of the open set $A_{k;t}$ lies either in the interior of E or in its exterior, and thus must lie in its interior. □

Subtracting (5.9) from (5.8), we obtain

Corollary 6.2. $\mu(\partial^* E) = \int_{\partial^* E} \big(\mathscr{F}_{\mathrm{e}}(y) - \mathscr{F}_{\mathrm{i}}(y)\big) \cdot \nu(y)\, d\mathcal{H}^{N-1}(y)$.

If we replicate the proof of Theorem 5.2 with E replaced by $\widetilde{E} := E^0 \cup \partial^* E$ then with

$$u_k(t) := \varphi_{1/k} * \chi_E = \fint_{B(1/k)} \chi_{A_{k;t}},$$

we obtain

$$\fint_{B(1/k)} \chi_E + \fint_{B(1/k)} \chi_{\widetilde{E}} = 1.$$

Hence, if we define

$$v_k(t) := \varphi_{1/k} * \chi_{\widetilde{E}}$$

we have

$$u_k(x) + v_k(x) = 1,$$

which implies $u_k^{-1}(t) = v_k^{-1}(1-t)$. Since $t \in (\frac{1}{2}, 1)$ if and only if $1 - t \in (0, \frac{1}{2})$, the following observation becomes evident.

Corollary 6.3. *The inner trace of $\boldsymbol{F}$ relative to E on $\partial^* E$ is the same as the outer trace of $\boldsymbol{F}$ relative to $\widetilde{E}$ on $\partial^* E$.*

We offer the following simple example to illustrate our result. This example also dramatically demonstrates the difference between the classical derivative and the weak (distributional) derivative.

Example 6.4. Consider the most elementary situation of $N = 1$ where $\Omega := [0, 2]$, $E := [0, 1]$ and f is a non-decreasing function defined on $[0, 2]$ which is continuous everywhere except at $x = 1$, at which point we assume right continuity.

(1) The case $\frac{1}{2} < s < 1$. Since f is BV, we know that $f' = \mu$, for some measure μ. Then, according to Theorem 5.2,

$$\mu(E^1) = \mu\big((0, 1)\big) := \int_0^1 f'\, dx = \mathscr{F}_{\mathrm{i}}(1) - \mathscr{F}_{\mathrm{i}}(0),$$

where $\mathscr{F}_{\mathrm{i}}(1) = \lim_{x \to 1^-} f(x)$ and $\mathscr{F}_{\mathrm{i}}(0) = \lim_{x \to 0^+} f(x)$. Indeed, the sets $A_{k;s}$, with fixed $\frac{1}{2} < s < 1$, form a nested family of open intervals contained in $[0, 2]$. The measures σ_k correspond to f evaluated on point masses located at x_k; thus, as in (4.2), $f(x_k)$ converges to a limit, $\mathscr{F}_{\mathrm{i}}(1)$.

(2) The case $0 < s < \frac{1}{2}$. Here the sets $A_{k;s}$, $0 < s < \frac{1}{2}$, form a nested family of open intervals containing $[0,1]$. Similar to above, we have

$$\mu(E) = \mu([0,1]) := \int_0^1 f' \, \mathrm{d}x = \mathscr{F}_{\mathrm{e}}(1) - \mathscr{F}_{\mathrm{e}}(0),$$

the measures σ_k correspond to f evaluated on point masses located at x_k and thus $\mathscr{F}_{\mathrm{e}}(1) = \lim_{x\to 1^-} f(x)$ and $\mathscr{F}_{\mathrm{e}}(0) = \lim_{x\to 0^+} f(x)$.

Since E has a uniform Lipschitz boundary, for each $x \in \partial E$ there is a finite cone, C_x, with vertex x that completely lies in the complement of E. Each cone C_x is assumed to be congruent to a fixed cone C. We note this implies that the hypothesis of Lemma 6.1 are satisfied and therefore there exists $0 < T < 1$ such that $u_k(x) < T$ for all k and all $x \in \partial E$.

7. Another interpretation of the inner and outer trace

In Theorem 5.2, we have defined the inner (resp. outer) normal trace of $\boldsymbol{F} \in \mathcal{DM}^\infty(\mathbb{R}^N;\mathbb{R}^N)$, $\mathscr{F}_{\mathrm{i}}$, (resp. $\mathscr{F}_{\mathrm{e}}$) over the boundary of a set of finite perimeter E. We know that

$$\mathscr{F}_{\mathrm{i}} = 2\overline{\boldsymbol{F}\cdot\nabla u_E}. \tag{7.1}$$

In order to obtain the normal trace of $\boldsymbol{F}$ on $\partial^*\widetilde{E}$, where $\widetilde{E} := E^0 \cup \partial^m E$, we reproduce the proof of Theorem 5.2 and apply it to $\widetilde{E}$. Therefore this trace, that we will denote by $\mathscr{F}_-$, is obtained using level sets $B_{k,s} = \{v_k > s\}$ for some $s \in (\frac{1}{2}, 1)$, where v_k is a sequence of mollifiers for $\chi_{\widetilde{E}}$. We note that for all $x \in (\mathbb{R}^N \setminus \partial E)$,

$$\rho_\varepsilon * \chi_E(x) + \rho_\varepsilon * \chi_{\widetilde{E}} = 1,$$

and therefore

$$v_k^{-1}(s) = u_k^{-1}(1-s),$$

where $1 - s \in (0, \frac{1}{2})$. Since $-\nu := \nu_{\mathrm{i}}$ is the inner unit normal to $\widetilde{E}$, we have

$$\begin{aligned} \int_{\partial^* E} \mathscr{F}_-(y)\, \mathrm{d}\mathcal{H}^{N-1}(y) &= -\lim_{k\to\infty} \int_{\partial B_{k,s}} \boldsymbol{F}\cdot\nu \\ &= -\lim_{k\to\infty} \int_{\partial A_{k,1-s}} \boldsymbol{F}\cdot\nu = -\lim_{k\to\infty} \sigma_{k,1-s}(\mathbb{R}^N). \end{aligned} \tag{7.2}$$

Corollary 4.2 implies that, for almost every $t \in (0, \frac{1}{2})$,

$$\mu(A_{k,t}) \to \mu(E). \tag{7.3}$$

Using (7.3) in the proof of Theorem 5.2 but replacing $1 - s \in (0, \frac{1}{2})$ with $s \in (\frac{1}{2}, 1)$ we obtain

$$\lim_{k\to\infty} \sigma_{k,1-s}(\mathbb{R}^N) = 2\overline{\chi_E \boldsymbol{F}\cdot\nabla u_E} - \mu(\partial^* E),$$

and therefore

$$\mathscr{F}_- = -2\overline{\chi_E \boldsymbol{F}\cdot\nabla u_E} + \mu(\partial^* E) = \mathscr{F}_{\mathrm{e}}. \tag{7.4}$$

From (7.1) and (7.4) we obtain the following result.

Theorem 7.1. *Let E and $\boldsymbol{F}$ be as in Theorem 5.2. Then the traces $\mathscr{F}_{\mathrm{i}}$ and $\mathscr{F}_{\mathrm{e}}$ are the traces of $\boldsymbol{F}$ on $\partial^* E$ corresponding to E and $\widetilde{E}$ respectively. Thus, because $-\nu := \nu_{\mathrm{i}}$ is the exterior normal to E, we see that $\nu := \nu_{\mathrm{e}}$ is the exterior normal to $\widetilde{E}$. Therefore, we have*

$$\mu(\partial^* E) = \operatorname{div} \boldsymbol{F} \llcorner \partial^* E = \int_{\partial^* E} \mathscr{F} \cdot \nu_{\mathrm{i}} + \int_{\partial^* E} \mathscr{F}_{\mathrm{e}} \cdot \nu_{\mathrm{e}}$$

from (7.1).

Acknowledgments. This paper was the subject of a lecture given at a conference on "Workshop on Singularities in PDE and the Calculus of Variations," which was held at the Centre of recherche mathématiques, Université de Montréal, July 17–21, 2006. The author wishes to thank the organizers, Stan Alama, Lia Bronsard, and Peter Sternberg, for the opportunity to present this work.

Total bibliography[2]

1. L. Ambrosio, *Some fine properties of sets of finite perimeter in Ahlfors regular metric measure spaces*, Adv. Math. **159** (2001), no. 1, 51–67.
2. ______, *Fine properties of sets of finite perimeter in doubling metric measure spaces: Calculus of variations, nonsmooth analysis and related topics*, Set-Valued Anal. **10** (2002), no. 2-3, 111–128.
3. L. Ambrosio, N. Fusco, and J. E. Hutchinson, *Higher integrability of the gradient and dimension of the singular set for minimisers of the Mumford–Shah functional*, Calc. Var. Partial Differential Equations **16** (2003), no. 2, 187–215.
4. L. Ambrosio, N. Fusco, and D. Pallara, *Functions of bounded variation and free discontinuity problems*, Oxford Math. Monogr., Oxford Univ. Press, New York, NY, 2000.
5. L. Ambrosio and B. Kirchheim, *Rectifiable sets in metric and Banach spaces*, Math. Ann. **318** (2000), no. 3, 527–555.
6. G. Anzellotti, *Dirichlet problem and removable singularities for functionals with linear growth*, Boll. Un. Mat. Ital. C (5) **18** (1981), no. 1, 141–159.
7. ______, *Pairings between measures and bounded functions and compensated compactness*, Ann. Mat. Pura Appl. (4) **135** (1983), 293–318.
8. G. Anzellotti, S. Delladio, and G. Scianna, BV *functions over rectifiable currents*, Ann. Mat. Pura Appl. (4) **170** (1996), 257–296.
9. G. Anzellotti and M. Giaquinta, *Funzioni* BV *e tracce*, Rend. Sem. Mat. Univ. Padova **60** (1978), 1–21.
10. A. S. Besicovitch, *On the definition of the area of a surface by means of inscribed polyhedra*, J. London Math. Soc. **19** (1944), 138–141.
11. ______, *A general form of the covering principle and relative differentiation of additive functions*, Proc. Cambridge Philos. Soc. **41** (1945), 103–110.
12. ______, *A general form of the covering principle and relative differentiation of additive functions*. II, Proc. Cambridge Philos. Soc. **42** (1946), 1–10.
13. W. W. Bledsoe and A. P. Morse, *Some aspects of covering theory*, Proc. Amer. Math. Soc. **3** (1952), 804–812.
14. S. Bloom, W. B. Jurkat, and G. Sampson, *Two weighted* (L^p, L^q) *estimates for the Fourier transform*, Rocky Mountain J. Math. **23** (1993), no. 2, 451–482.
15. Ju. S. Burago and V. G. Maz′ja, *Certain questions of potential theory and function theory for regions with irregular boundaries*, Zap. Naučn. Sem. Leningrad. Otdel. Mat. Inst. Steklov. (LOMI), vol. 3, 1967 (Russian); English transl. in *Potential theory and function theory for irregular regions*, Seminars in Mathematics, V. A. Steklov Mathematical Institute, Leningrad, vol. 3, Consultants Bureau, New York, NY, 1969.

[2]The number of references given here is much more than the number cited. This is for the interested reader who may wish to see related material.

16. Ju. S. Burago, V. G. Maz′ja, and V. D. Sapožnikova, *On the theory of potentials of a double and a simple layer for regions with irregular boundaries*, Problems of Mathematical Analysis. Boundary Value Problems and Integral Equations (V. I. Smirnov, ed.), Izdat. Leningrad. Univ., Leningrad, 1966, pp. 3–34 (Russian).
17. G. Q. Chen and C. M. Dafermos, *The vanishing viscosity method in one-dimensional thermoelasticity*, Trans. Amer. Math. Soc. **347** (1995), no. 2, 531–541.
18. G. Q. Chen and H. Frid, *Divergence-measure fields and hyperbolic conservation laws*, Arch. Ration. Mech. Anal. **147** (1999), no. 2, 89–118.
19. ———, *Large-time behavior of entropy solutions of conservation laws*, J. Differential Equations **152** (1999), no. 2, 308–357.
20. ———, *Extended divergence-measure fields and the Euler equations for gas dynamics*, Comm. Math. Phys. **236** (2003), no. 2, 251–280.
21. G. Q. Chen, H. Frid, and Y. Li, *Uniqueness and stability of Riemann solutions with large oscillation in gas dynamics*, Comm. Math. Phys. **228** (2002), no. 2, 201–217.
22. G. Q. Chen and M. Torres, *Divergence-measure fields, sets of finite perimeter, and conservation laws*, Arch. Ration. Mech. Anal. **175** (2005), no. 2, 245–267.
23. G. Q. Chen, M. Torres, and W. P. Ziemer, *Measure-theoretic analysis and nonlinear conservation laws*, Pure Appl. Math. Q. **3** (2007), no. 3, 1–39.
24. ———, *Gauss–Green theorem for weakly differentiable vector fields, sets o finite perimeter, and balance laws*, available at `arXiv:0709.3673`.
25. C. M. Dafermos, *Large time behavior of periodic solutions of hyperbolic systems of conservation laws*, J. Differential Equations **121** (1995), no. 1, 183–202.
26. ———, *Balance laws in continuum physics*, Advanced Topics in Theoretical Fluid Mechanics (Paseky and Jizerou, 1997), Pitman Res. Notes Math. Ser., vol. 392, Longman, Harlow, 1998, pp. 89–117.
27. ———, *Hyperbolic conservation laws in continuum physics*, Grundlehren Math. Wiss., vol. 325, Springer, Berlin, 2000; 2nd ed., 2005.
28. E. De Giorgi, *Sulla proprietà isoperimetrica dell'ipersfera, nella classe degli insiemi aventi frontiera orientata di misura finita*, Atti Accad. Naz. Lincei. Mem. Cl. Sci. Fis. Mat. Natur. Sez. Ia (8) **5** (1958), 33–44.
29. ———, *Complementi alla teoria della misura $(n-1)$-dimensionale in uno spazio n-dimensionale*, Seminario di Matematica della Scuola Normale Superiore di Pisa, 1960–61, Editrice Tecnico Scientifica, Pisa, 1961.
30. ———, *Frontiere orientate di misura minima*, Seminario di Matematica della Scuola Normale Superiore di Pisa, 1960–61, Editrice Tecnico Scientifica, Pisa, 1961.
31. E. De Giorgi, F. Colombini, and L. C. Piccinini, *Frontiere orientate di misura minima e questioni collegate*, Scuola Normale Superiore, Pisa, 1972.
32. T. De Pauw and W. F. Pfeffer, *The Gauss–Green theorem and removable sets for PDEs in divergence form*, Adv. Math. **183** (2004), no. 1, 155–182.
33. L. C. Evans and R. F. Gariepy, *Measure theory and fine properties of functions*, Stud. Adv. Math., CRC Press, Boca Raton, FL, 1992.
34. H. Federer, *The Gauss–Green theorem*, Trans. Amer. Math. Soc. **58**, 44–76.
35. ———, *A note on the Gauss–Green theorem*, Proc. Amer. Math. Soc. **9** (1958), 447–451.
36. ———, *Curvature measures*, Trans. Amer. Math. Soc. **93** (1959), 418–491.
37. ———, *The area of a nonparametric surface*, Proc. Amer. Math. Soc. **11** (1960), 436–439.
38. ———, *Currents and area*, Trans. Amer. Math. Soc. **98** (1961), 204–233.
39. ———, *Some theorems on integral currents*, Trans. Amer. Math. Soc. **117** (1965), 43–67.
40. ———, *Two theorems in geometric measure theory*, Bull. Amer. Math. Soc. **72** (1966), 719.
41. ———, *Some properties of distributions whose partial derivatives are representable by integration*, Bull. Amer. Math. Soc. **74** (1968), 183–186.
42. ———, *Geometric measure theory*, Grundlehren Math. Wiss., vol. 153, Springer, New York, NY, 1969.
43. ———, *Slices and potentials*, Indiana Univ. Math. J. **21** (1971/72), 373–382.
44. H. Federer and W. H. Fleming, *Normal and integral currents*, Ann. of Math. (2) **72** (1960), 458–520.
45. H. Federer and W. P. Ziemer, *The Lebesgue set of a function whose distribution derivatives are pth power summable*, Indiana Univ. Math. J. **22** (1972/73), 139–158.
46. B. Fuglede, *On a theorem of F. Riesz*, Math. Scand. **3** (1955), 283–302.

47. ______, *Extremal length and functional completion*, Acta Math. **98** (1957), 171–219.
48. ______, *Extremal length and closed extensions of partial differential operators*, Jul. Gjellerups Boghandel, Copenhagen, 1960.
49. E. Giusti, *Minimal surfaces and functions of bounded variation*, with notes by G. H. Williams, Notes on Pure Math., vol. 10, Dept. of Pure Mathematics, Australian National Univ., Canberra, 1977.
50. C. Goffman, C. J. Neugebauer, and T. Nishiura, *Density topology and approximate continuity*, Duke Math. J. **28** (1961), 497–505.
51. C. Goffman and D. Waterman, *Approximately continuous transformations*, Proc. Amer. Math. Soc. **12** (1961), 116–121.
52. M. E. Gurtin and L. C. Martins, *Cauchy's theorem in classical physics*, Arch. Rational Mech. Anal. **60** (1975/76), no. 4, 305–324.
53. M. E. Gurtin and W. O. Williams, *An axiomatic foundation for continuum thermodynamics*, Arch. Rational Mech. Anal. **26** (1967), 83–117.
54. ______, *On the first law of thermodynamics*, Arch. Rational Mech. Anal. **42** (1971), 77–92.
55. M. E. Gurtin, W. O. Williams, and W. P. Ziemer, *Geometric measure theory and the axioms of continuum thermodynamics*, Arch. Rational Mech. Anal. **92** (1986), no. 1, 1–22.
56. C. A. Hayes Jr. and A. P. Morse, *Convexical blankets*, Proc. Amer. Math. Soc. **1** (1950), 719–730.
57. E. J. Howard and W. F. Pfeffer, *Luzin's theorem for charges*, Proc. Amer. Math. Soc. **132** (2004), no. 3, 857–863.
58. W. B. Jurkat and D. J. F. Nonnenmacher, *The general form of Green's theorem*, Proc. Amer. Math. Soc. **109** (1990), no. 4, 1003–1009.
59. ______, *An axiomatic theory of non-absolutely convergent integrals in* $\mathbb{R}^n$, Fund. Math. **145** (1994), no. 3, 221–242.
60. ______, *A generalized n-dimensional Riemann integral and the divergence theorem with singularities*, Acta Sci. Math. (Szeged) **59** (1994), no. 1-2, 241–256.
61. ______, *A theory of non-absolutely convergent integrals in* $\mathbb{R}^n$ *with singularities on a regular boundary*, Fund. Math. **146** (1994), no. 1, 69–84.
62. ______, *The fundamental theorem for the* ν_1*-integral on more general sets and a corresponding divergence theorem with singularities*, Czechoslovak Math. J. **45(120)** (1995), no. 1, 69–77.
63. ______, *A Hake-type property for the* ν_1*-integral and its relation to other integration processes*, Czechoslovak Math. J. **45(120)** (1995), no. 3, 465–472.
64. J. L. Lewis, *Uniformly fat sets*, Trans. Amer. Math. Soc. **308** (1988), no. 1, 177–196.
65. ______, *Some applications of Riesz capacities*, Complex Variables Theory Appl. **12** (1989), no. 1-4, 237–244.
66. J. Malý, D. Swanson, and W. P. Ziemer, *The co-area formula for Sobolev mappings*, Trans. Amer. Math. Soc. **355** (2003), no. 2, 477–492.
67. O. Martio and W. P. Ziemer, *Lusin's condition* (N) *and mappings with nonnegative Jacobians*, Michigan Math. J. **39** (1992), no. 3, 495–508.
68. A. P. Morse, *Perfect blankets*, Trans. Amer. Math. Soc. **61** (1947), 418–442.
69. W. Noll, *The foundations of classical mechanics in the light of recent advances in continuum mechanics*, The Axiomatic Method. With Special Reference to Geometry and Physics (Berkeley, CA, 1957/58) (L. Henkin, P. Suppes, and A. Tarski, eds.), Stud. Logic Found. Math., North-Holland, Amsterdam, 1959, pp. 266–281.
70. ______, *A new mathematical theory of simple materials*, Arch. Rational Mech. Anal. **48** (1972), 1–50.
71. ______, *Lectures on the foundations of continuum mechanics and thermodynamics*, Arch. Rational Mech. Anal. **52** (1973), 62–92.
72. ______, *Continuum mechanics and geometric integration theory*, Categories in Continuum Physics (Buffalo, NY, 1982) (F. W. Lawvere and S. H. Schanuel, eds.), Lecture Notes in Math., vol. 1174, Springer, Berlin, 1986, pp. 17–29.
73. ______, *Finite-dimensional spaces. Vol.* I: *Algebra, geometry, and analysis*, Mech. Anal., vol. 10, Nijhoff, Dordrecht, 1987.
74. W. Noll and E. G. Virga, *Fit regions and functions of bounded variation*, Arch. Rational Mech. Anal. **102** (1988), no. 1, 1–21.

75. W. F. Pfeffer, *The Stokes theorem for the generalized Riemann integral*, Real Anal. Exchange **26** (2000/01), no. 2, 623–636.
76. ———, *Divergence theorem for vector fields with singularities*, New Integrals (Coleraine, 1988) (P. S. Bullen, P. Y. Lee, J. L. Mawhin, P. Muldowney, and W. F. Pfeffer, eds.), Lecture Notes in Math., vol. 1419, Springer, Berlin, 1990, pp. 150–166.
77. ———, *The Gauss-Green theorem*, Adv. Math. **87** (1991), no. 1, 93–147.
78. ———, *Derivatives and primitives*, Sci. Math. Jpn. **55** (2002), no. 2, 399–425.
79. ———, *The Gauss-Green theorem in the context of Lebesgue integration*, Bull. London Math. Soc. **37** (2005), no. 1, 81–94.
80. ———, *Distributions for which* $\operatorname{div} v = f$ *has a continuous solution*, Comm. Pure Appl. Math., to appear.
81. P. Sternberg and W. P. Ziemer, *The Dirichlet problem for functions of least gradient*, Degenerate Diffusions (Minneapolis, MN, 1991) (W.-M. Ni, L. A. Peletier, and J. L. Vásquez, eds.), IMA Vol. Math. Appl., vol. 47, Springer, New York, NY, 1993, pp. 197–214.
82. D. Swanson and W. P. Ziemer, *Sobolev functions whose inner trace at the boundary is zero*, Ark. Mat. **37** (1999), no. 2, 373–380.
83. ———, *The image of a weakly differentiable mapping*, SIAM J. Math. Anal. **35** (2004), no. 5, 1099–1109.
84. A. I. Vol′pert, *On the index and normal solvability of boundary-value problems for elliptic systems of differential equations on the plane*, Trudy Moskov. Mat. Obšč. **10** (1961), 41–87 (Russian).
85. ———, *Differentiation and quasi-linear operators in the space of functions whose generalized derivatives are measures*, Dokl. Akad. Nauk SSSR **171** (1966), 1027–1029 (Russian); English transl., Soviet Math. Dokl. **7** (1966), 1586–1589.
86. ———, *Spaces* BV *and quasilinear equations*, Mat. Sb. (N.S.) **73(115)** (1967), 255–302.
87. A. I. Vol′pert and S. I. Hudjaev, *The Cauchy problem for second order quasilinear degenerate parabolic equations*, Mat. Sb. (N.S.) **78(120)** (1969), 374–396 (Russian).
88. ———, *The Cauchy problem for composite systems of nonlinear differential equations*, Mat. Sb. (N.S.) **87(129)** (1972), 504–528.
89. H. Whitney, *On the topology of differentiable manifolds*, Lectures in Topology, Univ. of Michigan Press, Ann Arbor, MI, 1941, pp. 101–141.
90. ———, *Geometric integration theory*, Princeton Univ. Press, Princeton, NJ, 1957.
91. W. K. Ziemer, *A note on the support of a Sobolev function on a k-cell*, Proc. Amer. Math. Soc. **132** (2004), no. 7, 1987–1995.
92. W. P. Ziemer, *Cauchy flux and sets of finite perimeter*, Arch. Rational Mech. Anal. **84** (1983), no. 3, 189–201.
93. ———, *Weakly differentiable functions: Sobolev spaces and functions of bounded variation*, Grad. Texts in Math., vol. 120, Springer, New York, NY, 1989.
94. ———, *Functions of least gradient and* BV *functions*, Nonlinear Analysis, Function Spaces and Applications. Vol. 6 (Prague, 1998), Acad. Sci. Czech Repub., Prague, 1999, pp. 270–312.

Department of Mathematics, Indiana University, Rawles Hall, 831 East Third St. Bloomington, IN 47405-7106 USA
URL: http://www.indiana.edu/~ziemer
E-mail address: ziemer@indiana.edu

Titles in This Series

44 **Stanley Alama, Lia Bronsard, and Peter J. Sternberg, Editors,** Singularities in PDE and the calculus of variations, 2008

43 **Andrew Granville, Melvyn B. Nathanson, and Jozsef Solymosi, Editors,** Additive combinatorics, 2007

42 **Donald A. Dawson, Vojkan Jakšić, and Boris Vainberg, Editors,** Probability and mathematical physics: a volume in honor of Stanislav Molchanov, 2007

41 **André Bandrauk, Michel C. Delfour, and Claude Le Bris, Editors,** High-dimensional partial differential equations in science and engineering, 2007

40 **V. Apostolov, A. Dancer, N. Hitchin, and M. Wang, Editors,** Perspectives in Riemannian geometry, 2006

39 **P. Winternitz, D. Gomez-Ullate, A. Iserles, D. Levi, P. J. Olver, R. Quispel, and P. Tempesta, Editors,** Group theory and numerical analysis, 2005

38 **Jacques Hurtubise and Eyal Markman, Editors,** Algebraic structures and moduli spaces, 2004

37 **P. Tempesta, P. Winternitz, J. Harnad, W. Miller, Jr., G. Pogosyan, and M. Rodriguez, Editors,** Superintegrability in classical and quantum systems, 2004

36 **Hershy Kisilevsky and Eyal Z. Goren, Editors,** Number theory, 2004

35 **H. E. A. Eddy Campbell and David L. Wehlau, Editors,** Invariant theory in all characteristics, 2004

34 **P. Winternitz, J. Harnad, C. S. Lam, and J. Patera, Editors,** Symmetry in physics, 2004

33 **André D. Bandrauk, Michel C. Delfour, and Claude Le Bris, Editors,** Quantum control: Mathematical and numerical challenges, 2003

32 **Vadim B. Kuznetsov, Editor,** The Kowalevski property, 2002

31 **John Harnad and Alexander Its, Editors,** Isomonodromic deformations and applications in physics, 2002

30 **John McKay and Abdellah Sebbar, Editors,** Proceedings on moonshine and related topics, 2001

29 **Alan Coley, Decio Levi, Robert Milson, Colin Rogers, and Pavel Winternitz, Editors,** Bäcklund and Darboux transformations. The geometry of solitons, 2001

28 **J. C. Taylor, Editor,** Topics in probability and Lie groups: Boundary theory, 2001

27 **I. M. Sigal and C. Sulem, Editors,** Nonlinear dynamics and renormalization group, 2001

26 **J. Harnad, G. Sabidussi, and P. Winternitz, Editors,** Integrable systems: From classical to quantum, 2000

25 **Decio Levi and Orlando Ragnisco, Editors,** SIDE III—Symmetries and integrability of difference equations, 2000

24 **B. Brent Gordon, James D. Lewis, Stefan Müller-Stach, Shuji Saito, and Noriko Yui, Editors,** The arithmetic and geometry of algebraic cycles, 2000

23 **Pierre Hansen and Odile Marcotte, Editors,** Graph colouring and applications, 1999

22 **Jan Felipe van Diejen and Luc Vinet, Editors,** Algebraic methods and q-special functions, 1999

21 **Michel Fortin, Editor,** Plates and shells, 1999

20 **Katie Coughlin, Editor,** Semi-analytic methods for the Navier-Stokes equations, 1999

19 **Rajiv Gupta and Kenneth S. Williams, Editors,** Number theory, 1999

18 **Serge Dubuc and Gilles Deslauriers, Editors,** Spline functions and the theory of wavelets, 1999

17 **Olga Kharlampovich, Editor,** Summer school in group theory in Banff, 1996, 1998

16 **Alain Vincent, Editor,** Numerical methods in fluid mechanics, 1998

15 **François Lalonde, Editor,** Geometry, topology, and dynamics, 1998
14 **John Harnad and Alex Kasman, Editors,** The bispectral problem, 1998
13 **Michel C. Delfour, Editor,** Boundaries, interfaces, and transitions, 1998
12 **Peter C. Greiner, Victor Ivrii, Luis A. Seco, and Catherine Sulem, Editors,** Partial differential equations and their applications, 1997
11 **Luc Vinet, Editor,** Advances in mathematical sciences: CRM's 25 years, 1997
10 **Donald E. Knuth,** Stable marriage and its relation to other combinatorial problems: An introduction to the mathematical analysis of algorithms, 1997
9 **D. Levi, L. Vinet, and P. Winternitz, Editors,** Symmetries and integrability of difference equations, 1996
8 **J. Feldman, R. Froese, and L. M. Rosen, Editors,** Mathematical quantum theory II: Schrödinger operators, 1995
7 **J. Feldman, R. Froese, and L. M. Rosen, Editors,** Mathematical quantum theory I: Field theory and many-body theory, 1994
6 **Guido Mislin, Editor,** The Hilton Symposium 1993: Topics in topology and group theory, 1994
5 **D. A. Dawson, Editor,** Measure-valued processes, stochastic partial differential equations, and interacting systems, 1994
4 **Hershy Kisilevsky and M. Ram Murty, Editors,** Elliptic curves and related topics, 1994
3 **Rémi Vaillancourt and Andrei L. Smirnov, Editors,** Asymptotic methods in mechanics, 1993
2 **Philip D. Loewen,** Optimal control via nonsmooth analysis, 1993
1 **M. Ram Murty, Editor,** Theta functions: From the classical to the modern, 1993